Applied Agriculture

Springer-Verlag Berlin Heidelberg GmbH

A.S. Perry† I. Yamamoto I. Ishaaya R. Perry

Insecticides in Agriculture and Environment

Retrospects and Prospects

With 57 Figures and 45 Tables

Springer

A.S. Perry†

I. Yamamoto
Tokyo University of Agriculture
Department of Agricultural Chemistry
1-1 Sakuragaoka-1, Setagaya-ku, Tokyo
Japan

Prof. Dr. I. Ishaaya
Department of Entomology
Agricultural Research Organization
The Volcani Center, Bet Dagan 50250
Israel

R. Perry
Rehov Sold 48/2, Ramat Hasharom 47225
Israel

ISSN 1433-7576
ISBN 978-3-662-03658-7 ISBN 978-3-662-03656-3 (eBook)
DOI 10.1007/978-3-662-03656-3

Library of Congress Cataloging-in-Publication Data

Insecticides in Agriculture and environment: retrospects and prospects/A.S. Perry... [et al.].
p. cm. — (Applied agriculture)
Includes bibliographical references and index.

1. Insecticides. 2. Insecticides—Environmental aspects.
I. Perry, A.S. (Albert S.), 1915–1992. II. Series.
SB951.5.1638 1997
632' .9517—dc21 97-28341

Originally published by Springer-Verlag Berlin Heidelberg New York in 1998.
Softcover reprint of the hardcover 1st edition 1998

Cover design & production GmbH, Heidelberg

Typesetting: Replika Press Pvt Ltd, Delhi-110 040 (India)

SPIN: 10501707 31/3137/5 4 3 2 1 0

In memory of the Late Professor Albert S. Perry

Albert S. Perry

Foreword

Professor Albert S. Perry passed away suddenly on February 18, 1992, leaving behind his grieving family, friends and colleagues. It was his aspiration to produce a comprehensive work on insecticides to summarize his lifelong dedication to the field of entomology and public health.

On the day before his operation, he expressed his desire with the following words: "I am coming out of this surgery and will recuperate from it as soon as possible for the sake of my boy (then aged three) and the book". He also told me that he would like to add a chapter on IPM (Integrated Pest Management) and suggested that we write it together. The sad reality is that none of this took place the way he had planned and these became his last words. On my own, I found it difficult to proceed with the writing of the IPM chapter, since several chapters are required to cover this subject and, in fact, several books are already devoted to IPM. There was even an IPM article written in a journal (Awake 1983) for a general audience to which he commented that he would like to use it someday because it was well written for laymen, thus providing the readers a wide selection of journals and books to choose from.

I could not have accomplished the formidable task of completing the book alone and am grateful to the other two co-authors and to Professor Bruno Yaron, the chief editor of the editorial board of the *Advanced Series in Agricultural Sciences* of Springer-Verlag, for his generous support.

This volume of *Advanced Series in Agricultural Sciences* is dedicated to Professor Albert Perry's memory. In the name of the team of co-authors I would like to thank Dr. Dieter Czeschlik, Ursula Gramm and Springer-Verlag for agreeing to include the tribute to Professor Perry in the Series. This book provides the readers both retrospects and prospects of insecticide toxicology in agriculture and environment. I am convinced that this comprehensive work on insecticides, their mode of action and impact on the agricultural and other ecosystems will advance understanding of the role of insecticides, in the past, present and future, in the field of crop protection and public health, and will motivate the younger generation to pursue the development of selective insect control agents fitting well in IPM programs for the benefit of man and the environment.

Rica Perry

Preface

In general, the expert is knowledgeable in his own area of science, but for most agricultural scientists, students and teachers, it is a difficult and time-consuming task to gain a general understanding of the area of insecticides, which is under continuous development and progress. Taking as an example the mode of action of pyrethroids, one will be inundated by scientific papers and reviews, and only after enormous effort will one be able to make the correct conclusions from the diverse information available. Old-fashioned insecticides which are, in many cases, no longer in use, are not mentioned sufficiently in modern textbooks, and older editions are not always available. However, when found, the reader may again be swamped by a mass of literature.

The history of pesticide development periodically evolved new ideas and directions for research; however, this aspect is usually neglected by most textbooks. On the other hand, many novel compounds appear one by one, year after year, and they must be given the right loci in an updated textbook. The present book intends to provide all you ever wanted to know about insecticides and related control agents but never had the time to look up.

The authors have tried to review and collate the widespread literature in a concise and consolidated form. Condensing several books on a certain subject, each having several hundred pages, into a few pages, has simplified the complexity of the subject but retained as much as possible of its scientific strength. Such a condensation is necessary to provide suitable background information, the present state of the art, and a glance into the future.

The original draft of this book was written by the late Professor Albert Perry of Tel Aviv University, Israel. Thereafter, Izuru Yamamoto, Isaac Ishaaya and Rica Perry completed and updated the original manuscript.

This book contains the history, chemistry, mode of action, metabolism, toxicity and environmental problems of insect controlling agents as well as of nematicides, molluscicides, avicides, and rodenticides.

During the preparation of the manuscript, we were grateful for stimulating discussions, valuable suggestions, and criticism provided for the various chapters by many friends and colleagues including: M. Asada (Nippon Soda, Japan), K.R.S. Ascher (ARO, Israel), J.E. Casida (University of California, Berkeley), E. Cohen (Faculty of Agriculture, Israel), E. Dunkelblum (ARO, Israel), M. Elliott (Rothamsted Experimental Station, UK), Morifusa Eto (Miyakonojo National College of Technology, Japan), Masayoshi Eto (Himeji Women's College, Japan),

K. Hashimoto (Sumitomo, Japan), Y. Higashino (Mitsubishi, Japan), A.R. Horowitz (ARO, Israel), T. Jojima (Sankyo, Japan), N. Kyomura (Mitsubishi, Japan), Y. Margalit (Ben Gurion University Israel), T. Miyamoto (Tokyo University of Agriculture, Japan), T. Miyazaki (Tokyo Metropolitan Research Laboratory, Japan), Lea Muszkat (ARO, Israel), F. Matsumura (University of California, Davis), T. Narahashi (Northwestern University, USA), I. Silman (Weitzmann Institute, Israel), J. Sekizawal (National Institute of Health Sciences, Japan), and B. Sneh (Tel Aviv University, Israel).

We are most thankful to Mrs. Svetlana Kontsedalov for her patience in typing and organizing the various sections of this book and to Mr. Uri Lemberger for drawing the figures. Our gratitude also goes to the Entomological Society of America; the American Chemical Society; Pesticide Science Society of Japan; Plenum Publishing Corporation; John Wiley & Sons, Inc.; Taylor & Francis; CRC Press, Inc.; Academic Press; McGraw Hill Companies; Annual Reviews, Inc.; World Health Organization; Current Biology Limited; Buneido Publishing Co., Ltd.; International Atomic Energy Agency; Marcel Dekkar, Inc.; Elsevier Science/Pergamon; Thieme Verlag; MacMillan Magazine Ltd.; and to the publishing board of Phytoparasitica for granting us permission to publish part of the figures and tables presented in this book.

This book would have remained an unfinished manuscript if it were not for Prof. Bruno Yaron of ARO, Israel. At a crucial point, guidance from Prof. Ephraim Cohen of the Hebrew University at Rehovot was much appreciated. Deep appreciation goes to Mr. P. Gibson, Mrs. T. Nersesian, Mrs. G. Ramos, Mrs. Yuko Sato and Mr. and Mrs. H. Ward who also provided much practical support, as well as to many other personal friends who gave a helping hand from time to time. We are grateful to Mr. Zohar Zuk-rimmon of the Institute of Nature Conservation of the Tel Aviv University, who worked with Al Perry and supported him in writing the first draft, as well as to many other friends and colleagues of the Institute of Nature Conservation of the Tel Aviv University.

Contents

CHAPTER 1

Introduction

Since the dawn of civilization, insects have been plaguing human habitats as we see from the vivid descriptions of the plagues of insects recorded in ancient literature, such as the Bible. The Bible mentions plagues of locusts (Exodus 10: 12–15; Joel 1: 4; 2: 1–11), gnats (Exodus 8: 16, 17), gadflies (Exodus 8: 21–24), the olive fruit fly, the grape berry moth (Deuteronomy 28: 39, 40), and the clothes moth (Isaiah 50: 9; James 5: 2).

In order to combat the situation, man has come up with various methods, from magic and quackery to alchemy. Notwithstanding, from the early ages, rational pest control using crude chemical preparations has also been devised and put to use. Therefore, the history of insecticides may go back as far as 2500 B.C., when Sumerians already knew sulfur to have acaricidal and insecticidal properties. In China, about 1200 B.C., chalk and wood ash were used to control insects in enclosed spaces, and plant extracts were used for treatment of stored grain. In addition, arsenic sulfide was used to control human lice.

It is widely accepted that the ancient Greeks and Romans utilized sulfur, fumigants, oil sprays, oil and bitumen sticky bands, oil and ash, and other preparations for insect control. Progress with insecticides came with the introduction of the botanicals, such as pyrethrum, derris, quassia, and tobacco leaf infusion around the 16th century, but this may have been the extension of a usage which may have originated thousands of years earlier. In Japan, whale oil was used to control insects in rice paddies.

An inventory of insecticides used in the 19th century would include sulfur, arsenicals, fluorides, soaps, kerosene and various botanicals, of which nicotine, rotenone, pyrethrum, sabadilla, and quassia appear to have been most widely used. The first four decades of the 20th century saw significant progress in the synthesis of insecticides and the standardization of techniques such as the bioassay and the exploration of the relationship between chemical structure and biological activity. The insecticidal properties of DDT were discovered by Paul Müller of J.R. Geigy Company in 1939. Its first important use was to control malaria and typhus by the Western Allies during World War II. This marked the era of the chlorinated hydrocarbon insecticides, with the subsequent synthesis of hexachlorocyclohexane (HCH) and the cyclodiene compounds. These chlorinated hydrocarbon insecticides were generally persistent and had long residual properties. They were welcomed at the beginning, but their stability and hydrophobicity resulted in the contamination of the environment and bioconcentration in the body of many animals; thus they weve restricted in use or banned later on.

The Germans, on the other hand, experimented with the synthesis of organophosphorus compounds to replace nicotine. Three of these compounds, HETP, parathion and Schradan, attributed to Gerhard Schräder, were subsequently used on a worldwide scale despite their high toxicity to mammals. They were not persistent, and, in attempts to lower the mammalian toxicity and increase the efficacy, hundreds of other organophosphorus insecticides have since been synthesized.

The existence of another class of insecticides, the carbamates, was foreshadowed by Swiss workers in 1940, but the first major success was the introduction of the American insecticide carbaryl in 1950. It was followed by many other carbamates. The period between 1949 and the early 1970s saw the development of a number of synthetic pyrethrin analogs. The initial synthetic pyrethroids, represented by allethrin and resmethrin, were only for household use due to their low stability under field conditions. Between 1970 and 1977 a group of light-stable pyrethroids, represented by permethrin and deltamethrin, were introduced for agricultural use, and recent developments have added a variety of useful compounds.

Insecticide development has been guided mostly by chemorational design, but gradually the biorational design, based on a basic understanding of physiology and ecology of insects and crops, has been introduced and new types of compounds with different modes of action have been added to the plethora of insecticides. These include the insect growth regulators (IGR), such as chitin synthesis inhibitors, cyromazines, juvenile hormone mimics, ecdysone agonists, pymetrozine, and other novel agents such as imidacloprid, formamidines, avermectins, ***Bacillus thuringiensis***, and pheromones. Some of them are not conventional biocides and are variously named as insect growth regulators, insectistatics and behavioral regulators. As a whole, pesticides might be renamed bioregulators or bioregulants.

Since 1976 herbicides have represented the largest volume of pesticide production, followed by insecticides and fungicides. Current world consumption of pesticides is 43% herbicides, 32% insecticides, 19% fungicides, with the remaining 6% divided between growth regulators and miscellaneous agrochemicals. It is estimated that a steady increase in the use of crop protection chemicals will continue, so that two times the present level of consumption will be reached by the end of this century. There is no doubt that man has come to depend heavily on chemical insecticides, thus ignoring other approaches (biological and other nonchemical) to insect control. Together with other technologies, insecticides have contributed to the explosive growth of the human population from 2 billion in 1930 to about 6 billion at present. It is further estimated that a two fold increase in food production will be needed by the end of this century to feed the world's expanding population, which will probably require a large increase in pesticide use unless alternative methods, such as Integrated Pest Management (IPM), and other strategies are instituted on a global scale.

Whereas IPM methods can be evaluated on a cost/benefit basis for pest control in agriculture, when the objective of environmental protection is added, economic considerations in the abatement of insect pests related to human health become meaningless. In an epidemic, the control of insect vectors becomes crucial; however, it is not possible to express the effect of insecticides in cost/ benefit terms where human lives are concerned. Today, pesticides are used with regulations and limits to insure the unnecessary contamination of foodstuffs, the environment, and to protect human health. The use of insecticides is not an ecological sin if used judiciously and with common sense. It is the abuse of these chemicals that creates the hazard.

CHAPTER 2

Function and Classification

The term pesticide is an all-inclusive word meaning killer of pests (the ending "cide" comes from the Latin "cida", meaning killer). Pesticides are legally classed as economic poisons and are defined as any substance used for controlling, preventing, destroying, repelling, or mitigating any pest. According to this classification, many chemicals such as attractants, repellents, chemosterilants, hormonal agents, etc., are also designated as pesticides, although from a toxicological viewpoint they may not be directly involved in killing the pest (Table 2.1). Recently, terms like agrochemicals or bioregulators have been proposed to describe pesticides. The main classes of pesticides are shown in Table 2.2.

Insecticides may be classified in several ways:

1. According to their method of entry into the body, such as stomach poisons, contact poisons, and inhalation poisons or fumigants. This older classification has some limitations since, at present, there are many compounds which can act both as stomach and contact poisons, and others which can act in all categories.
2. According to their mode of action.
3. Based on their chemical structures.

In this book the classification of insecticides will be based mostly on their chemical structure (Table 2.3) and toxicity; mode of action and metabolism will be discussed under each group.

Table 2.1. Some chemicals legally classed as insecticides, but not ending in -cide and not necessarily killing the pest

Chemical	Function
Attractants	Attract insects
Chemosterilants	Sterilize insects, birds, or rodents; prevent reproduction
Insect growth regulators	Stimulate or retard growth of insects; interrupt normal development
Pheromones	Released by one individual and affect the physiology or behavior of another individual of the same species (e.g., females emit it to call males for copulation)
Repellents	Repel insects, mites and ticks, or pest vertebrates such as dogs, rabbits, birds, deer; prevent damage by rendering the commodity unattractive, unpalatable or offensive

Table 2.2. List of pesticide classes, their function, and derivation

Pesticide Class	Function	Derivation[a]
Acaricide	Kills mites	Gr. *akari*, mite or tick
Algicide	Kills algae	L. *alga*, seaweed
Avicide	Kills birds	L. *avis*, bird
Bactericide	Kills bacteria	L. bacterium; Gr. a staff
Fungicide	Kills fungi	L. fungus; Gr. mushroom
Herbicide	Kills weeds	L. herba, annual plant
Insecticide	Kills insects	L. insectum, divided into segments
Molluscicide	Kills snails and slugs	L. mollusca; soft or thin-shelled
Nematicide	Kills nematodes	L. nematoda; Gr. nema, thread
Ovicide	Kills eggs	L. ovum, egg
Piscicide	Kills fish	L. pisces, fish
Rodenticide	Kills rodents	L. rodere, to gnaw

[a]Gr., Greek; L., Latin.

Table 2.3. Classification of insecticides and related pesticides

Group	Examples
Organochlorines	DDT group such as DDT, DDD (TDE), methoxychlor; HCH; toxaphene; cyclodienes such as aldrin, dieldrin, heptachlor, heptachlor epoxide, isodrin, endrin, chlordane, mirex
Organophosphates	Phosphates such as dichlorvos, monocrotophos, naled, tetrachlorvinphos; phosphorothionates such as parathion, methyl parathion, fenitrothion, diazinon, chlorpyrifos, pirimiphos-methyl, fenthion; phosphorothiolates such as profenophos; phosphorothiolothioates such as malathion, dimethoate, prothiophos, azinphos-methyl; phosphonates, phosphonothionates and phosphonothiolothionates such as trichlorfon, EPN, fonofos; phosphoroamidates such as monitor, acephate; salithion
Carbamates	Aryl carbamates such as carbaryl, propoxur, carbosulfan, aminocarb, BPMC; oxime carbamates such as aldicarb, methomyl; N, N-dimethyl carbamates such as pirimicarb
Pyrethroids	Natural pyrethrins; synthetic pyrethroids such as allethrin, tetramethrin, resmethrin, phenothrin, permethrin, cypermethrin, deltamethrin, fenvalerate, fluvalinate, etophenprox, tefluthrin
Nitrogenous	Nicotine; neonicotinoids such as imidacloprid, nitenpyram, acetamiprid; cartap, bensultap, thiocyclam; chlordimeform, amitraz
Botanicals	Nicotine; pyrethrins; rotenone; neem
Microbial toxins	BT toxins; avermectins, milbemycins; polynactins
Inorganics	Arsenic, sodium fluoride
GABAergics	Avermectins, milbemycins; fipronyl; HCH; cyclodienes
Respiratory inhibitors	Electron transport inhibitors such as rotenone, fenazaquin, pyridaben, fenpyroximate, tebufenpyrad, hydrogen cyanide, phosphine; uncouplers such as dinitrophenols, chlorfenapyr, PCP
Insect growth regulators	Juvenile hormone mimics (juvenoids) such as methoprene, pyriproxyfen; ecdysone regulators such as tebufenozide; benzoylphenyl ureas such as dimilin, chlorfluazuron; buprofezin; cyromazine

(Contd)

Group	Examples
Chemosterilants	Alkylating agents such as apholate, thiotepa, metepa; anti-metabolites such as 5-fluorouracil; antibiotics such as mitomycin, cycloheximide; colchicine; triphenyltin; urea, thiourea; s-triazine compounds
Attractants	Cue-lure, siglure, methyl eugenol; pheromones such as codlelure, looplure, grandlure, riblure, gossyplure
Antifeedants	Organotine compounds, neem, pymetrozine
Synergists	Piperonyl butoxide, sesamex; MGK-264, SKF-525A
Fumigants	Hydrogen cyanide, phosphine (from aluminum phosphide); methyl bromide, ethylene dibromide, ethylene oxide, chloropicrin, sulfuryl chloride, methyl isothiocyanate (from metam sodium)
Nematicides	Methyl bromide, ethylene dibromide, ethylene oxide, chloropicrin, D-D (a mixture of 1,3-dichloropropene and 1,2-dichloropropane), 1,2-dibromo-3-chloropropane (Nemagon), methyl isothiocyanate (from matham-sodium-dazomet or from ammonium *N*-methyldithio-carbamate); carbamates such as aldicarb, oxamyl; organophosphorus such as diamidafos (Nellite), fenamiphos (Nemacur), thionazin
Miticides	Diarylcarbinols such as dicofol; organic sulfur compounds such as chlorfenson, tetradifon; phenolics such as binapacryl; organotin compounds such as cyhexatin, fenbutatin oxide; antibiotics such as avermectins, milbemycins, tetranactin; hexythiazox; pyridaben, fenpyroximate, tebufenpyrad; miticidal pyrethroids such as halfenprox
Molluscicides	Metaldehyde, clonitralid, tiphenyltin acetate, PCP
Avicides	4-Aminopyridine, 3-chloro-*p*-toluidine, strychnine
Rodenticides	Red squill, strychnine, norbormide; phosphorus, barium carbonate, arsenic trioxide, zinc phosphide, thallium sulfate; anti-coagulants such as warfarin, indandione-type compounds (chlorophacinone); α-naphthyl thiourea (ANTU), pyriminyl

CHAPTER 3

Formulations of Insecticides

The majority of insecticides are applied as sprays, dusts, granules, microcapsules, vapors, aerosols, or seed dressings. After an insecticide is manufactured in a relatively pure form (technical grade), it must be formulated before it can be applied. Formulation is the processing of a compound by such methods that will improve its properties of storage, handling, application, effectiveness and safety to the applicator and the environment, and profitability. Formulation is the final physical condition in which the insecticide is sold commercially. In most cases, it must be diluted according to the formulator's instructions before use. The price for a given weight of chemical depends largely on the type of formulation, the most expensive being the pressurized aerosol.

3.1 Common Formulations

Emulsifiable Concentrates (EC). More than 75% of all insecticide formulations are applied as sprays. The majority of these are water emulsions prepared from emulsifiable concentrates, which are solutions of the technical-grade material in an appropriate organic solvent with enough emulsifier added to allow the concentrate to mix readily with water for spraying. When an emulsifiable concentrate is added to water, the emulsifier causes the oil to disperse uniformly throughout the water phase, giving an opaque or milky appearance when agitated. This oil-in-water suspension is a normal emulsion. A few formulations are invert emulsions, that is, water-in-oil suspensions. These are opaque and thick, concentrated emulsions resembling salad dressing or thick cream, and are employed almost exclusively as herbicide formulations, because they result in little drift.

Emulsifiable concentrates, if properly formulated, should remain stable without further agitation for several days after dilution with water. If a precipitate forms after 24 h, a small amount of detergent or emulsifier can be added (about four tablespoons per liter of concentrate) and mixed thoroughly to improve its quality.

Most insecticides sold to the homeowner are formulated as emulsifiable concentrates and have a shelf life of approximately 3 years.

Water-Miscible Liquids. The technical-grade material is miscible with water or alcohol. These resemble emulsifiable concentrates in viscosity and color, but they do not form milky suspensions when diluted with water. Very few insecticides are formulated in this manner. They are not safe to use and should be applied only by experienced persons.

Wettable Powders (WP). Wettable powders are concentrated dusts containing a wetting agent to facilitate mixing the powder with water before spraying. The technical material is added to an inert diluent, such as talc or fine clay, in addition to a wetting agent or surfactant and mixed thoroughly in a ball mill. Without the wetting agent, the powder will float when added to water and the two phases will not mix. Because wettable powders contain 50–75% talc or clay, they tend to settle rather quickly to the bottom of the spray tank unless the mixture is agitated constantly.

Advantages of wettable powders are:

1. They do not cause damage to materials or places that are sensitive to organic solvents.
2. They do not dissolve washers and rubber hoses in spray machines or spray tanks.
3. They leave effective residues in cracks and crevices where sprays cannot penetrate.
4. They are not phytotoxic when sprayed on foliage even at high concentrations, in contrast to oil-containing sprays, which can cause foliage burning at temperatures above 32 °C.

Disadvantages of wettable powders:

1. They leave spots in areas sensitive to water spots, especially when they contain dyes.
2. They cause corrosion of valves, nozzles and pumps in sprayers.
3. They require constant agitation.

Water-Soluble Powders. Water-soluble powders contain a finely ground water-soluble solid which dissolves readily upon the addition of water. They may contain a small amount of wetting agent to assist their solution in water. Unlike wettable powders, they do not require constant agitation and form no precipitate. There are relatively few insecticides of this kind.

Oil Solutions. Oil solutions are formulated by dissolving the insecticide in an organic solvent for direct use in insect control. They are rarely used on crops because they can cause severe burning of foliage. They are used effectively on livestock, as weed sprays along roadsides, in standing pools for mosquito-larvae control, and in fogging machines for adult mosquito control. The concentrated solutions may be diluted with kerosene or diesel oil before application.

Flowable Powders. Some insecticides are soluble in neither oil nor water. In this case the technical material is wet-milled with a clay diluent and water with a suspending agent, a thickener and an anti-freeze compound thus forming a thick creamy pudding-like mixture which mixes well with water but needs frequent agitation.

Aerosols. Aerosols are the most common of all formulations for home use. To produce an aerosol, the active ingredients must be soluble in volatile petroleum oil under pressure. The pressure is provided by a propellant gas. When the solvent is atomized, it evaporates readily, leaving behind small droplets of the insecticide suspended in air. In the past, fluorocarbon was used as a propellant gas, but because of its effect on the ozone layer, it has been replaced by carbon dioxide and other gases. Aerosols are used for the knockdown and control of flying insects and cockroaches, but they provide no residual effect. Total release aerosols have been introduced in the last few years, which discharge the entire content of the container in a single application while the occupants of the house leave for a few hours. Upon their return, all windows must be opened to aerate the building. Aerosols produce droplets well below 10 μ, which are readily absorbed by alveolar tissue in the lungs. Therefore, caution must be taken when they are used.

Granulars. Granulated formulations consist of small pellets (20–80 mesh) containing 5–25% of the active ingredient sprayed onto a clay and allowing the solvent to evaporate. These formulations are used mainly in agriculture for soil applications as systemic insecticides to be absorbed by the roots of a plant and transported to other parts.

Fumigants. Chemical fumigants are gases or volatile liquids of low molecular weight which readily penetrate the material to be protected. In most cases they are formulated as liquids under pressure and are stored in cans or tanks. When released in open air, the liquid changes back into gas. Fumigants are used for the control of insects in stored products but are also important for soil sterilization as a means of killing soil insects, nematodes, weed seeds, and fungi.

Ultra-Low Volume Concentrates (ULV). These usually contain a technical product dissolved in a minimum amount of solvent. They are applied without further dilution by aerial or ground equipment in volumes of 0.6 l to a maximum of 4.7 l/ha in very small spray droplets of 1–15 μ. The advantage of ULV sprays is that the small droplets can better penetrate thick vegetation and other barriers. These formulations are sold only for commercial use where insect control is desired over large areas and a high volume of water would constitute a technical difficulty.

Fogging Concentrates. These are used in the control of adult flies and mosquitoes and sold for public health use to pest control operators. Fogging machines generate droplets of 1–10 μ. There are two types of fogging machines: (1) Thermal fogging machines, which utilize flash heating of the oil solvent to produce vapor or fog; and (2) ambient fogger-type machines, which atomize a small jet of liquid into a venturi tube, through which passes an ultra-high velocity air stream.

Thermal foggers utilize oil only, usually deodorized kerosene. Ambient foggers use emulsions, water-base or oil-base formulations.

Dusts. Dusts are the simplest of all formulations and the easiest to apply. An example is sulfur dust, which is used on ornamental plants. Usually the toxic dust is mixed with an inert diluent such as talc or pyrophylite. Dusts are neither very effective and nor economical. They are rarely used in public health pest control.

Impregnating Materials. These are chemicals used in the treatment of woolens for moth-proofing and timber against wood-destroying organisms.

Poison Baits. Formulated baits contain low levels of toxicants incorporated into materials such as foodstuffs, sugar, molasses, etc. that are attractive to the target pest. These can be purchased commercially or they can be formulated by individuals or by pest control operators.

Slow-Release Insecticides. These formulations are relatively new and only a few are available commercially. An example is the Shell-No-Pest Strip, which incorporates dichlorvos (DDVP) into strips of polychlorovinyl resin. The insecticide volatilizes slowly, killing flying and crawling insects over a long period of time.

Other forms of slow-release insecticides, such as microencapsulated diazinon, methyl parathion, fenitrothion, pyrethrins plus synergists and permethrin, have recently been introduced. Other types of formulations include a lacquer preparation of chlorpyrifos, which can be applied with a paint brush; adhesive tapes containing propoxur, diazinon or chlorpyrifos; and briquettes containing the insect growth regulator methoprene for mosquito larva control. Other slow-release formulations are introduced from time to time as the technology of producing such formulations improves.

The method of slow release involves the incorporation of the insecticide in a permeable covering, microcapsules or small spheres with diameters ranging from 15–50 μ. The insecticide escapes through the small sphere wall at a slow rate over an extended period of time, thus preserving its effectiveness much longer.

CHAPTER 4

Toxicology of Insecticides

Toxic interaction of a chemical with a given biological system is dose-related. Hence, toxicology can be termed the *science of doses.* The toxicity of an insecticide to an organism is usually expressed in terms of the LD_{50} (*lethal dose*), that is, the amount of poison per unit weight of the organism required to kill 50% of the test population. The LD_{50} is usually expressed in milligrams per kilogram of body weight (mg/kg). Under certain conditions, the term micrograms per insect (μg/insect) may be used when the chemical is applied topically to the insect.

The LC_{50} (*lethal concentration*) is the concentration of the chemical in the external medium required to kill 50% of the test population. This value is used when the exact dose given to the insect cannot be determined. For example, the toxicity of an insecticide to mosquito larvae or to fish is usually assessed by the concentration of the insecticide in water that will result in 50% mortality of the organisms which are exposed for a specified period of time. This is expressed as the percent of active ingredient of the chemical in the medium or as parts per million (ppm).

The term LT_{50} (*lethal time*) represents the time required for 50% mortality of the test organisms at a specified dose or concentration. This method is used when the number of test individuals is limited and is often employed in field tests where it is impossible to collect sufficient numbers of insects for testing.

In some instances, the rate of knockdown rather than kill is measured as a criterion of toxicity. In such cases, the KD_{50} (*knockdown dose*) or the KT_{50} (*knockdown time*) is used.

In other situations, killing or knockdown does not constitute the desired criterion. For example, in testing insect growth regulators or chemosterilants, the object is not to kill the organisms but rather to affect their developmental stages, fecundity, and egg viability. In such cases, the ED_{50} (*effective dose*) or EC_{50} (*effective concentration*) is used to describe the results of such tests.

CHAPTER 5

Methods of Testing Chemicals on Insects

There are several ways of administering a chemical to an insect. A commonly employed method is topical application, where the insecticide is dissolved in a relatively nontoxic solvent, such as acetone, and small, measured droplets are applied at a chosen location on the body surface. Topical application as is practiced today was made possible by the invention of the micrometer-driven syringe (Trevan 1922) and the use of continuous carbon dioxide anesthesia (Williams 1946). The advantages of this method are:

1. The high degree of precision and reproducibility that can be attained.
2. The large number of tests that can be performed in a relatively short time.
3. The small number of insects (10–20) required per replication.
4. The simple and inexpensive equipment needed.
5. The very small amount of chemicals and solvents used.
6. The fact that the LD_{50} values obtained for any species are reasonably constant and reproducible from laboratory to laboratory, provided that identical conditions of testing are maintained.

The results are expressed as micrograms of active ingredient per insect (μg AI/insect) or micrograms AI per gram insects (μg AI/g insects).

5.1 Injection Method

When knowledge of the exact amount of insecticide inside the body of the insect is needed, the injection method is used. For this, very fine stainless steel needles of 27 or 30 gauge (0.41 or 0.30 mm in diameter) are used. Small glass needles of 0.1–0.16 mm in diameter may be used for injection to small insects. The insecticide is commonly dissolved in propylene glycol or peanut oil and injection is made intraperitoneally (into the body cavity). Care must be taken to avoid bleeding by the insect.

5.2 Dipping Method

This method is employed when topical application or injection are impractical, for example, with small plant-feeding insects, stored-product insects, housefly larvae, insect eggs, red spiders, etc. The insects are dipped in aqueous solutions, emulsions, or suspensions of the chemical for short periods of time. In this case, the LC_{50} is used to express the results.

5.3 Contact or Residual Method

The insecticide in a volatile solvent is applied to a glass container such as a vial or a jar. The solvent is allowed to evaporate by rotating the container so that the insecticide is spread evenly over the entire surface leaving a residual film. Alternatively, the insecticide is applied evenly on a glass, filter paper, wood panel or other types of building materials and allowed to dry before exposing the insects to the residual deposits. The deposits are expressed as milligrams or grams of active ingredient per square meter (mg or g AI/m^2).

5.4 Feeding and Drinking Method

These methods are used to evaluate the toxicity of ingested chemicals. They may be classified as unlimited availability of food or drink or as limited dose feeding. Unlimited feeding includes:

1. Textile tests for moth-proofing.
2. Treated flour or grain.
3. Media for plant-boring insects.
4. Media for fly larvae.
5. Sprayed or dusted foliage.
6. Poison baits.

Limited dose feeding include: coated leaves (disks, sandwiches, squares, strips) and pellets. Drinking methods include: sugar syrups and drinking through membranes (such as plant juices or blood). In addition, there are methods for screening chemical attractants and repellents; for screening animal sprays, dusts, dips and dressings; and techniques for evaluating systemic insecticides against livestock insects.

CHAPTER 6

Evaluation of Toxicity in Insects

The susceptibility of an insect population to a poison is assessed by constructing a dosage-mortality curve in which the dosage is plotted against the percentage mortality at a given period of time. Such a plot produces a sigmoid curve whose asymptotic approaches at the regions of zero and 100% mortality are difficult to define without extensive testing. The use of probit transformation in which the sigmoid curve is converted to a straight line by plotting the logarithm of the dosage against the probit value of percent mortality (Log-Probit paper is available commercially for making these plots) is recommended. This method of computation yields a straight line which greatly facilitates the determination of the LD_{50} and other values on the plot (Fig. 6.1). Detailed discussions of this method are

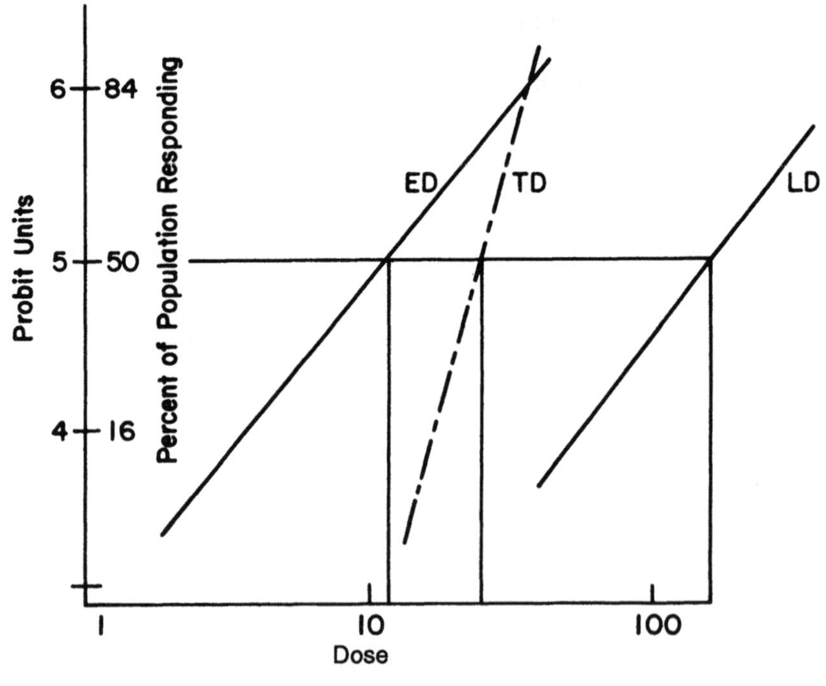

Fig. 6.1. Probit transformation. Comparison of effective dose (*ED*), toxic dose (*TD*), and lethal dose (*LD*). Plot of log dose versus percent of population responding and a translation into probit units. (Casarett and Doull 1975)

available in specialized books (Finney 1949, 1952; Swaroop and Uemura, 1956; Busvine 1971).

If there is natural mortality in the controls, adjusted mortality is used according to Abbott's formula (Abbott 1925) as follows:

$$\text{Corrected Mortality } \% = \frac{P - P_0}{100 - P_0} \times 100$$

where P is the percent mortality of treated insects and P_0 is the percent mortality of insects in the untreated control. This adjusted value is permissible when mortality in the controls does not exceed 20% or when mortality is based on a large number of replications.

CHAPTER 7

Evaluation of Toxicity in Higher Animals

Evaluation of toxicity in higher animals is different from that of insects in that the number of available animals for testing usually is limited. While the process for determining LD_{50} is identical, greater emphasis is placed on qualitative rather than quantitative aspects of poisoning. Another characteristic of toxicological tests in higher animals is that, in most cases, the overriding concern is the evaluation of safety for man.

Selection of the test animal is usually based on convenience and cost. For ordinary testing of LD_{50} values rats or mice are the animals of choice. The animals should be healthy and of acceptable genetic homogeneity. Factors influencing toxicity include: duration of exposure; route of administration; species; individual variation; age; sex; population density temperature; and nutrition (Hayes and Laws 1991).

7.1 Acute Toxicity

This toxicity is defined as the toxicity that causes injury or death from a single or multiple exposures over a short period of time, usually 24 h. The most common route of administration of toxicants is the oral route. There are two methods of oral administration: one is to add the insecticide to the animal's diet; the other is to give the chemical directly through a stomach tube or syringe equipped with a 19-gauge stainless steel needle about 10 cm long with a spherical bead of silver solder about 2 mm in diameter at the tip to prevent perforation of the esophagus. The vehicle used for the administration of a toxicant with a stomach tube is important and should cause no ill effects when given to control animals. Vegetable oil and polyethylene glycol are the solvents of choice, and the volume should not exceed 2–3% of the animal's body weight. When many oral doses are to be given, it is preferable to mix the chemical into the animal's feed in such concentration that they will consume it at the desired rate. Generally, a volatile solvent and vegetable oil are used in this process. Sometimes, prior fasting is used to promote ingestion of food, but this method is not recommended for experimental work.

Other methods of administration of toxicants include intraperitoneal, intravenous, intramuscular and subcutaneous injections. The intravenous route is the fastest acting, introducing the chemical directly into the blood stream, thus eliminating the process of absorption. The volume of the injected material should not exceed 0.1–0.5 ml for rodents.

The intraperitoneal route results in rapid absorption of the toxicant due to the rich blood supply of the peritoneal cavity and the large surface area. Compounds administered by this route are absorbed primarily through the portal circulation and, therefore, must pass through the liver before reaching other organs (Casarett and Doull 1975). Toxicants administered intramuscularly and subcutaneously are absorbed at a slower rate, especially if given in an oil solution.

From the standpoint of the safety of pesticides to pest control operators and agricultural workers, the acute dermal toxicity and acute inhalation toxicity are the most important toxicity tests. Dermal toxicity tests are usually performed on rabbits. A rabbit weighing 2–3 kg is shaved around the abdomen and the back, and the chemical is painted over the area, which is then covered with a rubber sleeve or cotton gauze held in place with wire screen. Toxicity studies of this sort with organochlorine, organophosphorus, and carbamate insecticides have been reported by Gaines (1960), Hayes et al. (1964), and others.

Acute inhalation toxicity tests can be conducted by using either static or dynamic air flow systems. In the static test, the animal is kept in a closed chamber and is exposed to one or more spray or aerosol applications. In the dynamic system, the animal is kept in a chamber which is continuously supplied with an air current containing a constant amount of the insecticide. The different devices used in the inhalation tests have been described by Gage (1970), and methods of vaporization, atomization, etc. have been discussed by Lehman et al. (1964).

The respiratory volume for a man doing light work is estimated at 36 m^3/day. Hartwell et al. (1964) found the ratio of parathion toxicity to man to be: inhalation 10, oral 3, and dermal 1.

Inhalation toxicity is usually expressed as milligrams of active ingredient per cubic meter of air space (mg/m^3). Threshold limit values (TLVs) refer to airborne concentrations of substance and represent conditions under which it is believed that nearly all workers may be repeatedly exposed, day after day, without adverse effect. The time-weighted average TLVs for a normal 8-h workday and a 40-h workweek are given in Table 7.1. Listed substances followed by the designation "skin" refer to the potential contribution to the overall exposure by the cutaneous route including mucous membranes and eye, either by airborne or more particularly, by direct contact with the substance.

7.2 Chronic Toxicity

Chronic toxicity involves a multi-dose exposure over a prolonged period of time, usually 6–18 months for tests with larger animals, or exposure for the duration of the lifespan of smaller test animals, such as mice. In true chronic toxicity tests, the purpose is to find any undesirable delayed harmful effects of the chemical. The doses given are generally high, and clinical signs of illness and gross physiological and pathological changes are carefully watched for and recorded.

Chronic toxicity may lead to secondary harmful effects such as carcinogenicity, mutagenicity, and teratogenicity.

Table 7.1. Adopted threshold limit values for selected pesticides and other compounds (American Conference of Governmental Industrial Hygienists 1989–1990)

Substance	TLV (mg/m^3)	Substance	TLV (mg/m^3)
Aldrin—skin	0.25	Fonofos—skin	0.1
Amitrole	0.2	Heptachlor—skin	0.5
Arsenic and soluble compounds, as As	0.2	Hexane	176
Azinphos-methyl—skin	0.2	Hydrogen chloride	7.5
Benomyl	10	Hydrogen cyanide—skin	11
Bromacil	11	Lead arsenate as $Pb_3(AsO_4)_2$	0.15
Bromine	0.66	Lindane—skin	0.5
Camphor	12	Malathion—skin	10
Captafol—skin	0.1	Methomyl	2.5
Captan	5	Methoxychlor	2.5
Carbaryl	5	Methyl bromide—skin	19
Carbofuran	0.1	Methyl demeton—skin	0.5
Carbon disulfide—skin	31	Methyl isothiocyanate—skin	0.047
Carbon tetrachloride—skin	31	Methyl parathion—skin	0.2
Chlordane—skin	0.5	Mevinphos—skin	0.09
Chloropicrin	0.67	Naled—skin	3
Chlorpyrifos—skin	0.2	Naphthalene	52
Cyanides as CN—skin	5	Nicotine—skin	0.5
Cyhexatin	5	Paraquat, respirable sizes	0.1
2, 4-D	10	Parathion—skin	0.1
DDT	1	Pentachlorophenol—skin	0.5
Demeton—skin	0.11	Phorate—skin	0.05
Diazinon—skin	0.1	Phosphine—skin	0.42
1,2-Dichloroethylene	793	Picloram	10
Dichlorovos—skin	0.90	Pindone	0.1
Dicrotophos—skin	0.25	Pyrethrum	5
Dieldrin—skin	0.25	Ronnel	10
Dinitro-o-cresol—skin	0.2	Rotenone (commercial)	5
Diquat	0.5	Sodium fluoroacetate—skin	0.05
Disulfoton	0.1	Starch	10
Diuron	10	Sucrose	10
Endosulfan—skin	0.1	Sulfotep—skin	0.2
Endrin—skin	0.1	Sulfuric acid	1
EPN—skin	0.5	2, 4, 5-T	10
Ethion—skin	0.4	Temephos	10
Ethylene dibromide—skin	2	TEPP—skin	0.047
Ethyl ether	1210	Thiram	5
Fenamiphos—skin	0.1	Toxaphene—skin	0.5
Fensulfothion	0.1	Vegetable oil mists	10
Fenthion—skin	0.2	Warfarin	0.1
Ferbam	10		

7.3 Carcinogenicity

The course of neoplasia is unknown, but the incidence can be increased by radiation, viral infection, defective immune response, and some chemicals. A number of pesticides have been reported to induce tumors in mice and rats in laboratory tests (International Agency for Research in Cancer 1979; Hayes 1982; Hayes and Laws 1991). DDT and other chlorinated hydrocarbon insecticides have been shown to cause marked changes in the liver of various rodents, and these changes progressed to tumor formation in some species, notably the mouse. There is disagreement as to whether the mouse tumors are malignant, but it is certain that they are peculiar to rodents and, therefore, interpretation of their significance for man and useful animals is difficult. However, DDT failed to produce detectable tumors in pesticide industry workers who absorbed DDT for 19 years or more at rates hundreds of time higher than those found in the general population (Laws et al. 1967).

7.4 Mutagenicity

For a long time man has introduced chemical compounds into the environment in ever increasing variety and quantity without knowing their potential hazard. Some of these chemicals, including pesticides, may have adverse effects on human genetic material.

Two major factors are responsible for the concern over genetic effects. The first is the concern for the protection of the human gene pool. The second is connected with oncogenesis. The ultimate relationship between the tumorigenic and genotoxic properties of chemicals makes genetic testing mandatory for chemicals of unknown oncogenic potential.

Another important factor in assessing the consequences of genetic effects is the location of the affected cell. If mutations occur in cells which are not part of the reproductive system (somatic cells), the resultant effect is only on the individual exposed to the genotoxic agent. If alteration occurs in gametes (sperm or ova), the change may affect subsequent generations (Brusick 1980). The mutation induced in the gametes of an exposed individual may show expression immediately if it is dominant, or it may show expression after several generations if it is recessive.

There are a large number of assay tests for mutagenesis, such as point mutations in bacteria, yeast and mold, mammalian cells in vitro, tests for chromosomal aberrations, etc. The Ames Test with *Salmonella typhimurium* is a popular assay procedure (Ames 1974; Ames et al. 1975) and is used extensively by many investigators and chemical industries to test the safety of new products before their introduction to the market.

7.5 Teratogenecity

The word teratogenesis is derived from the Greek *teras*, a monster, and *genesis*, origin. By general agreement, the word refers to deformities resulting from abnormalities of embryonic development. Such deformities can also be observed in rodent fetuses by direct inspection or with the aid of a microscope. Teratogenesis does not include toxic injury to organs after they are fully formed. A detailed review of teratology is given by Wilson (1972).

Another important aspect of the chronic toxicity tests is the NOEL (no observed effect level), which is defined as the maximal or near-maximal dosage level at which no difference from untreated or solvent-treated controls is detected. After considering all available data on the test compound, NOEL is determined for human safety by regulatory government agencies, including international organizations such as the World Health Organization (WHO) and the Food and Agricultural Organization (FAO) of the United Nations.

The NOEL is not too far below any harmful dose level as to be insignificant. For this reason, considering human safety, this level is divided by 100 or more; but in some cases this factor can be smaller and is expressed as the ADI (*Acceptable Daily Intake*), which means that such values should reflect the daily doses which can be taken during an entire lifetime and which appear to be without any appreciable risk. Hence, this term is employed in connection with long-term studies, and assessment is made mainly through food ingestion. More recently, NOAEL (no observed adverse effect level) has been used rather than NOEL.

Other criteria for establishing acceptable daily intake levels include:

1. Toxicity data for the most sensitive system in the most sensitive species to establish minimum-effect levels. Man is considered as equivalent to the most sensitive species.
2. Human data, if available, take precedence over any other animal data.
3. Minimum-effect levels are divided by 100 (the safety factor) to arrive at the ADI values.

Criteria for *effects* as recommended by Fitzhugh (1965) include:

1. Acute LD_{50} and LC_{50} data on experimental animals.
2. Chronic toxicity data on two or more species, for example, rat, 2 years; dog, 5 years.
3. Metabolism and storage studies, and determination of toxic and non toxic metabolites.
4. Biochemical studies, such as effects on enzymes and enzyme induction.
5. Reproduction studies, recommended for at least three generations.
6. Human studies where applicable, such as measurement of inhibition of cholinesterase by organophosphorus and carbamate insecticides.

In addition, delayed neurotoxicity, hypersensitivity, and allergic reactions must also be considered. Several examples of NOAEL and ADI levels of a few insecticides are shown in Table 7.2.

Table 7.2. Examples of ADI and NOAEL of some selected insecticides[a]

Common name	ADI (mg/kg body wt.)	Animal	NOAEL (mg/kg body wt.)	Exposure period
Organochlorines				
DDT	0.02	Rat	6.25	Life span
		Monkey	10	7 Years
		Human	0.25	25 Years
Dieldrin	0.0001	Rat	0.025	2 Years
		Dog	0.025	68 Weeks
Endosulfan	0.006	Rat	0.6	2 Years
		Mouse	0.84	2 Years
		Dog	0.57	1 Year
Lindane	0.008	Rat	0.75	13 Weeks
		Dog	1.6	2 Years
Organophosphorus				
Azinphos-methyl	0.005	Rat	0.86	2 Years
		Rat	0.48	2 Generations
		Mouse	0.88	2 Years
		Dog	0.74	52 Weeks
		Human	0.3	30 Days
Diazinon	0.002	Rat	0.07	99 Weeks
		Rat	0.5	Reproduction
		Dog	0.02	1 Year
		Rabbit	25	Teratology
		Human	0.025	34–36 Days
Fenitrothion	0.005	Rat	0.5	Embryo toxicity/ teratogenecity
		Dog	1.25	1 Year
		Human	0.08	
Malathion	0.02	Rat	5	2 Years
		Human	0.2	47 or 88 Days
Parathion	0.004	Human	0.05	3 Weeks
		Rat	0.25	2 Years
		Rat	0.05	Reproduction
		Rat	1	Developmental toxicity
		Mouse	7.5	90 Days
		Dog	0.008	6 Months
Pirimiphos-methyl	0.03	Rat	0.5	2 Years
		Rat	5	3 Generations
		Mouse	0.5	80 Weeks
		Dog	10	2 Years
		Human	0.25	28 Days
Carbamates				
Carbaryl	0.003	Rat	10	2 Years
		Rat	2	1 Year
		Dog	1.8	1 Year
		Human	0.06	6 Weeks
Carbofuran	0.002	Rat	1	2 Years
		Mouse	2.8	2 Years

(Contd)

Common name	ADI (mg/kg body wt.)	Animal	NOAEL (mg/kg body wt.)	Exposure period
Methomyl	0.03	Rat	2.5	90 Days
		Mouse	8.7	2 Years
		Dog	3.1	2 Years
Propoxur	0.02	Rat	10	2 Years
		Mouse	300	24 Months
		Dog	5	1 Year
		Human	0.2	
Pyrethroids				
Bioresmethrin	0.03	Rat (male)	3	104 Weeks
		Rat (female)	4	104 Weeks
		Dog	80	90 Days
Fenvalerate	0.02	Rat	7.5	104 or 119 Weeks
		Mouse	3.5	87 or 91 Weeks
Permethrin	0.05	Rat	5	2 Years
Others				
Buprofezin	0.01	Rat	0.9	24 Months
		Mouse	1.82	104 Weeks
		Dog	2	2 Years
Diflubenzuron	0.02	Rat	2	2 Years
		Mouse	2.4	91 Weeks
		Dog	2	1 Year
Methoprene	0.1	Rat	25	3 Generations
		Dog	12.5	90 Days

[a]Selected from Sekizawa (1997).

CHAPTER 8

The Nervous System as Insecticide Target

Most insecticides exert their lethal action by virtue of their effects on the nervous system. The reason for this lies in the special sensitivity of the nervous system, which shows irreversible damage, more so than any other tissue in the body. Poisons which affect other organs of the body commonly produce their ultimate effect on the nervous system. For example, atropine (a heart poison) and carbon monoxide (which blocks the oxygen carrying capacity of the blood) exert their lethal effects by depriving the brain's essential oxygen requirements, causing brain damage. In order to understand how insecticides exert their toxic effects, it is essential to have some fundamental understanding of the physiology and biochemistry of the mammalian and the insect nervous systems, their similarities and their basic differences. The general features of these systems are depicted in Fig. 8.1. Both systems have two important subdivisions in common, namely, the central nervous system and the peripheral nervous system.

The basic element in the nervous system is the nerve cell or neuron. This consists of a cell body containing the nucleus and long cytoplasmic projections (the axons), which extend to make contact with other neurons or muscle fibers. Nerve impulses are conducted from one cell to the next along the axons. Part of each neuron is specialized for the reception of stimuli. This part, which is called the dendrite, may arise directly from the cell body or may represent the distal endings of the axon. The junction with other neurons is called a synapse, while the junction with muscle fiber is called a neuromuscular junction or end plate. Aggregations of neurons form the ganglia.

8.1 The Mammalian Nervous System

The central nervous system is comprised of the brain and the spinal cord. It is composed of millions of nerve cells which are connected to each other by junctions called synapses. The brain is the integrating center of all body activities. The peripheral nervous system consists of the somatic and the autonomic systems.

The Somatic System. The somatic system handles that part of movement characterized by reactions to environmental stimuli and the corresponding muscle response. This system is composed of incoming and outgoing pathways. Detection

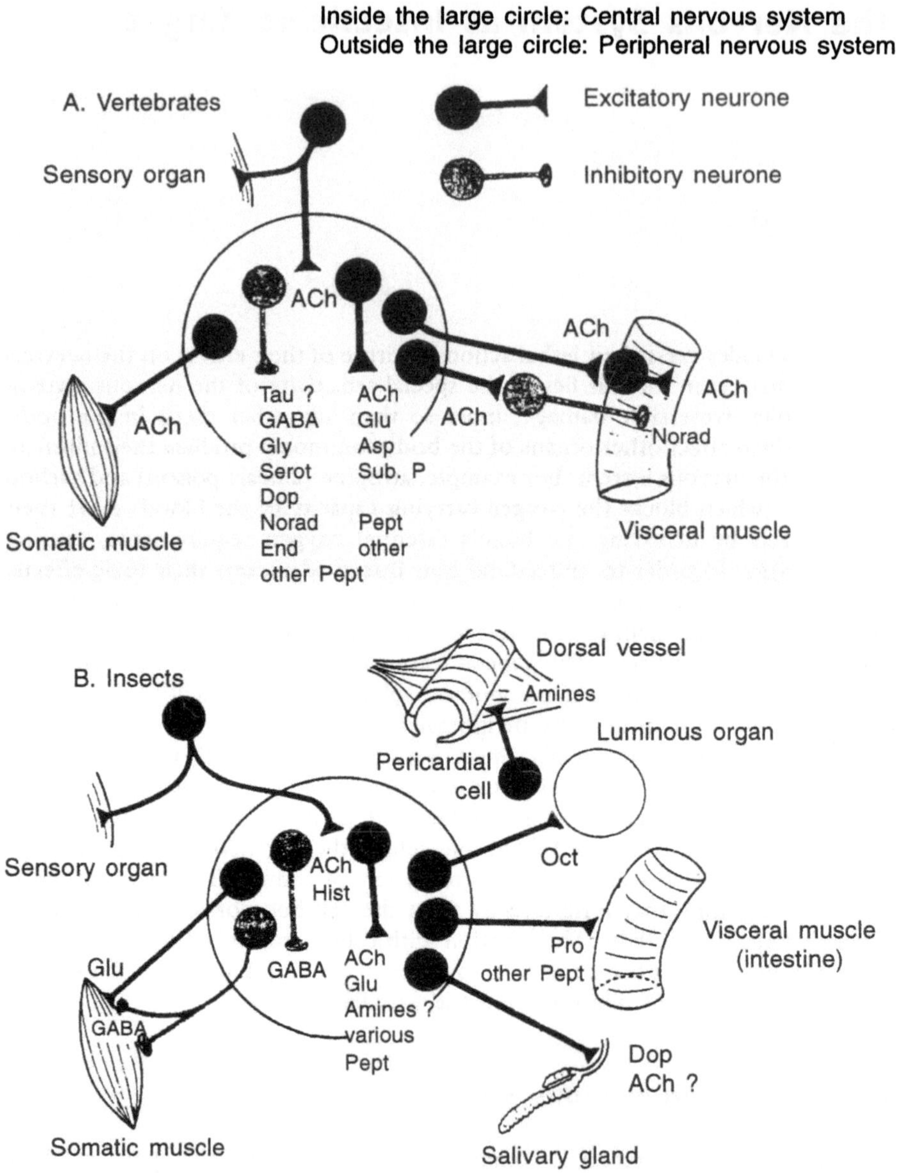

Fig. 8.1. Comparison of nervous systems and neurotransmitters between **A** vertebrates and **B** insects. *ACh* acetylcholine; *Asp* aspartic acid; *Dop* dopamine *End* endorphin; *GABA*, γ-aminobutyric acid; *Glu* glutamic acid; *Gly* glycine; *Hist* histamine; *Noradr* noradrenaline; *Oct* octopamine; *Pept* peptides; *Pro* Proctolin; *Serot* serotonin; *Sub P* substance P; *Tau* taurine. Prepared especially for this chapter by Dr. Daisuke Yamamoto, Mitsubishi Life Science Institute, Yokohama, Japan

of environmental changes is accomplished by sensory receptor cells which are directly connected to the central nervous system.

The stimulus received by a sensory cell can be transmitted either directly to a responding motor nerve (a simple reflex arc), or it can first be analyzed by the brain, which sends an order to the voluntary muscle through the motor nerve. Transmission of the stimulus across the synapse and at the neuromuscular junction is mediated by a chemical transmitter, acetylcholine. This is known as the cholinergic system. It is assumed that the transmitter combines with its receptor on the far side of the synapse to produce a configurational change, which alters the sodium ion (Na^+) permeability of the nerve membrane, thus triggering an action potential or an appropriate response. In order to restore the sensitivity of the synapse, the transmitter acetylcholine must be eliminated so that the receptor can return to its resting condition. At cholinergic junctions this is accomplished by the enzyme acetylcholine esterase, which breaks down acetylcholine to the inactive components, choline and acetate. With the help of other enzymes, choline acetylase and acetyl Coenzyme A, choline and acetate combine again to form new acetylcholine and the process is repeated.

The Autonomic System. The autonomic division of the nervous system innervates glands, cardiac muscle, all smooth muscle (involuntary muscle), including that of the digestive, respiratory and urogenital systems, blood and lymph vessels, and skin. In other words, this division innervates all the effectors of the body except skeletal muscle, which is voluntary muscle. Practically all effectors except striated muscle have a double nerve supply. On this basis, the autonomic system is subdivided into the sympathetic and parasympathetic systems, which are antagonistic in their action. For example, the sympathetic system accelerates the heart whereas the parasympathetic system retards it.

The sympathetic nervous system has a large chain of ganglia outside the central nervous system. The axon connecting the central nervous system to a ganglion is called ***preganglionic*** and the axon connecting the ganglion to a gland or organ is called ***postganglionic***. At the terminal of the postganglionic axon is a synapse which is bridged to a gland or organ. Transmission of the stimulus across this synapse is mediated by another chemical, noradrenaline (norepinephrine), and is called the adrenergic system.

In the parasympathetic system, though resembling the sympathetic system, the ganglia are usually located right beside the organ, and the junction between the postganglionic axon and the organ is cholinergic.

Although the cholinergic and adrenergic systems are the most important in the peripheral nervous system, there are many other chemicals which act as transmitters in the central nervous system. γ-Aminobutyric acid is one of them.

8.2
The Insect Nervous System

The basic elements of the insect nervous system are nerve cells forming long processes, or axons, along which nerve impulses are conducted. The bodies of nerve cells are aggregated to form ganglia.

Insect neurons, with the exception of sensory cells, are grouped together to form the brain and a series of ganglia. All ganglia are joined by a pair of connectives into a ventral chain that extends the length of the body. From this chain, nerves run to the peripheral sense organs, the muscle systems, and a stomatogastric system, which innervates the anterior alimentary canal.

Surrounding the central nervous system is a blood-brain interface that can maintain a constant internal ionic environment against the fluctuating one of the blood (Treherne and Pichon 1972). Unlike vertebrates and some other arthropods, no blood channels penetrate the insect's central nervous system. Instead, gaseous exchange is mediated by an extensive array of tracheae and tracheoles within the ganglia which, in turn, are supplied by a direct pathway from the spiracles (small openings on the surface of the cuticle). It has been suggested that insecticides applied to the surface of the cuticle penetrate the nervous system by diffusion through the tracheal route (Gerolt 1972).

The insect central nervous system is analogous to that of mammals. Acetylcholine was established as the transmitter at the insect central nervous system synapses, although γ-aminobutyric acid (GABA), glutamic acid, glycine and biogenic amines, such as dopamine, norepinephrine, serotonin, and tryptamine, do occur in the insect central nervous system (Gerschenfeld 1973; Kehoe and Marder 1976; Evans 1978) as in the mammalian central nervous system. Octopamine is unique to the insect system. Monoamine oxidase and diamine oxidase are also found in a variety of insects (Brown and Nestler 1985).

There are, however, some notable differences. For example: (1) no cholinergic system is involved in the insect's peripheral nervous system, but glutamic acid stimulates and GABA suppresses muscle contraction as transmitters at the neuromuscular junctions; (2) no distinct autonomic system exists; (3) insect ganglia are located mainly in the central nervous system and form the insect's brain.

There are also some morphological and histological differences. In contrast to the vertebrate nerve, insect nerve has no distinct myelination. The insect neurone exists in a protected environment and possesses an extremely effective blood-brain barrier, the nerve sheath (Hoyle 1953), which prevents penetration of water-soluble or polar compounds across the nerve membrane but allows the passage of compounds with favorable lipid/water partition coefficients, such as contact insecticides.

8.3 Conduction of Nerve Impulses Along the Axon

The axon is part of the neuron, which is specialized for carrying nerve impulses or action potentials rapidly and without any change in size or pattern of the impulses as it moves along. The extracellular fluid surrounding the axonal membrane contains a high concentration of sodium ions (Na^+) and a low concentration of potassium ions (K^+). The reverse is true for the inside of the nerve cell. At the resting stage (not conducting an impulse), its membrane is relatively permeable to K^+, but has very little permeability to Na^+. This situation makes the inside of the cell negative with respect to the outside, and the difference in potential is usually around-60 mV (resting potential).

A stimulation causes the axonal membrane to become permeable to Na^+ (sodium channel opening), which flows into the axon according to the electrochemical gradient and causes the inside to become transiently positive (rising phase of the action potential) (Fig. 8.2). However, the sodium channel starts closing quickly, usually within 1 msec (sodium inactivation), and the channel becomes substantially permeable to K^+ as a result of the opening of potassium channels, causing an efflux of K^+ according to their electrochemical gradient. This sodium inactivation and potassium activation generate the falling phase of the action potential. Thus, the axon is ready to generate another impulse. The action potential generates a local circuit current, which in turn stimulates the (depolarized) neighboring region of the axon to propagate impulse. This event passes down the axon. Thus, the impulse takes the form of a wave of depolarization along the axon.

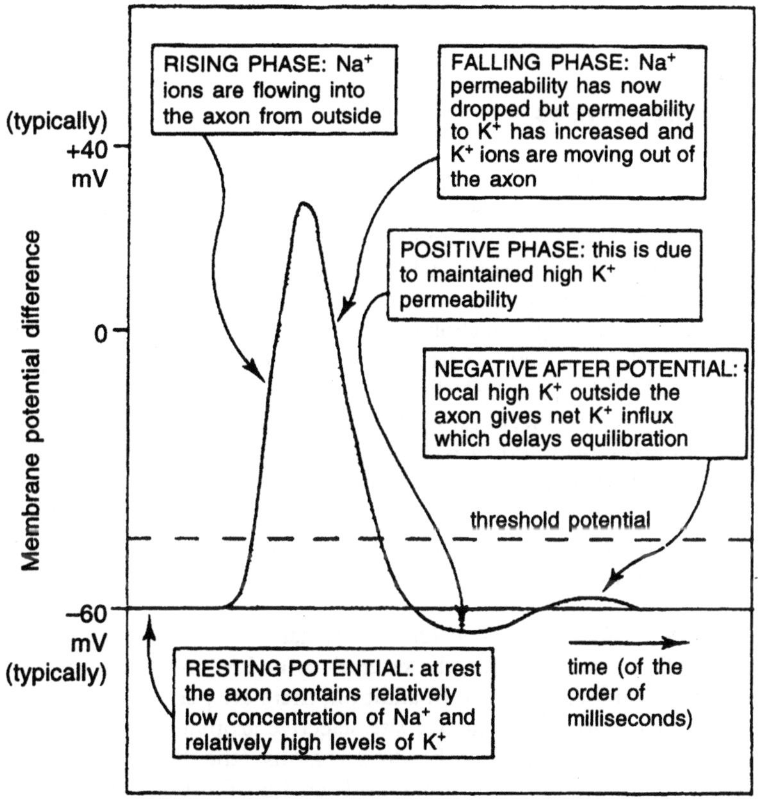

Fig. 8.2. An idealized action potential. Negative after potential used in this figure equal to depolarizing after potential mentioned in the text. (Corbett et al. 1984)

The ability to perform all these events depends upon the maintenance of ion gradients across the membrane. This is done by a metabolism-driven Na^+–K^+

pump, which maintains the gradients in the first place, and compensates for the leakage that occurs during impulse transmission. DDT and pyrethroids affect the axonal transmission.

8.4 Synaptic Transmission of the Impulse

When an impulse has passed along an axon, it must cross a synapse in order to stimulate another neuron or trigger a response. Unlike transmission along the axon, which is electrical, transmission across the synapse involves a chemical (transmitter), which is stored in vesicles at the end of the axon. The transmitter becomes attached at receptor sites to the postsynaptic membrane and through a change in ion permeability, directly or indirectly through the adenylate cyclase system, thereby causing the membrane to be depolarized and to generate an action potential.

8.5 Cholinesterases

The acetylcholine is the most important neurotransmitter in the nervous system. In order to restore the sensitivity of the postsynaptic membrane, the chemical transmitter must be eliminated. At cholinergic junctions this is accomplished by the enzyme cholinesterase, which breaks down acetylcholine into its inactive components, choline and acetate.

There are two types of cholinesterase. One is true cholinesterase, or acetylcholine esterase, which occurs in vertebrate erythrocytes, the electric organs of eels and fishes and is the principal cholinesterase of the vertebrate nervous system and the insect central nervous system. This is the enzyme which is the target of organophosphorus and carbamate insecticide poisoning. The other type is called pseudocholinesterase (or serum or plasma cholinesterase) and occurs primarily in vertebrate blood plasma and to some extent in brain and other tissues. It has no known biological function and can be totally inhibited without showing toxic consequences.

Active Site of Acetylcholinesterase. It has been recognized that the active site of AChE has two subsites, the esteratic site (ES) and the anionic site or anionic substrate-binding site (AS). In addition, there are peripheral anionic site(s) (PAS), the phenyl binding site (active site-selective aromatic cation-binding site, ACS) and putative hydrophobic binding regions (the hatched areas) as shown in Fig. 8.3A. Imidazole and hydroxyl side chains of histidine and serine, respectively, function at the ES. It is generally believed that, within the AS, carboxylate ions are present, which help to bind the positively charged quaternary group of ACh. Also, the trimethyl group on the quaternary nitrogen atom seems to have hydrophobic interaction with AS. ACS seems important to bind aromatic ring-containing ligands, such as organophosphorus or carbamate inhibitors. However, the exact contributions of such bonding to this site is not certain.

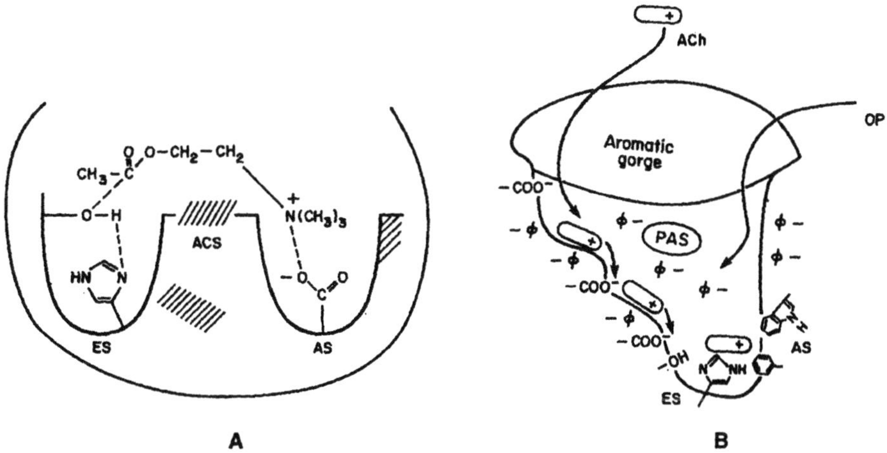

Fig. 8.3. Diagrams of the acetylcholine (*ACh*) binding sites on acetylcholinesterase (*AChE*): **A** former view and **B** modern view. Prepared by Izuru Yamamoto based on Sussman and Silman (1992)

Recently, the concept on the active site of AChE has advanced drastically (Sussman and Silman 1992, Fig. 8.3B). The X-ray structure reveals that AChE has an ellipsoid shape, into which a deep and narrow gorge penetrates. This cavity named the ***active-site gorge*** or ***aromatic gorge*** contains the ES consisting of serine and histidine residues near the bottom, and its wall and bottom are lined with 14 aromatic amino acid residues (ϕ). Only a few acidic residues ($-COO^-$) are located atop the gorge, about half way down and near the bottom. The AS, which functions to bind the cationic heads of ACh and other ligands, is uncharged and hydrophobic and consists of at least two aromatic residues [tryptophane and phenylalanine (Torpedo) or tyrosine (mammals)] in contrast to the classical model. The nature of the binding is not through the formerly conceived Coulombic interaction with carboxylate ions. On the other hand, dipole-dipole interaction between the π electrons of the aromatic rings and the quaternary nitrogen of the ACh plays a key role in ligand binding. Tryptophane residue, near the top of the gorge, is a part of PAS and is involved in the substrate inhibition.

The cationic ligands may be guided by an array of negative charges and slide down along the hydrophobic wall toward the active site via a series of low-affinity sites. There appears to be an electrostatic field around AChE which draws the positively charged ACh into and down the gorge.

8.6 GABA Receptor Complex

In the insect's peripheral nervous system, no distinct autonomic system exists and no cholinergic system is involved, but L-glutamic acid stimulates muscle, while GABA suppresses muscle contraction as transmitters. GABA is also known to function in the central nervous system of insects and mammals. GABA is released from presynaptic endings of stimulated inhibitory neurons and diffuse across the synaptic cleft to bind at postsynaptic sites on the GABA receptor complex, which increases chloride permeability. This results in hyperpolarization, thereby deactivating the postsynaptic cell.

Therefore, activation of GABA receptors produces inhibition at a variety of sites within the vertebrate and invertebrate central nervous system as well as invertebrate neuromuscular junctions, while inhibition of GABA-induced chloride ion permeability causes excessive release of acetylcholine at presynaptic sites or stimulates glutamate action, which may account for the stimulant and convulsant effects of the inhibitors.

The GABA receptor complex or GABA receptor chloride ionophore is a single complex protein having at least three distinct interacting components both in mammals and insects: one that binds GABA, another that binds benzodiazepine, and an ion channel which binds picrotoxinin, bicyclophosphates, bicycloorthobenzoates, dithianes and barbiturates. These sites interact allosterically and affect each other's bindings. In mammals there are at least two subtypes of GABA receptors, $GABA_A$ and $GABA_B$, but such subclassification of GABA receptors has not been extended to insects. The $GABA_A$ receptor in the mammalian brain is a primary target for the toxic action of chlorinated cyclodienes and HCH and insect GABA receptor is a molecular target for their insecticidal action (Eldefrawi et al. 1985).

CHAPTER 9

The Organochlorine Insecticides

9.1
DDT Group

DDT is the common name given to the technical mixture of compounds in which the major component is *p,p'*-DDT [dichlorodiphenyl trichloroethane; 2,2-bis (p-chlorophenyl)-1,1,1-trichloroethane). Pure *p,p'*-DDT is a white, tasteless, almost odorless crystalline solid. It is practically insoluble in water but soluble in many organic solvents such as acetone, benzene, carbon tetrachloride, kerosene, diesel oil, etc. It is slightly soluble in alcohol (about 2%). For field application, DDT has been formulated as solutions, emulsifiable concentrates, wettable powders, dusts, and aerosols.

9.1.1
History

DDT was first synthesized by Othmar Zeidler in 1874. However, the insecticidal properties of DDT were first discovered in 1939 by Paul Müller of the J.R. Geigy Company, for which he obtained a Swiss patent in 1942 and the Nobel Prize in Medicine in 1948. The early dust formulations containing 5 or 10% DDT proved to be very effective against a wide variety of insect pests of agricultural crops, in particular in controlling infestations of the Colorado potato beetle which threatened the Swiss potato crop in 1939. However, the most remarkable impact of DDT in the early years after its discovery was in the field of public health. The first full-scale use of DDT was during World War II in Naples, Italy. Here, in December 1943 a typhus epidemic (a rickettsial disease transmitted by body lice) broke out in the overcrowded civilian population, which was louse-ridden. A mass disinfestation was undertaken by the Allied Forces by dusting people's clothing with 10% DDT powder. Records show that over 2.5 million people were deloused, bringing the typhus epidemic completely under control within a few weeks (West and Campbell 1950; Cushing 1957). This successful operation was even more remarkable in that a typhus epidemic was controlled in winter for the first time in history (Simmons 1959).

The control of insects affecting man and animals was revolutionized by the application of DDT. Aside from its favorable toxicity to a wide range of insects, its lack of color, odor, and skin irritancy made it an ideal chemical for the control of ectoparasites. It proved to be an excellent agent against larvae of mosquitoes and blackflies, against adults of all the important biting flies (such as blackflies, tsetse flies, sandflies, horseflies, deerflies, etc.), and against houseflies, bedbugs, fleas, ticks, and other insects of public health importance. Direct spraying of DDT on dairy cattle was avoided due to its accumulation in body fats and milk. Also, its toxicity to certain species of mites, ticks, and dipterous larvae was found to be inferior to other insecticides.

Dramatic effects were also obtained in the control of anopheline mosquitoes, vectors of malaria, by DDT. A worldwide program of malaria eradication undertaken by the World Health Organization (WHO) in 1955 resulted in striking reductions of malaria cases in many areas of the world, particularly in Africa. By 1970, the number of people living in areas where malaria had been eradicated amounted to 710 million, or 39.4% of the total population of the originally malarious areas (Wright et al. 1972). DDT drastically changed antimalaria strategy because of its persistence, low cost, economy of application, and relative safety to man.

Many examples could be given to show the impact of DDT on the general health of populations throughout the world. Sri Lanka can be taken as an example. Following the introduction of a countrywide spraying operation with DDT, the number of malaria cases dropped from 2.8 million in 1946 to 110 in 1961, and deaths from malaria cases dropped from 12,587 to almost zero in the same period. Parallel to this, the general death rate fell from 20.3 to 8.6/1000 and the infant mortality rate from 141 to 57/1000. When spraying with DDT was ceased in 1964, the remarkably improved situation in that country reverted and malaria began to increase, reaching more than 2 million cases in 1968–1969. Many such examples were reported in other countries (Wright et al. 1972).

The health improvement, economic gains, and social benefits resulting from malaria control have been tremendous. DDT has probably saved tens of millions of human lives and may have prevented hundreds of millions of human illnesses (Knipling 1953).

It is appropriate to mention one important aspect of DDT usage in malaria control and in other areas where DDT applications are made indoors. In such instances DDT is applied as a residual spray inside buildings on walls and other surfaces. The amount of DDT escaping to the environment outside of the buildings is only a small fraction of the total amount of DDT used, unlike those amounts used in agriculture, which remain in the environment in their entirety. Hence, the use of DDT in malaria control by spraying the inner surfaces of houses has little chance of contaminating the environment and bringing any considerable amount of DDT in the food chain (Simmons 1959).

DDT proved its effectiveness in the control of other disease vectors such as yellow fever, dengue, filariasis, encephalitis, plague, and other scourges of mankind.

In the agricultural field the chemical control of phytophagous insects was revolutionized by the appearance of DDT. Commercial preparations of DDT were completely safe to plant foliage with the exception of cucurbits. DDT could persist on the foliage for over a month since neither heat nor light destroys it, rain does not dissolve it, and, due to its low vapor pressure, its loss by evaporation is very slow. Its use on feed for dairy cattle was prohibited from early times due to its accumulation in cow's milk.

DDT showed great promise for the control of a wide variety of insects, including the Colorado potato beetle, European corn borer, various species of leafhoppers, flea beetles, alfalfa weevils, pea aphids, Japanese beetles, codling moth, oriental fruit moth, and many others. However, DDT was found to be ineffective against heavily sclerotized insects, such as orthopterans, larger hemipterans, adult cicadids, etc., or against insects having rapid life cycles, such as aphids, psyllids, aleyrodids, coccids, and others.

DDT was especially ineffective in the control of acarina. Mites infesting plants present a peculiar problem in that their life cycle is so short that there is always a sufficient proportion of the population in the egg stage which is not susceptible to DDT. However, the mite predators lack this ample supply of eggs to replenish their populations. Therefore, when the destructive mites hatch from their surviving eggs they find that they have escaped from their biological control agents. Although DDT is effective against nymphs and adults of the two spotted mite *Tetranychus bimaculatus* and the red spider mite *T. telarius*, it does not kill their eggs, and in the absence of natural enemies, the mite population explodes and becomes more abundant than before. Thus, DDT has been found to encourage the development of infestations of several mite species in orchards, greenhouses and crop plants.

Although the use of DDT, especially in agriculture, has been banned in the developed countries due to bioaccumulability (1970 in Sweden; 1971 in Japan; 1972 in the U.S.A. and many other countries subsequently), it continues to be one of the most interesting insecticides ever discovered and is still being used to combat insect vectors of disease in some parts of the world.

9.1.2 DDT Analogs

The discovery of the insecticidal properties of DDT stimulated the search for analogous organochlorine compounds. Although hundreds of compounds were synthesized, only a few were found to be sufficiently active and of low cost to be suitable for commercial exploitation.

Some of the important analogs of DDT and their structural formulae are listed in Table 9.1.

None of the structural analogs of DDT shown in Table 9.1 and scores of other modifications proved to be as effective as DDT. The compound DDD (TDE) introduced in 1945 under the name of Rhothane is a non-systemic contact and stomach insecticide which does not possess the broad spectrum of insecticidal activity of DDT but is equally effective against larvae of some mosquito and lepidopteran species.

Methoxychlor. This is a biodegradable insecticide used as a substitute for DDT. While it is more expensive and somewhat less effective than DDT, it still provides effective control of many agricultural pests and of insects affecting man and animals. It does not persist in aquatic ecosystems to the same extent as DDT and has been successfully used for blackfly (*Simulium*) larva control. Methoxychlor is a moderately persistent compound in the soil (up to 14 months or more). It

Table 9.1. Chemical name and structural formulae of DDT and its analogs

Compound	Structure	Rat LD_{50} (mg/kg)	Comment
DDT	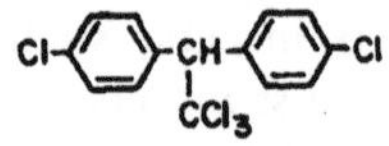	113–250	
DDD (TDE) (Rothane)	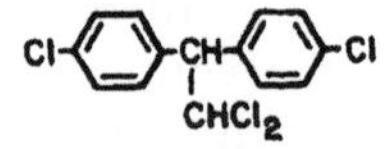	2500	
DFDT	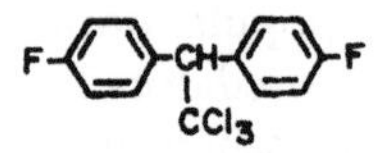	1120	
DNP (Prolan)	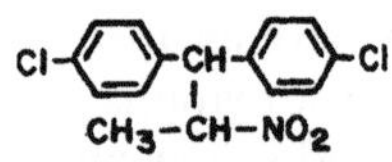	4000	A component of Dilan
DNB (Bulan)	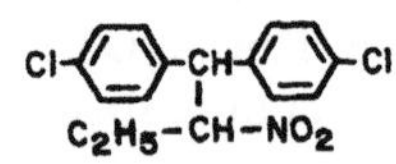	330	A component of Dilan
Ethylane (Perthane)	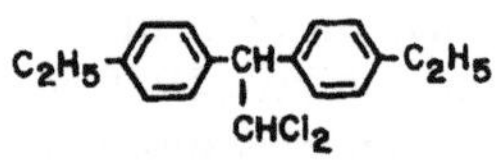	8170	
Methoxychlor	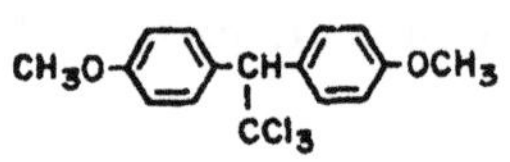	5000	
Dicofol (Kelthane)	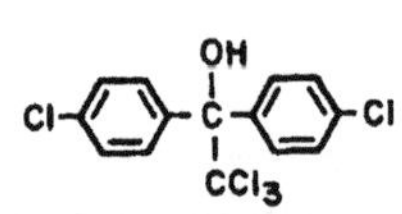	684–809	Miticide
Chlorfenethol (DMC)	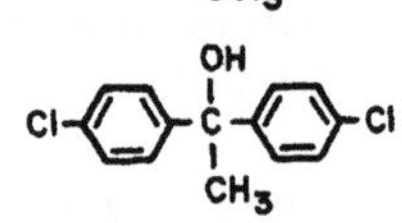	926–1391	Miticide
Chlorobenzilate	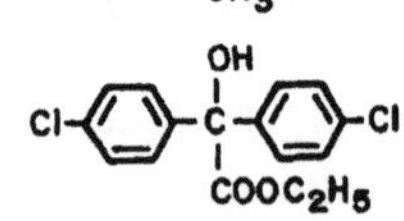	700–3100	Miticide
Chloropropylate	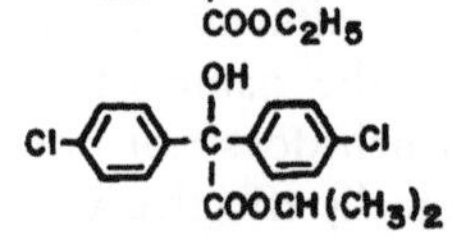	> 5000	Miticide

is susceptible to degradation by ultraviolet light (photolysis) and volatilizes from soil surfaces and aquatic suspensions in the same manner as DDT. Methoxychlor has some advantages over DDT from an environmental standpoint. The oral LD_{50} for rats is 5000 mg/kg, significantly lower than that of DDT, which is 113 mg/kg, and, unlike DDT, it does not accumulate in the body fats of animals and is not excreted in the milk. It is a good substitute for DDT in fly control in dairy barns and for control of other insect pests of public health importance. Methoxychlor is a safe insecticide to use since it is rapidly broken down in the mammals to nontoxic, water-soluble derivatives.

DFDT. This has a much shorter residual activity than DDT, acts more quickly, and is less toxic to mammals.

Dilan. This is a mixture of 53.3% Bulan, 27.6% Prolan and 20% related compounds. It is a nonsystemic insecticide with a similar range of action, but is of lower toxicity to insects than DDT.

Perthane. This is generally less toxic to insects than DDT and DDD but has found some application in the control of fruit and vegetable insects as well as against livestock pests, clothes moths, and carpet beetles. An advantageous feature is its low toxicity to mammals.

Dicofol. This has little or no insecticidal activity but is used to control mites on a wide variety of crops.

Chlorobenzilate and chloropropylate. These are nonsystemic acaricides recommended for use against phytophagous mites on fruits, nuts, vegetables, cotton, and ornamental plants.

Chlorfenethol (DMC). This is effective against both eggs and active stages of several mite species. It has a moderate residual effect, kills slowly, and produces semiparalysis of the mite. In laboratory experiments DMC was found to be a good synergist for DDT against DDT-resistant houseflies by its ability to inhibit the conversion of DDT to the nontoxic metabolite DDE (Perry et al. 1958).

9.1.3 Toxicity and Mode of Action

As mentioned earlier, the majority of insecticides owe their toxicity to their ability to interfere with the function of the nervous system. For some insecticides, such as the organophosphorus and carbamate insecticides, it is well established that their action on the nervous system is primarily due to inhibition of the enzyme cholinesterase. For the DDT group, however, a biochemical mechanism has not been elucidated.

DDT is rather slow acting. In the DDT-poisoned insect there are tremors throughout the body and the appendages, characteristically labeled "*DDT jitters*". According to Tobias and Kollros (1946), the sequence of symptoms of poisoning is as follows:

1. Hyperextension of the legs and uncoordinated movements.
2. Tremors throughout the body.
3. Ataxia (loss of coordinated motion).
4. Hyperactivity resulting from external stimuli.

5. Repeated falling on the back and righting efforts.
6. Disappearance of fast tremors.
7. Prostration with the heart still beating.
8. Finally, paralysis and death.

The insect dies in a desiccated condition with the body muscles contracted. During the tremoring period, the oxygen consumption of the insect rises appreciably; the insect rapidly loses weight due to tissue oxidation, water loss, glucose and glycogen consumption, and utilization of fat reserves (Ludwig 1946; Dahm and Kearns 1951). However, death occurs before the caloric reserves decrease to the level of starvation (Buck and Keister 1949). DDT is more toxic as the temperature is decreased, hence it is said to have a negative temperature dependence or coefficient. For the DDT group, no biochemical mechanism has been established, though the inhibition of the ATP-dependent portion of the Na/Ca exchange may be involved (Matsumura and Clark 1982). It is now well established that DDT acts primarily on neurons and interferes with the axonal and synaptic transmission. DDT prolongs the inward sodium current (delay of sodium channel closing; the prolongation of falling phase) and increases the depolarizing after potential. When the depolarizing after potential is increased to a certain level, a sudden burst of repetitive discharge, a train of impulses, can be provoked by a single stimulus. This leads to the hyperexcitability of the nervous system, resulting in tremor, paralysis, and eventually death of insects. DDT has been shown to cause the release of neurohormones which might be involved in its toxicity.

Toxicity to Mammals. The acute oral toxicity of DDT to mammals has been established as an LD_{50} of 113 mg/kg for rats (WHO 1988). There are, of course, species differences in the LD_{50} value, but rats seem to be more susceptible to DDT than other mammals. The toxic effect of DDT is enhanced if given in a fatty vehicle rather than in the dry crystalline form. For example, it is more dangerous if applied to the skin in an oil or a fat-soluble solvent than if applied as a water suspension or as powder. DDT-impregnated clothing does not pose a hazard to man.

Aerosols and mists containing DDT for use in the control of household pests are of negligible toxicity to animals unless a concentration of 20 ppm is reached. This is much higher than the concentration needed for insect control.

Symptoms of poisoning in animals include hyperexcitability, tremor, ataxia (loss of coordinated movements), and finally convulsions. Death is usually due to respiratory failure at the convulsive stage of poisoning. The mechanism of the toxic action of DDT is associated with its effects on the membranes of the nervous system as in insects. There are changes in the movement of sodium and potassium ions through the axonal membrane, a movement which is involved in the transmission of nerve impulses.

Apart from the nervous system, the liver is the only other organ significantly affected by DDT. Potentially fatal doses of DDT produce focal necrosis of liver cells. Another form of liver change is the stimulation of microsomal enzymes (enzymes involved in oxidation of foreign compounds and endogenous steroidal hormones) primarily in rodents.

Current available information indicates that DDT is neither mutagenic in bacterial test systems, nor teratogenic in experimental animals. Although liver tumors are produced in mice and possibly in rats, there is no evidence that DDT is carcinogenic to man.

9.1.4 Metabolism

DDT metabolism in insects, vertebrates and plants is shown in Fig. 9.1.

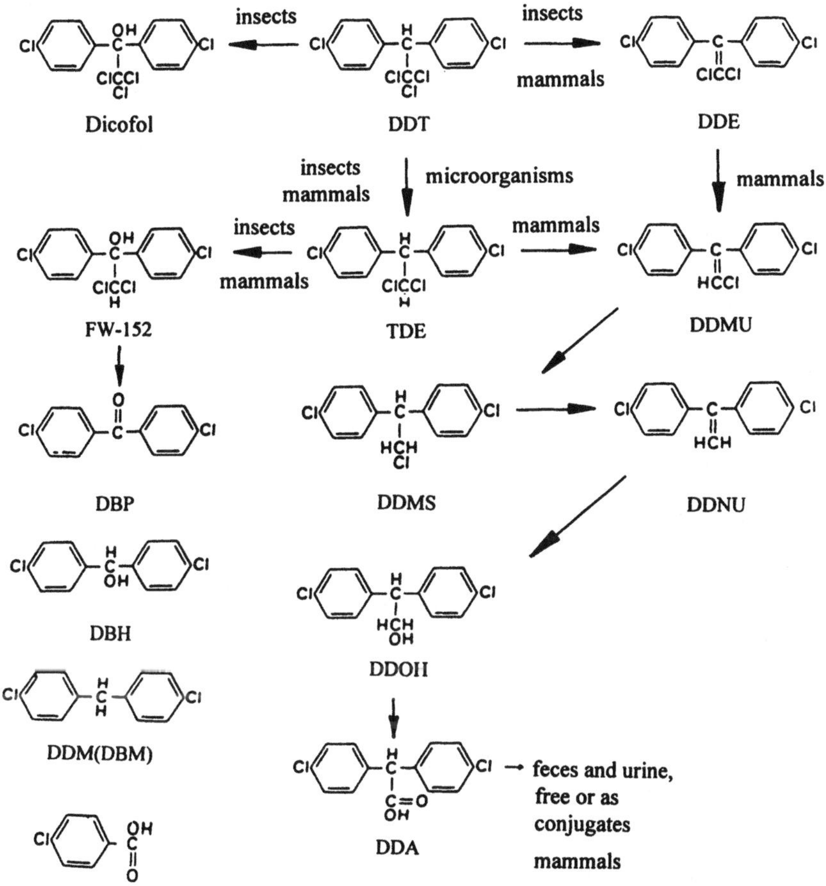

Fig. 9.1. DDT metabolism in insect, mammals, and microorganisms. *DDT* 2,2-bis (*p*-chlorophenyl)-1,1,1-trichloroethane; *TDE (DDD)* 2,2-bis(*p*-chlorophenyl)-1,1-dichloroethane; *DDE* 2,2-bis(*p*-chlorophenyl)-1,1-dichloroethylene; *Dicofol* 2,2-bis(*p*-chlorophenyl)-2-hydroxy-1,1,1-trichloroethane; *FW-152* 2,2-dichloro-1,1-bis(*p*-chlorophenyl) ethanol; *DBP p, p'*-dichlorobenzophenone; *DBH p, p'*-dichlorobenzhydrol; *DDM p, p'*-dichlorodiphenylmethane; *DDMU* 2,2-bis(*p*-chlorophenyl)-1-chloroethylene; *DDNU* 1,1-bis(*p*-chlorophenyl) ethylene; *DDMS* 2,2-bis(*p*-chlorophenyl)-1-chloroethane; *DDA* 2,2-bis(*p*-chlorophenyl) acetic acid. Modified from Matsumura 1975

Insects. In insects the most important route of DDT metabolism is through its conversion to DDE. However, there are other routes yielding as many as seven metabolites. The three most common routes of DDT are dehydrochlorination to DDE, hydroxylation to dicofol, and reductive dechlorination to TDE (DDD). Practically all insects studied accomplish the conversion of DDT to DDE, and the latter seems to be an important metabolic product in most other organisms studied. Insect resistance to DDT has been attributed largely to the greater ability of the resistant insect to convert DDT to DDE. Another mechanism is the decreased sensitivity of nerve targets to DDT. Other metabolites include FW-152, DBP, and probably *p*-chlorobenzoic acid. All the conversion products of DDT are less toxic to insects and other organisms. TDE is the only metabolite which is fairly toxic to a number of insects.

The conversion of DDT to DDE is mediated by the enzyme DDT-dehydrochlorinase, a glutathion-dependent enzyme, and the hydroxylation of DDT to dicofol is accomplished by the mixed-function oxidase system.

Mammals. Like most other species, mammals, including man, metabolize DDT to DDE, TDE, and DDA. In addition, DBP, DBH, DDM, DDMU, and DDMS have been identified in mammals fed or exposed to DDT. DDA is the principal water-soluble metabolite which is excreted in the urine and in the feces. Certain microorganisms effectively convert DDT to DDE. Both DDT and DDE are stored preferentially in body fat and in the fat content of other tissues and organs. In blood, DDT and related metabolites are carried mostly by proteins and less than 1% is carried by fat droplets in the blood.

DDT has been found in human milk in the range of 0.01 to 0.1 mg/l and together with other metabolites, especially DDE, in ranges twice as high. In a few countries average values for total DDT in human milk of 1 to 5 mg/l have been reported, the highest value observed being 12.2 mg/l (WHO 1979).

Generally, DDT analogs are persistent and bioaccumulable, except dicofol and methoxychlor. Conversion of DDT to dicofol occurs only in some insects. Methoxychlor is demethylated by the mixed-function oxidase system to form the polar metabolites:

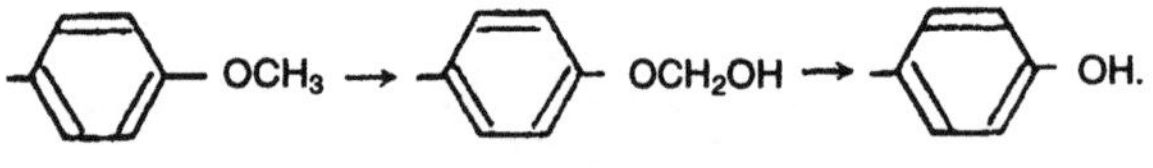

9.2 Hexachlorocyclohexane (HCH, BHC)

Hexachlorocyclohexane (previously called BHC, benzenehexachloride) is the oldest of the organochlorine insecticides, having been first prepared in 1825 by Faraday, who did not recognize then its insecticidal properties. There are theoretically many isomers of HCH, in which seven (α, β, γ, δ, ε, η, and θ) are known (Table 9.2). The α- and β-isomers were discovered by Meunier in the late 1880s. In 1912, Van der Linden confirmed the presence of these two isomers and reported the presence of two additional isomers, the γ and δ. The insecticidal properties of HCH were discovered in England and France in 1942. The toxic principle in HCH is the γ-isomer, named Lindane after its discoverer.

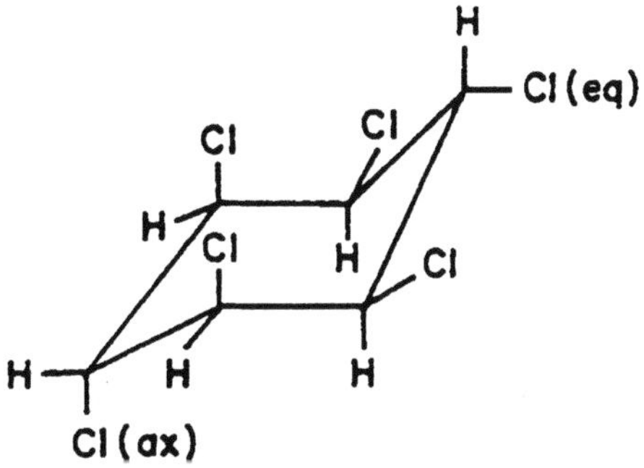

γ-HCH (lindane)

Three chlorine substituents take axial (ax or a) and the other three take equatorial (eq or e) conformations on the chair form of cyclohexane.

In general, the isomers of HCH are relatively stable to light, high temperature, hot water, and acid, but they are dechlorinated by alkali. Lindane is soluble in water to the extent of 10 ppm. It is 100 times more volatile than DDT and therefore has a fumigant action. This stability to heat, its low melting point, and its relatively high vapor pressure made it possible to use lindane as an insecticidal vapor dispensed by various types of dispensers. However, because of its relatively high mammalian toxicity, this practice was discontinued in areas where unprotected humans might be present.

The technical product of HCH consists of five isomers named α, β, γ, δ, ε in the following proportion: 55–70, 5–14, 10–18, 6–10, 3–4%, respectively. The crude material has a characteristic musty odor imparting an off-flavor to some edible crops which have been treated with this compound. Hence, this product is no longer used. The γ-isomer is practically odorless and present-day preparations for insect control contain 99% of this isomer registered as *Lindane* (or Gammexane, or one of many other trade names).

Lindane is outstanding for its speed of action as well as its high acute toxicity as a contact, stomach, or fumigant insecticide. The latter characteristic made it

Table 9.2. HCH isomers and toxicity

Isomer	Conformation	Melting point (°C)	Relative toxicity to German cockroach	Rat LD_{50} (mg/kg)
α	aaeeee	159–160[a]	3.6	500
β	eeeeee	309–310	–	> 6000
γ	aaaeee	112–113	100	125
δ	aeeeee	138–139	0.3	1000
ε	aeeaee	219–220	–	–
η	aeaaee	90	0.7	–
θ	aeaeee	124–125	–	–

[a]The racemate, which was resolved to the optical antipodes (m.p. 128–132 °C) (Eto 1979)

useful as a soil insecticide but its high volatility resulted in poor performance when sprayed on crops. Lindane has proved highly effective for the control of insects of public health importance, especially those which had become resistant to DDT. It is quite effective against mites infesting man, such as *Sarcoptes scabei* and chiggers.

9.2.1 Toxicity and Mode of Action

Only lindane (γ-isomer) has high toxicity toward insects and other organisms. It is five to ten times as effective as DDT against most insects and is a more acute nerve poison than DDT. The symptoms of poisoning in insects include tremors, ataxia (loss of coordinated movements), convulsions, falling on the back, prostration, paralysis, and death (Savit et al. 1946). There is a large increase in respiratory rate (larger than that caused by DDT) as a consequence of nervous and muscular activity (Harvey and Brown 1951). Houseflies poisoned with lindane show abnormal "*fanning*" movements of their wings, a symptom not seen in DDT poisoning (Busvine 1954). There is a negative relationship between temperature and degree of poisoning, but it is not as pronounced with lindane as it is with DDT (Guthrie 1950).

Lindane is a broad-spectrum insecticide with an LD_{50} of 125 mg/kg for rats and 1–10 mg/kg for insects. In mammals it stimulates the central nervous system, which causes a rise in blood pressure and a fall in heart beat (bradycardia). There is an increase in respiration followed by restlessness, coarse tremors of the whole body, salivation, grinding of the teeth, and convulsions; then, slowing of respiration and death from cardiac arrest.

In man, the symptoms include dizziness, headache, weakness, diarrhea and epileptiform attacks. Phenobarbital or pentobarbital counteracts the poisoning effect by suppressing convulsions. The convulsive effects produced by lindane are also antagonized by the α-, β-, and δ-isomers of HCH.

Lindane does not interact with any specific enzyme, although it is a better inhibitor of Na^+, K^+, and Mg^{+2}-ATPase than DDT.

Lindane causes the accumulation of acetylcholine in nerves of insects but it does not inhibit the enzyme cholinesterase. The same is true for mammals as evidenced by the ability of atropine to counteract the poisoning effect by preventing a fall in heart beat. The mechanism of action was established as blocking of the GABA-gated chloride channel (see Chap. 10).

9.2.2 Metabolism

Normally, lindane does not accumulate in the tissues of warm-blooded animals to the same extent as DDT does. However, residues of HCH have been found in human fat and milk. Although the α, β, and γ-isomers were all found as residues, the α- and γ-isomers are more rapidly metabolized, whereas the β-isomer accounts for most of the residue.

Lindane is metabolized in insects by progressive dechlorination to form several chlorinated benzene derivatives and water soluble glutathion conjugates (Fig. 9.2). Hydrolysis of the conjugates produce dichlorothiophenols.

In mammals, lindane is metabolized by dehydrochlorination, hydroxylation, and conjugation to form several chlorinated phenols and conjugates which are excreted in the urine. PCCH was not found as the metabolite, but PCCH gave 1,2,4-TB and the following derivative phenols (Fig. 9.3).

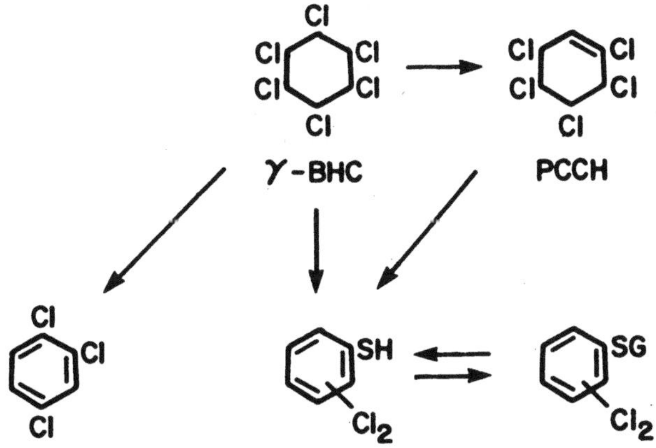

Fig. 9.2. Metabolism of lindane in insects. *SG* Glutathione moiety. (Matsumura 1972)

9.3 The Cyclodiene Group

Cyclodiene compounds are a group of highly active insecticides. However, many of them have long persistency and their use is restricted. They include chlordane, heptachlor, aldrin, dieldrin, isodrin, endrin, endosulfan (Thiodan), mirex, chlordecone (Kepone), and the chlorinated terpene, camphechlor (toxaphene). The oral toxicity to rats is summarized in Table 9.3.

Chlordane. Technical chlordane is a dark amber viscous liquid with a cedarlike odor, a multicomponent mixture of chlorinated hydrocarbons, representing more than 45 distinct chemical entities, of which at least 11 are major components. Table 9.4 shows the composition of technical chlordane and the structure of some of them are shown in Fig. 9.4.

Heptachlor. The compound is a white crystalline solid, four to five times more toxic to insects than technical chlordane. Heptachlor is stable to heat, light, moisture, air, and oxidizing agents. Both insects and mammals are able to convert heptachlor to heptachlor epoxide which is stored in tissues in that form (Fig. 9.5). Heptachlor epoxide is more toxic than heptachlor.

Aldrin. The compound is a white crystal almost insoluble in water (0.2 ppm) but soluble in most organic solvents. It is stable to dilute acids and to alkalis. It is readily converted in biological systems and in the soil to its epoxide, dieldrin. Metabolism of aldrin and dieldrin is shown in Fig. 9.6. Aldrin has a high vapor pressure, hence it is effective as a soil insecticide.

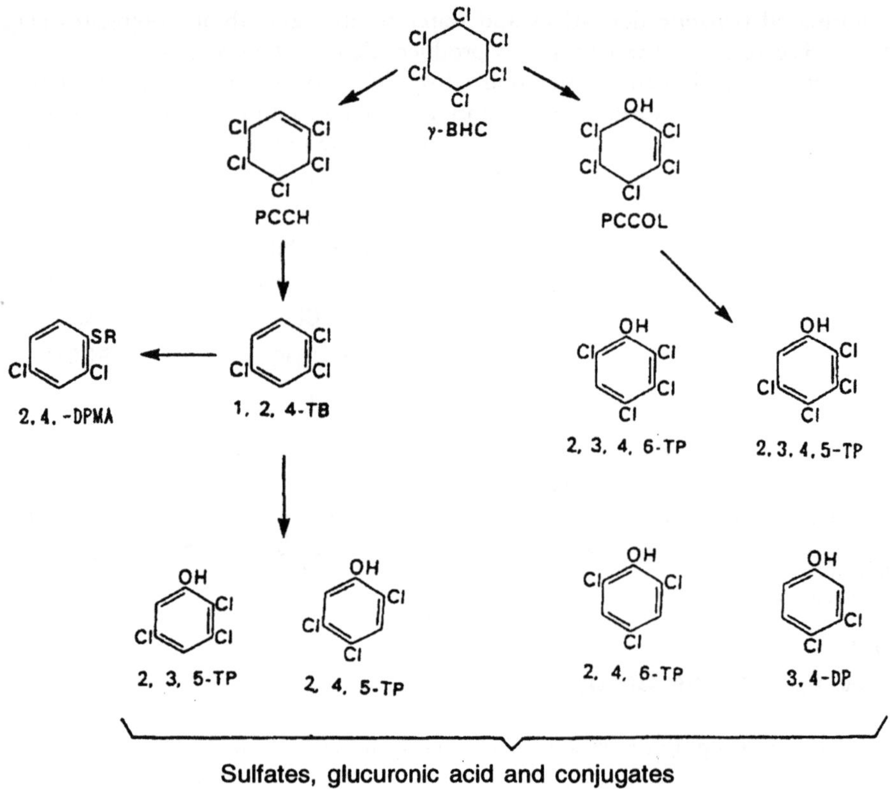

Fig. 9.3. Metabolism of lindane in rats. *PCCH* γ-2,3,4,5,6-pentachlorocyclohex-1-ene; *PCCOL* 2,3,4,5,6-pentachloro-2-cyclohexen-1-ol; *1,2,4-TB* 1,2,4-trichlorobenzene; *2,3,4,6-TP* 2,3,4,6-tetrachlorophenol; *2,3,4,5-TP* 2,3,4,5-tetrachlorophenol; *2,3,5-TP* 2,3,5-trichlorophenol; *2,4,5-TP* 2,4,5-trichlorophenol; *2,4,6-TP* 2,4,6-trichlorophenol; *3,4-DP* 3,4-dichlorophenol; *2,4-DPMA* 2,4-dichlorophenylmercapturic acid.

R = —CH_2—CH—COOH
                 |
            $NHCOCH_3$

(Modified from Matsumura 1975)

Table 9.3. Acute oral toxicity of cyclodienes to rats. (Büchel 1983)

	Rat LD_{50} (mg/kg)
Chlordane	283–590
Heptachlor	90–135
Aldrin	67
Dieldrin	40–87
Isodrin	12–17
Endrin	7.5–17.5
Endosulfan	100 (α, 76; β, 240)
Toxaphene	40–120
Mirex	235–702
Kepone	95–140

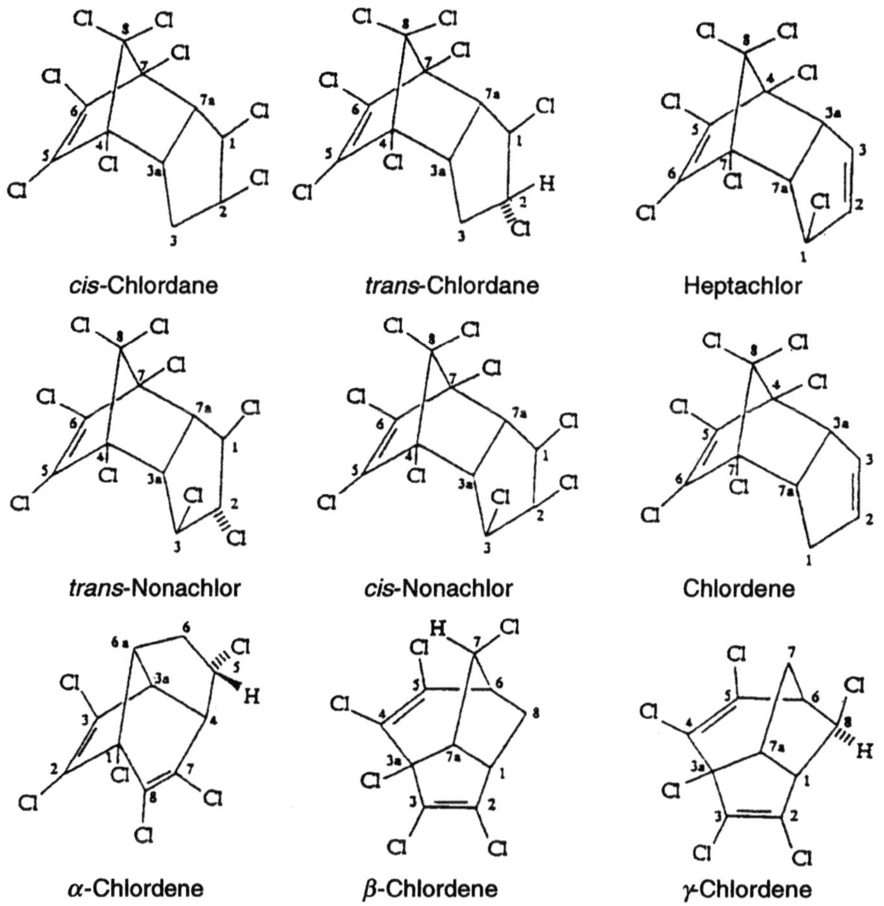

Fig. 9.4. Some components of technical chlordane. (Wilson and Sovocool 1977)

Table 9.4. Composition of technical chlordane. (Sovocool et al. 1977)

Component	Approximate (%)
α- or *cis*-Chlordane	19
γ- or *trans*-Chlordane	24
Heptachlor	7
trans-Nonachlor	7
"Isomer 2-Chlordene"	7.5
α-Chlordene	3
β-Chlordene	4
γ-Chlordene	9
Chlordene	1
Related compounds (including *cis*-nonachlor and compound K)	19.5

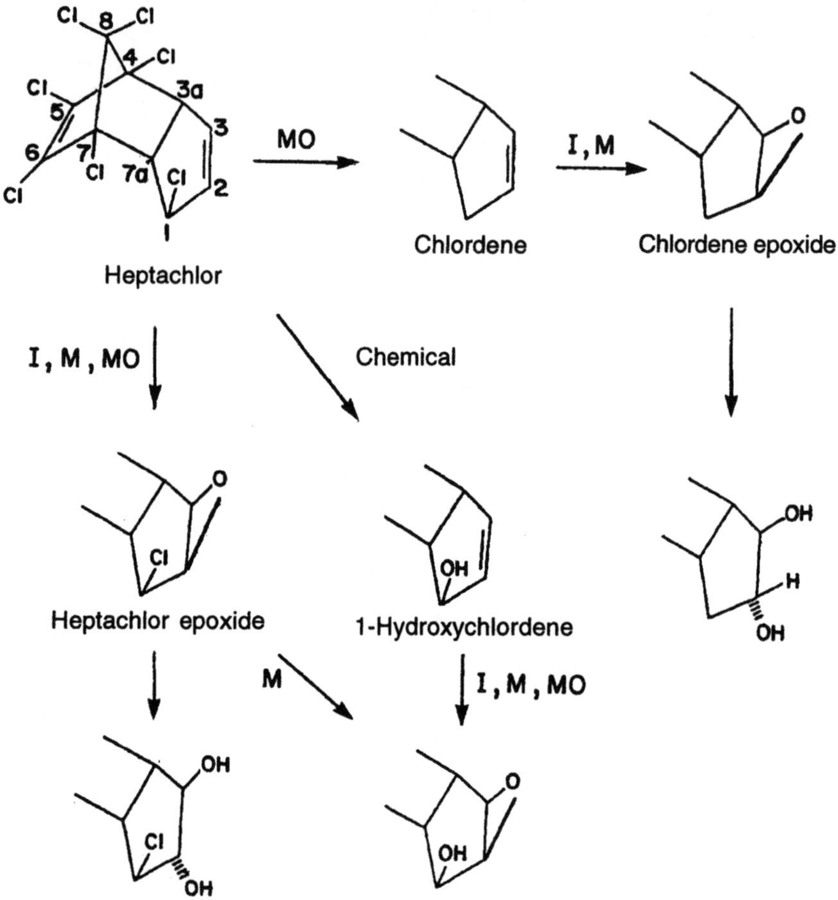

Fig. 9.5. Metabolism of heptachlor and chlordene in insect, mammals and microorganisms. *I* Insects; *M* Mammals; *MO* Microorganisms. (Brooks 1972)

Dieldrin. This compound is the epoxide of aldrin. It is an odorless, white crystalline compound, insoluble in water (0.25 ppm) but less soluble in organic solvents than aldrin. It is one of the most persistent of all insecticides and has been used for many years as a long-lasting residual insecticide. Only treatments with strong acids and long exposure to ultraviolet light can decompose it. Dieldrin can be absorbed through the skin.

The toxic action of aldrin and dieldrin resembles that of chlordane, heptachlor, and lindane. Although lindane does not belong to this group of compounds, it shares with them a common mode of action (Busvine 1964). Dieldrin and aldrin are the best known members of the cyclodiene group of insecticides and are named after Diels and Alder, the discoverers of this type of synthesis.

Isodrin and Endrin. Isodrin is an isomer of aldrin and is converted to its epoxide, endrin. Isodrin and endrin are less stable than aldrin and dieldrin and can easily

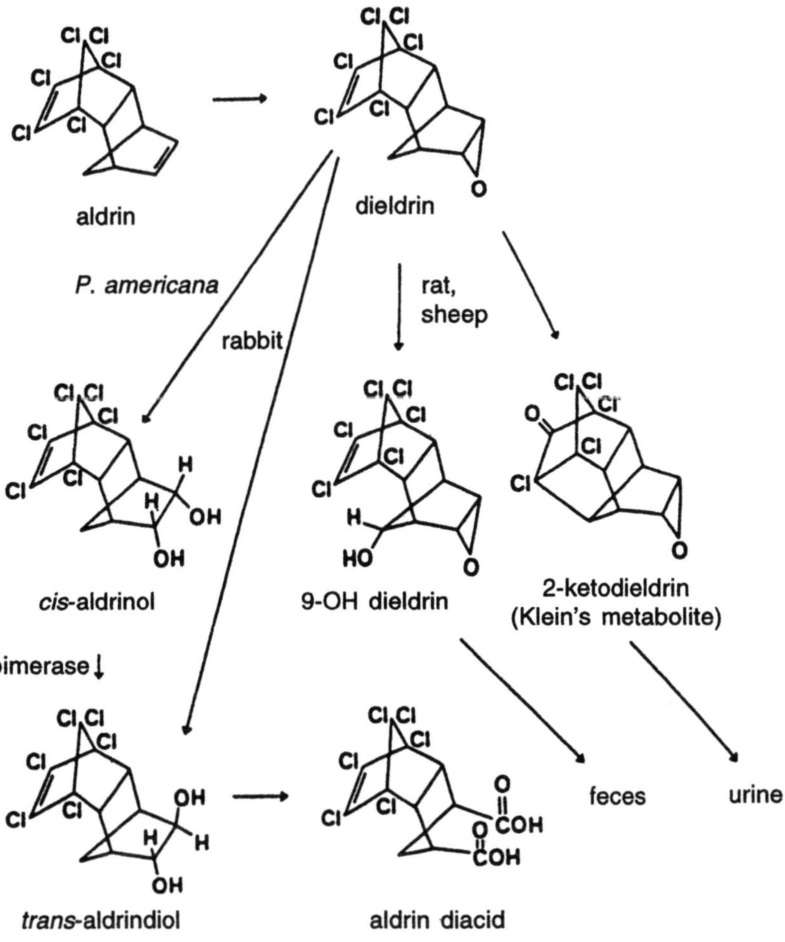

Fig. 9.6. Metabolism of aldrin and dieldrin in insects and mammals. (Matsumura 1975)

be degraded by heat and light, but their toxic effects are similar. Technical endrin is a light tan powder, but the pure compound is a white crystalline solid with a high melting point at 200 °C. Endrin, aldrin and dieldrin are excellent soil insecticides and have been used extensively for termite control. Isodrin has not been used commercically.

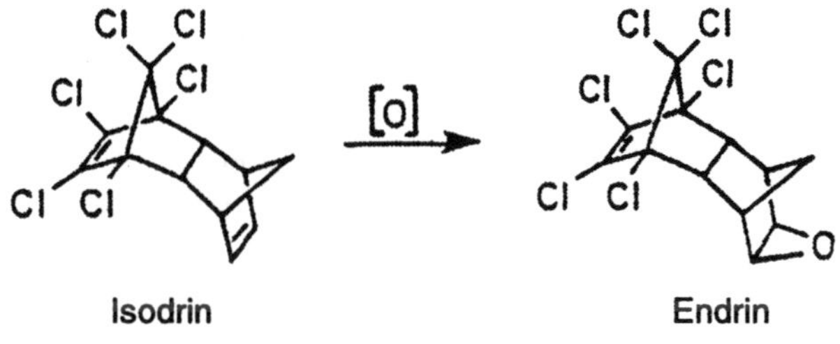

Endosulfan (Thiodan). The compound is both an insecticide and acaricide. It is a brownish crystalline solid consisting of two isomers. It is insoluble in water and moderately soluble in organic solvents. It is absorbed through the skin and very toxic to fish. Unlike most other cyclodiene insecticides it is low in persistency and still used. It undergoes oxidation in insects, mammals, and plants to form a primary insecticidal metabolite, endosulfan sulfate. Endosulfan and the sulfate are hydrolyzed to form endosulfan diol (Fig. 9.7).

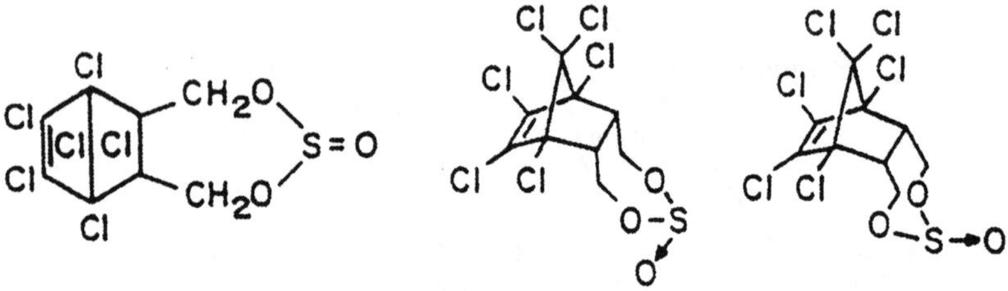

Endosulfan α-Endosulfan β-Endosulfan

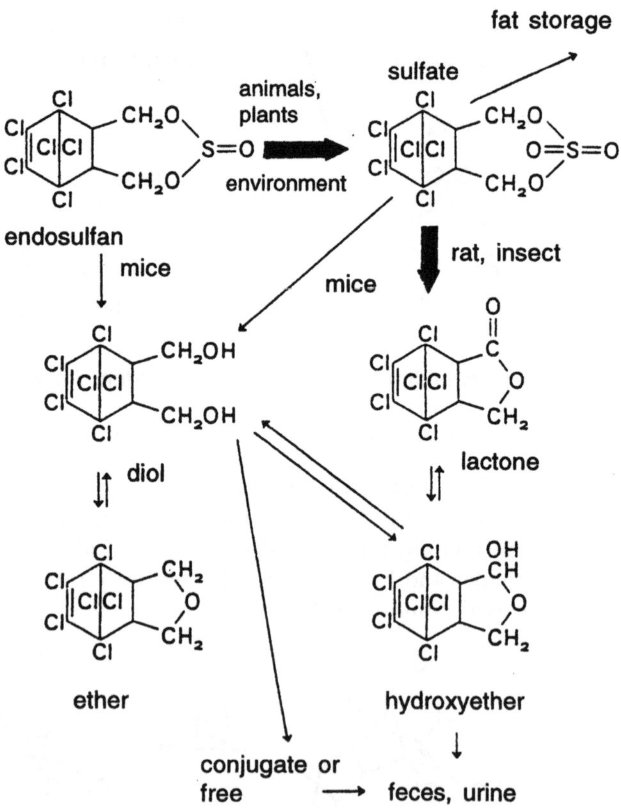

Fig. 9.7. Metabolism of endosulfan in insects, mammals, and plants. (Matsumura 1975)

Camphechlor (Toxaphene). Toxaphene is a highly chlorinated camphene containing 67–69% chlorine (5–12 chlorine atoms). It is a complex mixture of about 180 different components with yellow waxy appearance, slightly soluble in water (about 3 ppm), and soluble in most organic solvents. Toxaphene is a stable compound, except that it undergoes dehydrochlorination in the presence of alkali, sunlight, or heat above 155 °C.

Ten components (Fig. 9.8) are identified but approximately 75% of the components by weight remain unknown. The most toxic component and a major factor in toxaphene toxicity to mice, houseflies and goldfishes is 2,2,5-endo,6-exo,8,9,9,10-octachlorobornane (I) followed by II (Saleh and Casida 1979).

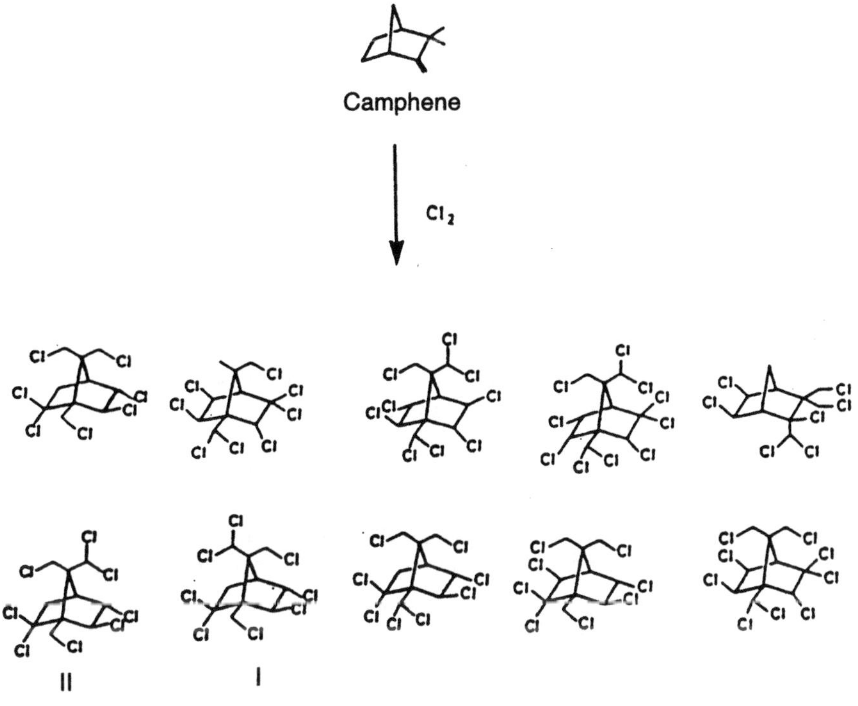

Fig. 9.8. Identified toxaphene components. (Saleh and Casida 1979)

Mirex and Chlordecone (Kepone). Mirex is a white crystalline solid, insoluble in water, and moderately soluble in organic solvents with high melting point (485 °C). It has exceptional chemical stability and is very persistent. Mirex is a stomach poison with little contact action and has been used in the form of small granules made from corncobs impregnated with vegetable oils containing mirex, which served as a bait for fire and control. In mammals, mirex is poorly absorbed by the gastrointestinal tract and more than half of a single administered dose of 6 mg/kg is excreted unchanged in the feces in 48 h.

Chlordecone is similar in structure to mirex. The technical material is a solid, slightly soluble in water, readily soluble in acetone but less soluble in other

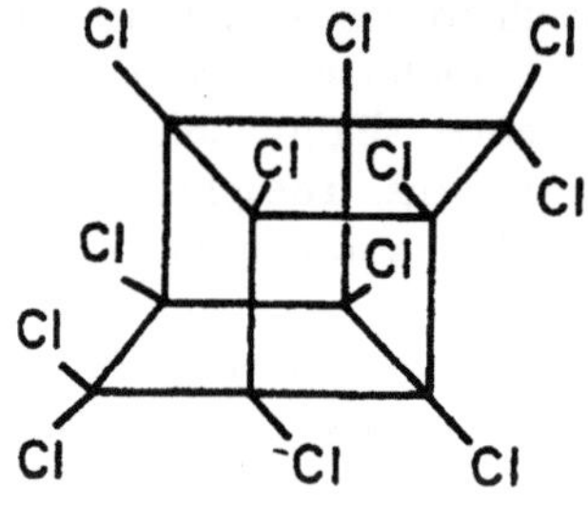

Mirex

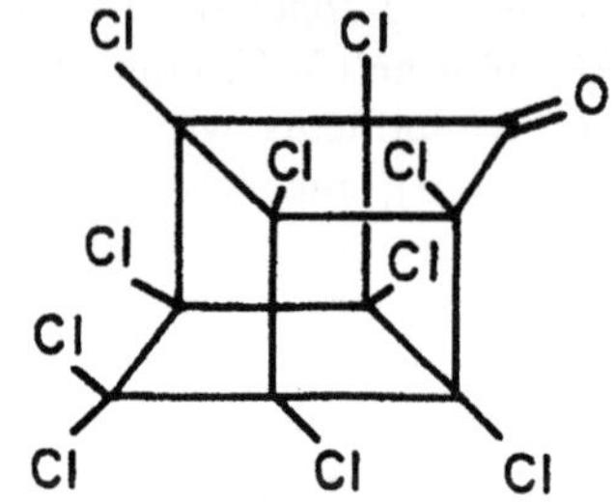

Chlordecone

organic solvents. Like mirex it is effective as a stomach poison with a weak contact activity and is used in bait. It is also a miticide. In mammals, it is poorly absorbed through the skin. After oral administration, chlordecone is found chiefly in blood plasma and appears to be bound to high density lipoproteins, the concentration decreasing by 50% after 8.5 days. Both mirex and chlordene are strong inducers of liver microsomal enzymes in rats and mice (Hayes 1982).

9.3.1
Toxicity and Mode of Action

As a group, the cyclodiene compounds, like most other chlorinated hydrocarbon insecticides, act as neurotoxicants. The primary mode of action is the blocking of the GABA-gated chloride channels like HCH, but not like DDT. An important difference between DDT and the chlorinated cyclodienes is that the cyclodienes are absorbed from the intact skin. As shown in Table 9.5, the difference between the oral and dermal LD_{50} values for the cyclodienes is much smaller than the difference for DDT, indicating that skin contact is more hazardous with the cyclodiene compounds.

In insects, the cyclodienes have a positive temperature relationship, that is, their toxicity is enhanced as the external temperature rises (the opposite effect

Table 9.5. Oral and dermal LD_{50} values for the chlorinated hydrocarbon insecticides

Compound	LD_{50} mg(AI)/kg[a]	
	Oral	Dermal
p,p'-DDT	113	2510
p,p'-DDE	880	–
DDA	740	–
Methoxychlor	6000	> 6000
Lindane	88	1000
Aldrin	39	98
Dieldrin	46	90
Endrin	18	
Heptachlor	100	195
Chlordane	335	840
Mirex	740	> 2000
Chlordecone	132	> 2000

[a]Values obtained in standardized tests in the same laboratory (Gaines 1969).

is seen with DDT and HCH). The symptoms of poisoning in insects resemble those of DDT, with strong tremors and short trains of impulses, but these effects appear 2 hours after treatment with dieldrin, 3 h with heptachlor and toxaphene, 4 h with aldrin, and 5–8 h with chlordane. This "lag period" between administration of the poison and the appearance of symptoms of poisoning is characteristic of the cyclodiene compounds.

The effect on the rate of respiration of the poisoned insect is much more pronounced with the cyclodienes than it is with DDT, but less so than with lindane (Harvey and Brown 1951). Here, too, the lag period is quite noticeable.

Cyclodienes cause an excessive release of acetylcholine at presynaptic sites, but they do not inhibit the enzyme cholinesterase. They do not affect any other enzyme systems; however, there is evidence that they interact with ATPases from nerve cord and muscle (Matsumura 1985). With houseflies and blowflies, a period of normal activity is followed by a more quiescent period, after which wing tremors become frequent, increasing in severity, and the insects show spasms of wing beat without flying. After a while this culminates in uncoordinated flight, loss of ataxic gait and falling on their backs. This process is irreversible with dieldrin and ends in death, but with related compounds that can be detoxified, recovery can occur even after 36 hours of immobilization; the symptoms resemble those of an anesthetic effect (O'Brien 1967).

In mammals, the most prominent symptoms of poisoning are convulsions. These symptoms can appear even before any other signs of toxicity are apparent. Convulsions may be accompanied by confusion, excitability, incoordination, and in some cases, coma. The toxicities of heptachlor, aldrin and isodrin are increased due to their conversion to their corresponding epoxides, heptachlor epoxide, dieldrin and endrin. Dieldrin is the most persistent among the cyclodienes, accumulating in fatty tissues and being retained for long periods of time. Endrin does not accumulate in fat and is metabolized to water-soluble metabolites, which are excreted.

Most of the cyclodiene compounds produce changes in liver cells, and some are considered carcinogenic to man. Because of their high oral and dermal toxicities, effects on liver cells, the potential for tumor production, and detrimental environmental effects, most of these insecticides have been banned from usage in agriculture and public health, except for termite control.

Aldrin, dieldrin, heptachlor and endrin undergo conversion in sunlight to form photoproducts which are more toxic and more persistent in the environment than the parent compounds.

9.4 Therapy

The usual treatment for poisoning by the chlorinated hydrocarbon insecticides is by administration of the anticonvulsants phenobarbital and pentobarbital. These barbiturates are also of value in promoting increased metabolism of the insecticide. Another drug, diazepam, is effective in the control of convulsions. For details on the toxicology and therapy of the chlorinated hydrocarbon insecticides in animals and man, see Hayes (1982).

CHAPTER 10

Insecticides Acting as GABA-Ergic Agents

The GABA receptor-chloride ionophore contains the GABA recognition site and binding sites for a variety of agonists and antagonists (Fig. 10.1).

10.1 Agonists

Avermectins and Milbemycins. These have excellent miticidal, insecticidal, and anthelmintic activities and are probably the agonists to the GABA recognition site (see Chap. 23).

10.2 Antagonists

Old insecticides, such as chlorinated cyclodienes and HCH, bind to the ion channel non-competitively and suppress the chloride ion permeability without affecting GABA binding. There are many compounds which share the same mode of action with the above (Casida 1993).

[^{35}S]TBPS (t-butylbicyclophosphorothionate), [^{3}H] TBOB [t-butylbicyclo-orthobenzoate]; [^{3}H] EBOB [1-(4-ethynylphenyl)-4-n-propylbicycloorthocarboxylate].

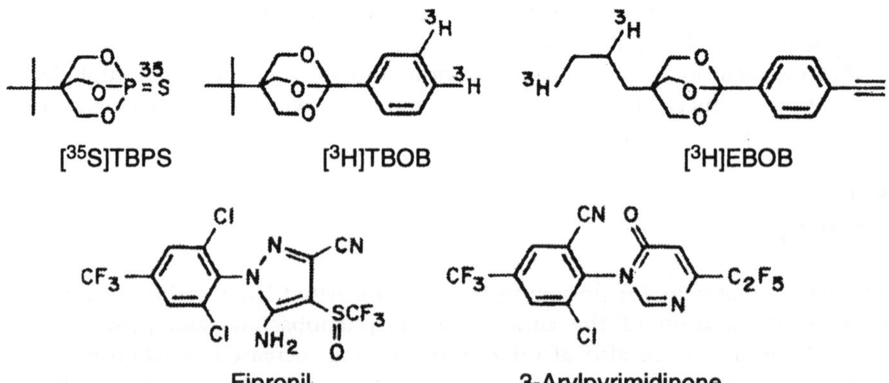

Fig. 10.1. GABA-ergic agents

These are excellent noncompetitive antagonists useful as radioligands and affinity probes: [^{35}S]TBPS and [^{3}H]TBOB are for the mammalian brain binding site and [^{3}H]EBOB for both the housefly and mammalian brain binding sites.

Arylpyrazoles. Recently, a series of arylpyrazoles as well as an aryltriazole have been developed as insecticides, which are noncompetitive antagonists. The most important one is fipronyl. It is effective to lepidopterous larvae and many other agricultural pests. Fipronyl and dieldrin give the same poisoning signs in houseflies and mice. The cross-resistance of the housefly is developed among them, but not in the avermectins. Such cross-resistance of insects is caused by the decreased binding on the receptor. Some of the arylpyrazoles, including fipronyl, are provided with low mammalian toxicity and high insecticidal activity (Cole et al. 1993).

Arylpyrimidinones. Several compounds belonging to the group shown in Fig. 10.1 show insecticidal activity on public health pests many times that of pyrethroids (Whittle 1994).

CHAPTER 11

Organophosphorus Insecticides

11.1 Structure and Toxicity

The organophosphorus (OP) compounds are one of the largest groups of insecticides in use today and they have largely replaced the organochlorine insecticides. They are neutral ester or amide derivatives of phosphorous acids [P(V) oxyacids] carrying a phosphoryl (P-O) or thiophosphoryl (P-S) group as shown in Fig. 11.1. They are generally the most toxic of all insecticides to insects and vertebrates (with some exceptions) by inhibiting cholinesterases.

$$\underset{\text{Cholinesterase}}{\text{ENZYME–OH}} + \underset{\text{OP compound}}{\text{Z–P–(O)(OR)}_2} \rightarrow \underset{\text{Inhibited enzyme}}{\text{ENZYME–O–P(O) (OR)}_2} + \text{ZH}$$

where Z denotes the leaving group.

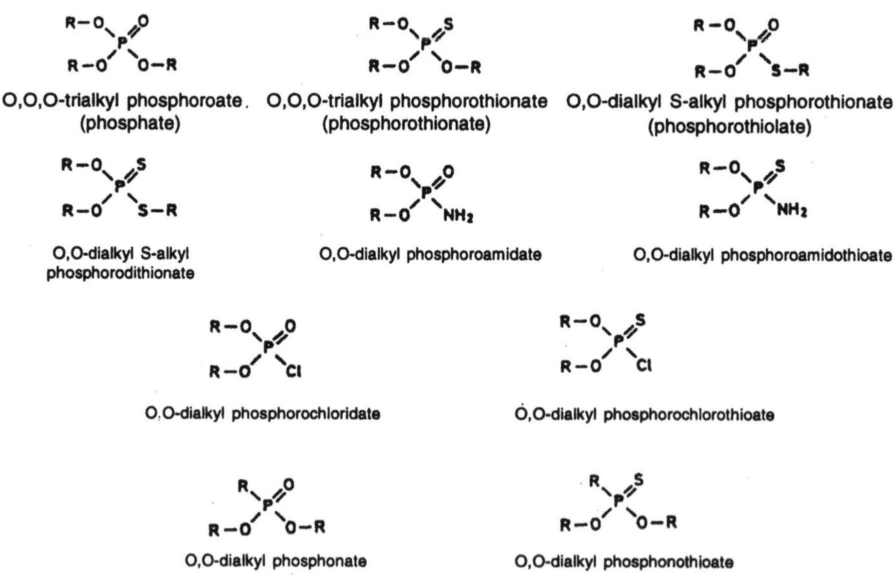

Fig. 11.1. Types of organophosphorus insecticides and related compounds

Because OP compounds are manufactured at high temperatures (150–200 °C) they commonly contain isomers or byproducts which are responsible for their unpleasant odor, and for their direct anticholinesterase activity. Organophosphorus insecticides are related in chemical structure and in mode of action to the toxic "nerve gases" of World War II, sarin, tabun, and soman (Fig. 11.2). The insecticidal action of OP compounds was discovered in Germany by Gerhard Schräder in the early 1940s. Three of these compounds, i.e., TEPP, parathion and Schradan, despite their high mammalian toxicity, were used as insecticides on a worldwide scale. Parathion is still being used in many areas of the world for the control of agricultural insect pests.

The OP compounds have two distinct features: firstly, they are acute toxicants of high level of activity, and, secondly, they are chemically unstable, or nonpersistent, and are quite biodegradable by most organisms and in the environment. These qualities make them useful as substitutes for the persistent organochlorines for the control of insect pests in agriculture and public health. Virtually thousands of OP compounds have been synthesized in the past 50 years, but not all have reached commercial application. About 100 OP insecticides (on active ingredient basis) are in commercial use throughout the world today. Of all the OP insecticides in use worldwide, 30 are classed as extremely hazardous (class Ia); 29 are highly hazardous (class Ib); 24 are moderately hazardous (class II), and 8 are slightly hazardous (class III). Only six OP insecticides supposedly present no hazard (WHO 1988). The OP insecticites in commercial use in agriculture, as well as those used in public health and historical ones are listed in Table 11.1. There are several review books on the OP compounds (Schräder 1963; Fest and Schmidt 1973; Eto 1974; Eto and Casida 1995).

11.2 Symptoms of Poisoning in Vertebrates

The toxicity of OP insecticides is due to their ability to inhibit the vital enzyme cholinesterase. Signs and symptoms of acute poisoning by OP insecticides are those of stimulation of acetylcholine receptors in the parasympathetic nervous system, such as sweating, salivation, urination, defecation, lachrymation, nausea, vomiting, abdominal cramps, constriction of the pupil (miosis), slowing of the heart (bradycardia), and drop in blood pressure (hypotension). These are called muscarinic effects from similar effects of muscarine produced by the housefly fungus *Amanita muscaria*. In addition, there are symptoms resulting from the

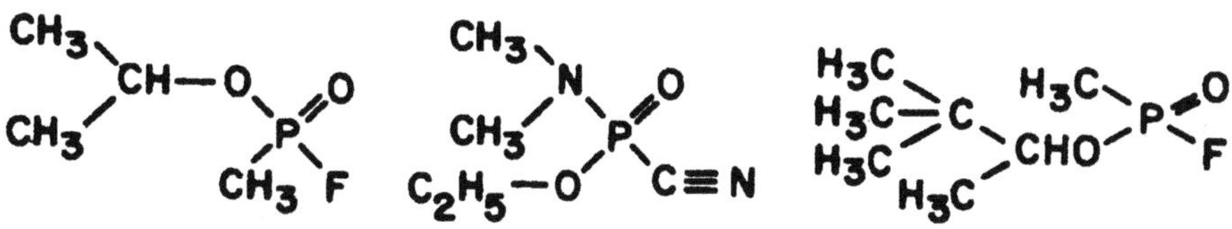

Sarin Tabun Soman

Fig. 11.2. Nerve gases

Table 11.1. Organophosphorous insecticides

Common name	Structure	Toxicity oral LD_{50}(mg/kg)	Comment
Phosphate			
Chlorfenvinphos	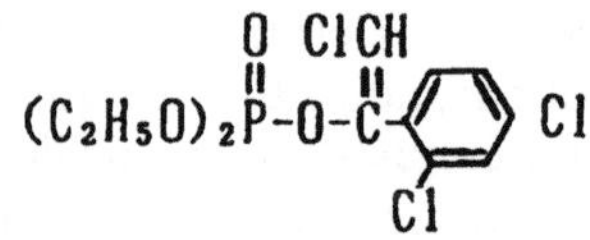 Z & E	117–200 mice 24–39 rats	
Dichlorvos[a]	$(CH_3O)_2P(=O)\text{-}OCH{=}CCl_2$	c. 50 rats	Fumigant Also miticidal
Mevinphos	$(CH_3O)_2P(=O)\text{-}OC(CH_3){=}CHCOOCH_3$ Z & E	7–18 mice 3–12 rats	Also miticidal Systemic
Monocrotophos	$(CH_3O)_2P(=O)\text{-}OC(CH_3){=}CHCONH(CH_3)$ E	18 male rats 20 female rats	Systemic
Naled[a]	$(CH_3O)_2P(=O)\text{-}OCH(Br)C(Br)Cl_2$	430 rats	Also miticidal
Phosphamidon	$(CH_3O)_2P(=O)\text{-}OC(CH_3){=}C(Cl)\text{-}CON(C_2H_5)_2$ Z & E	17.9–30 rats	Systemic
Propaphos	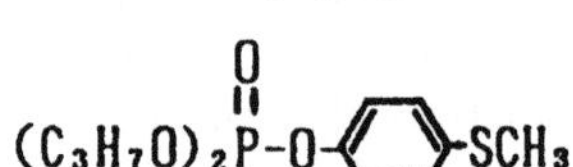	90 mice 70 rats	Systemic
Tetrachlorvinphos[a]	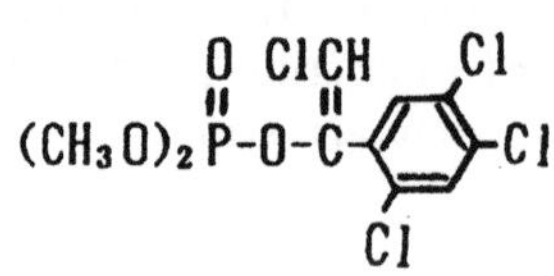 Z	2500–5000 mice 4000–5000 rats	
Phosphorothionate			
Chlorethoxyfos	$(C_2H_5O)_2P(=S)\text{-}OCH(Cl)CCl_3$	1–10 rats 20–50 mice	Soil insecticide

(Contd)

Common name	Structure	Toxicity oral LD_{50}(mg/kg)	Comment
Chlorpyrifos[a]	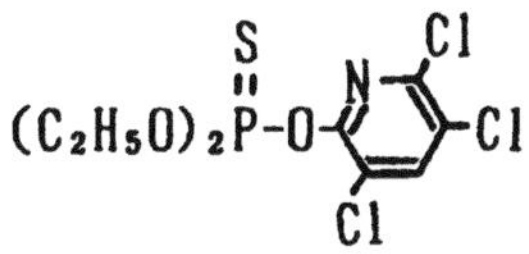	135–163 rats	
Chlorpyrifos-methyl[a]	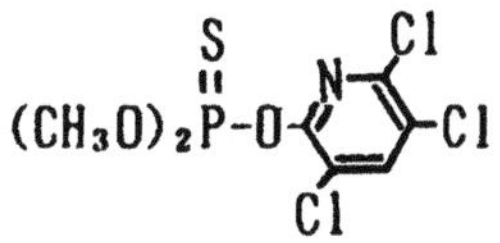	1100–2250 mice > 3000 rats	
Diazinon[a]	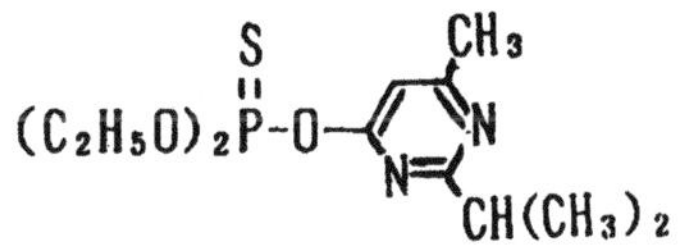	400 rats 80–135 mice	Also miticidal
Fenitrothion[a]	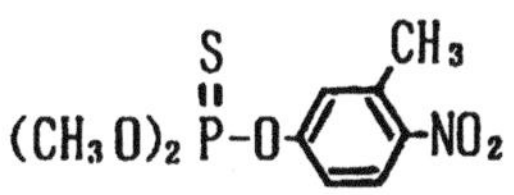	1700 male rats 1720 female rats	Also miticidal
Fenthion[a]	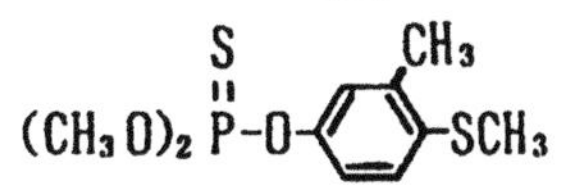	ca. 250 rats 245–615 female rats	Also miticidal
Isoxathion	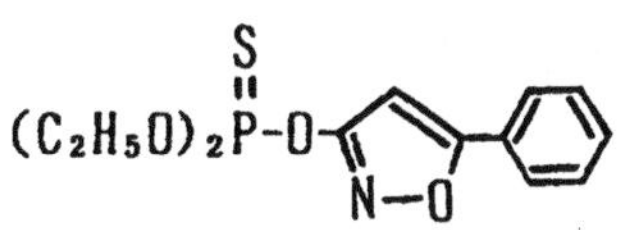	112 rats 98.4 mice	
Parathion	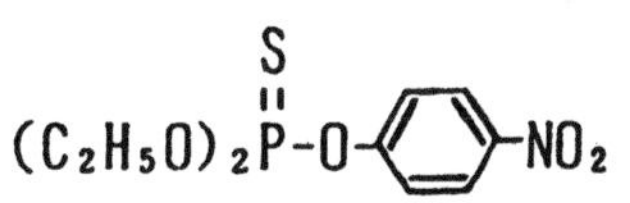	ca. 2 rats ca. 12 mice	Also miticidal
Parathion-methyl	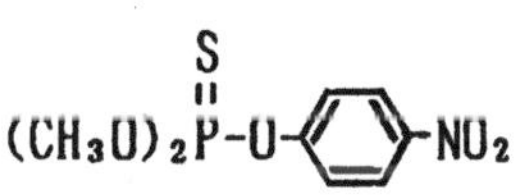	ca. 6 rats ca. 30 male mice	Also miticidal
Phoxim[a]	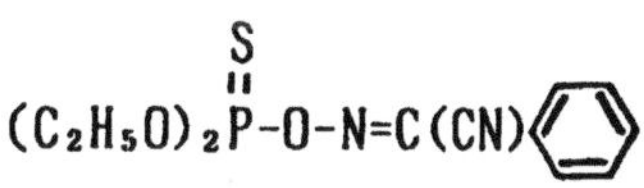	> 2000 rats	
Pirimiphos-ethyl	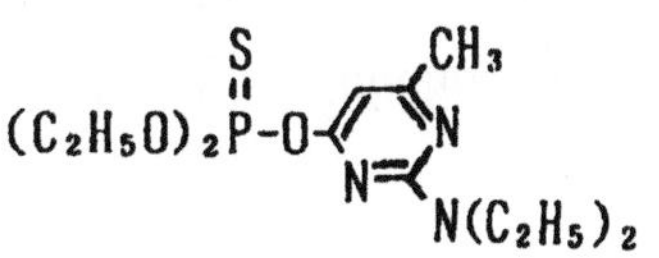	140–200 rats	
Pirimiphos-methyl	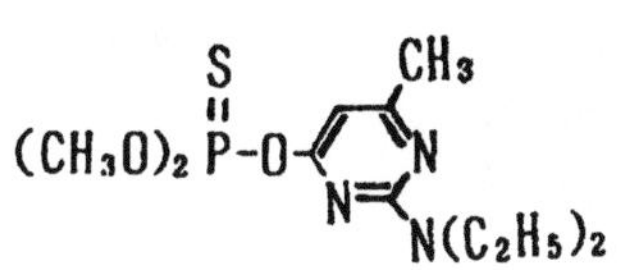	1180 mice 2050 rats	

(Contd)

Common name	Structure	Toxicity oral LD_{50}(mg/kg)	Comment
Quinalphos	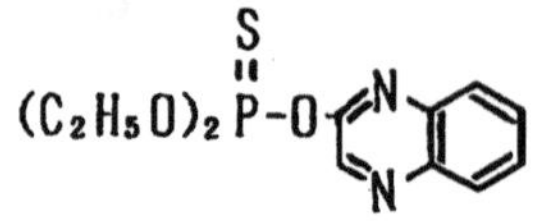	71 male rats	
Temephos[a]	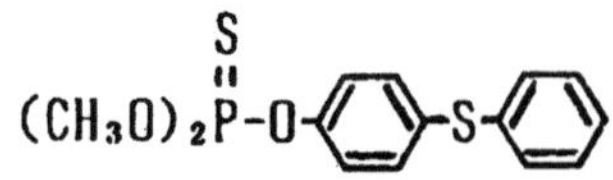	4204 male rats > 10000 female rats	
Phosphorothiolate			
Cadusafos	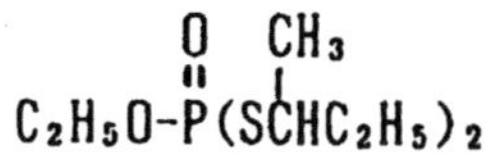	37.1 rats 71.4 mice	Nematicidal
Fosthiazate	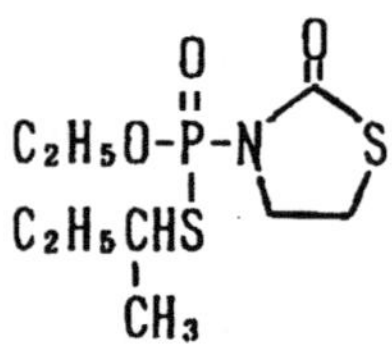	57 rats 91 mice	Systemic Nematicidal
Oxydemeton-methyl	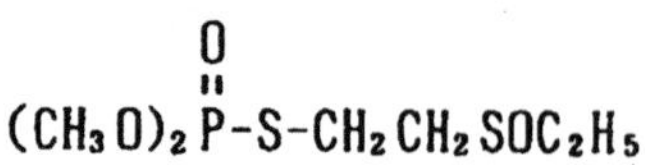	ca. 50 rats 75 female rats	Systemic Also miticidal
Oxydeprofos (ESP; Estox)	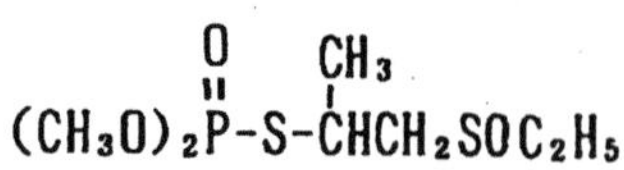	105 rats	Systemic
Profenofos	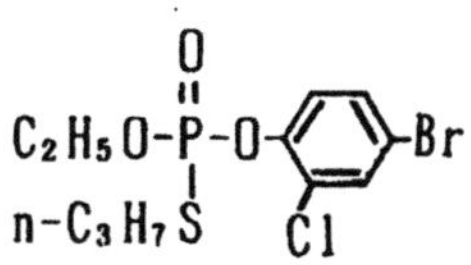	358 rats	
Pyraclofos[a]	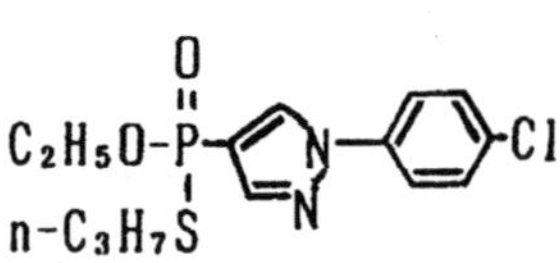	237 rats 575 male mice 420 female mice	Nematicidal
Vamidothion		100–105 male rats 64–67 female rats	Systemic Also miticidal
Phosphorodithioate			
Azinphos-methyl	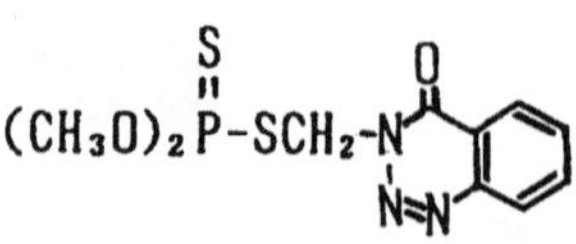	ca. 9 female rats[3]	Also miticidal

(Contd)

Common name	Structure	Toxicity oral LD_{50}(mg/kg)	Comment
Carbophenothion	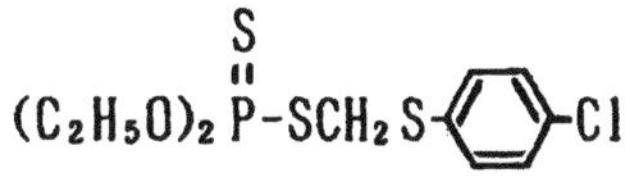	32.3 male rats	Also miticidal
Dimethoate[a]		290–325 rats 160 mice	Systemic Also miticidal
Disulfoton	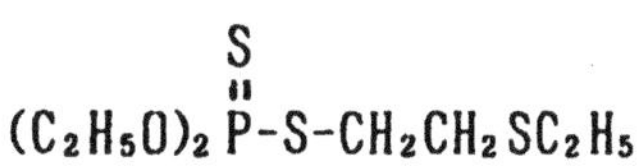	2–12 rats 7.5 mice	Systemic Also miticidal
Malathion[a]	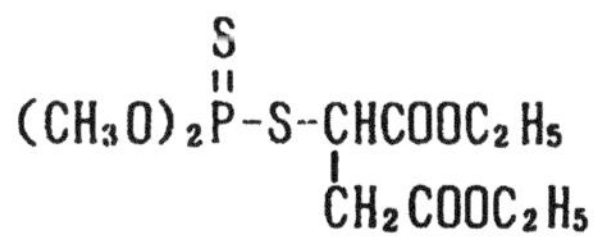	775 3320 mice 1375–2800 rats	Also miticidal
Methidathion	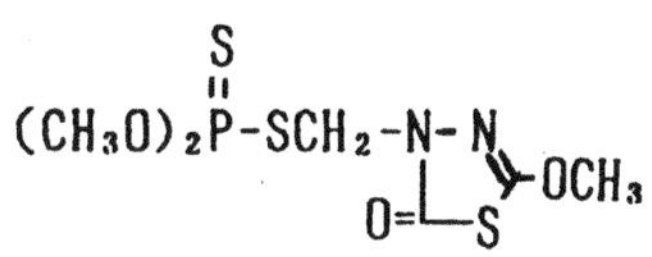	25–70 mice 25–54 rats	Also miticidal
Phenthoate	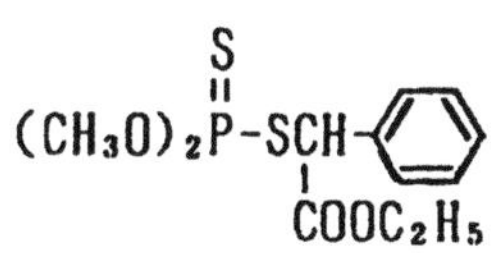	350–400 mice 300–400 rats	Also miticidal
Phorate	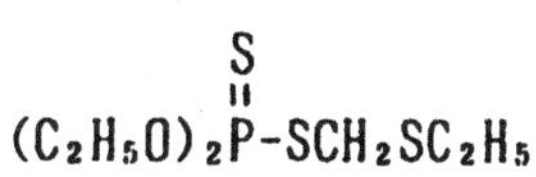	3.7 male rats 1.6 female rats	Systemic Also miticidal
Prothiophos	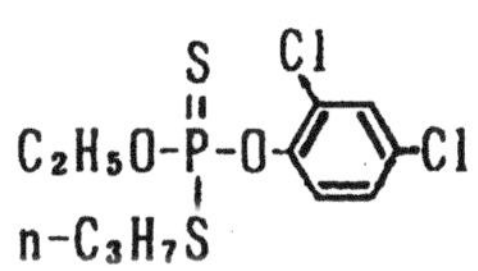	ca. 1500 rats ca. 2200 mice 420 female mice	
Terbufos	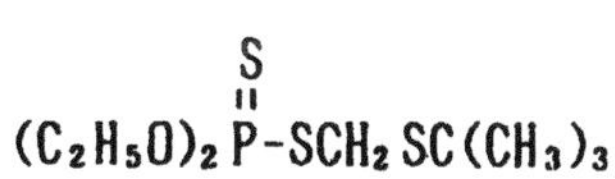	1.6 male rats 5.4 female mice	Soil insecticide Also nematicidal
Phosphonate, phosphonothionate, phosphonodithioate			
EPN	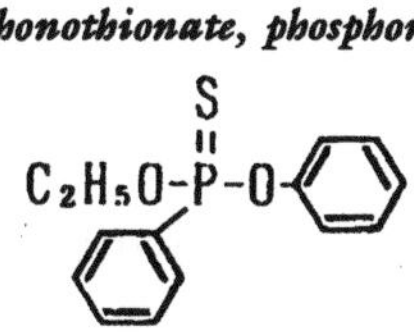	94.8 male mice 59.4 female mice 36 male rats 24 female rats	Also miticidal
Fonofos	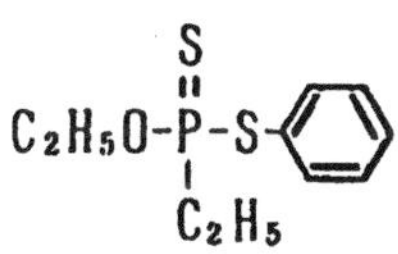	11.5 male rats 5.5 female rats	Soil insecticide

(Contd)

Common name	Structure	Toxicity oral LD_{50}(mg/kg)	Comment
Trichlorfon[a]	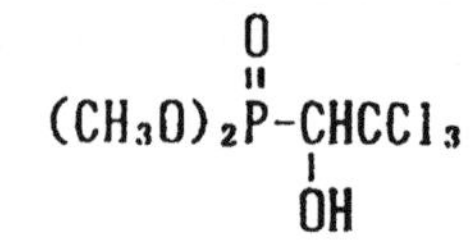	950 mice 630 male rats 560 female rats	
Phosphoroamidate,others			
Acephate	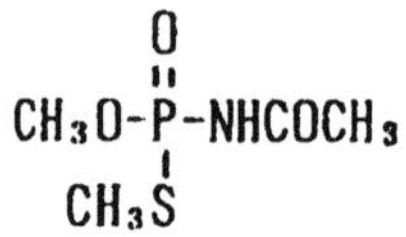	361 mice 945 male rats 866 female rats	Systemic
Fenamiphos	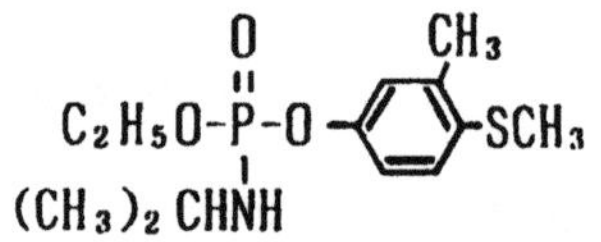	ca. 6 rats ca. 10 mice	Systemic Nematicidal
Isofenfos	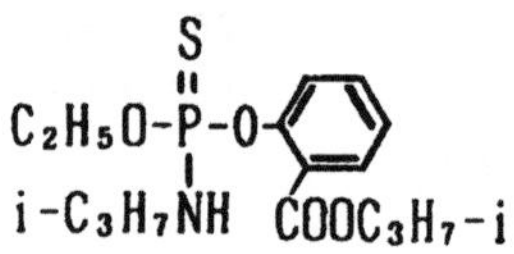	ca. 20 rats ca. 125 mice	
Methamidophos	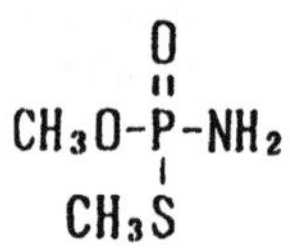	ca. 20 rats	Systemic Also miticidal
Phosphoran	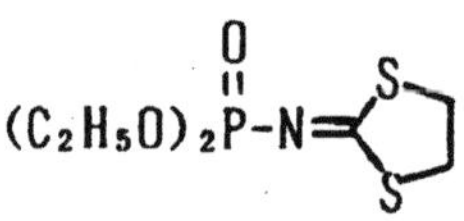	120–170 male rats 180 mice	Systemic
Propetamphos[a]	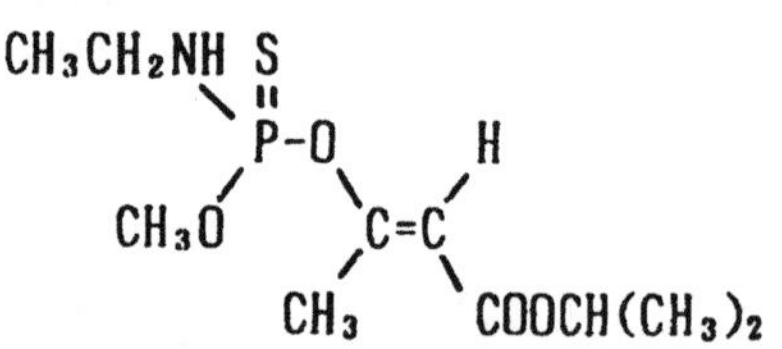	59.5 female rats	
Schradan (OMPA)	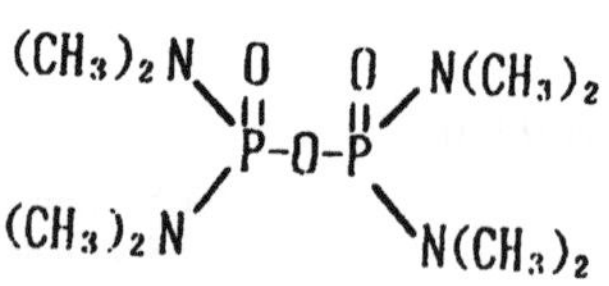	9.1 male rats 42 female rats	Systemic
TEPP	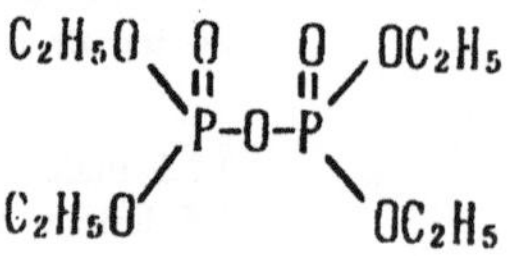	1.12 rats	Also miticidal Very hydrolyzable

[a]These compounds are also used in the control of insects in public health.

accumulation of acetylcholine at the junctions of motor nerves and skeletal muscle and in ganglia of the autonomic nervous system. These symptoms are called nicotinic, because of their similarity to the action of nicotine, and include involuntary twitching of the muscles, muscle spasms, weakness of respiratory muscles leading to dyspnea and cyanosis. Accumulation of acetylcholine in the central nervous system produces restlessness, anxiety, insomnia, neurosis, tremor, ataxia, convulsions, depression of respiratory centers, and coma. The immediate cause of death in fatal OP poisoning is asphyxia resulting from respiratory failure. The usual symptoms of poisoning in man include: headache, giddiness, nervousness, blurred vision, weakness, nausea, cramps, diarrhea, slowing of the heart, and drop in blood pressure. Signs of poisoning include: sweating, salivation, lachrymation, vomiting, constriction of the pupil, uncontrollable muscular twitches, and in severe cases convulsions, coma, loss of reflexes, and death from respiratory failure.

In addition to their direct toxic action a few OP compounds exert, in some vertebrates, a specific action which appears as a flaccid paralysis, especially of the legs (in man and fowl) or in the hind limbs of other animals, such as calf, cat, lamb, and rabbit. When seen under the microscope, the lesion appears to be as damage to the myelin sheath (a mixture of lipoid and proteinaceous material) surrounding the nerve. Symptoms usually appear within 2 weeks after poisoning. Recovery may be fast (1 day or more) or the paralysis may be prolonged or even become permanent. The more permanent type of neurotoxicity occurs with only a few compounds such as tri-ortho-cresyl phosphate (TOCP), di-isopropyl-fluorophosphate (DFP), mipafox, merphos, *p*-cresyl saligenin phosphate, and a few others. Several OP insecticides such as azinphos-methyl, carbophenothion-methyl, fenthion, malathion, parathion-methyl, chlorpyrifos, coumaphos, and others (see Hayes 1982 for an extended list of compounds) also produce these neurotoxic symptoms, but they last only a few days to 4–6 weeks, after which there is complete recovery. The mechanism of neurotoxic action of these compounds has been proposed as initial phosphorylation of "neurotoxic esterase or neuropathy target esterase (NTE)" followed by subsequent "aging", i.e., dealkylation, of the phosphorylated NTE (Johnson 1982).

The best test for assessing the degree of exposure to the OP insecticides is the inhibition of blood cholinesterase since it is easy to obtain samples and determine the degree of enzyme inhibition rather quickly. It has been calculated that a decrease of both plasma and red blood cell cholinesterase by 23–30% below a single preexposure value would be considered to be significant. Acetylcholinesterase inhibition of 50% of the preexposure value requires immediate action including temporary removal from further exposure and appraisal of the work situation (WHO 1975). Death usually occurs when 95% of brain acetylcholinesterase is inhibited.

11.3 Symptoms of Poisoning in Insects

Symptoms of poisoning in insects roughly follow the general pattern of nerve poisoning, i.e., restlessness, hyperexcitability, tremors (especially of the extremities), convulsions, paralysis, and death. The symptoms appear slowly and death may occur after several hours or up to 24 h. The ultimate cause of death is not certain but is regarded to be the result of cholinesterase inhibition by analogy with mammalian poisoning. A study of 17 OP insecticides tested on houseflies showed that, at least with 12 of them, death occurred at the time of maximum cholinesterase inhibition (Mengle and Casida 1958). In most insect species cholinesterase inhibition by OP compounds is not reversible.

11.4 Mode of Action

In order to understand the mechanism of toxic action of OP compounds it is essential to describe the interaction of acetylcholine with the enzyme acetylcholinesterase (AChE) (Fig. 11.3). AChE first produces E-OH • ACh, that is, the intermediate reversible complex between the enzyme and the substrate, which in turn forms acetylated AChE. Acetylated AChE is hydrolyzed to restore AChE. In the living organism these reactions proceed at a fast rate so that there is no accumulation of acetylcholine across the synapse or at the neuromuscular junction. The choline portion that is removed and the acetic acid combines again (with the help of other enzymes) to form new acetylcholine and the cycle is repeated.

OP compounds react with AChE in the same manner as the normal substrate acetylcholine (Fig. 11.4). In the first step the enzyme forms a reversible complex with the OP compound; next, the complex breaks down whereby a portion of the OP compound (the leaving group) is removed and the enzyme is phosphorylated (inhibited). The inhibition results from the long life of the phosphorylated enzyme. In the last step, hydrolysis takes place, the AChE is liberated along with a phosphate residue (such as dimethyl phosphoric acid)

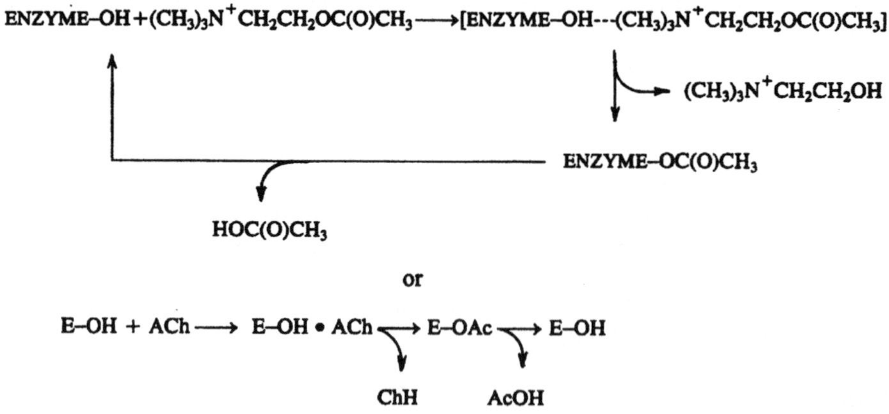

Fig. 11.3. Hydrolysis of acetylcholine by acetylcholinesterase. *E-OH* or *ENZYME-OH* acetylcholinesterase (AChE); *ACh* acetylcholine; *ChH* choline; *AcOH* acetic acid; *ENZYME-OC(O)CH₃* or *E-OAc* acetylated *AChE*

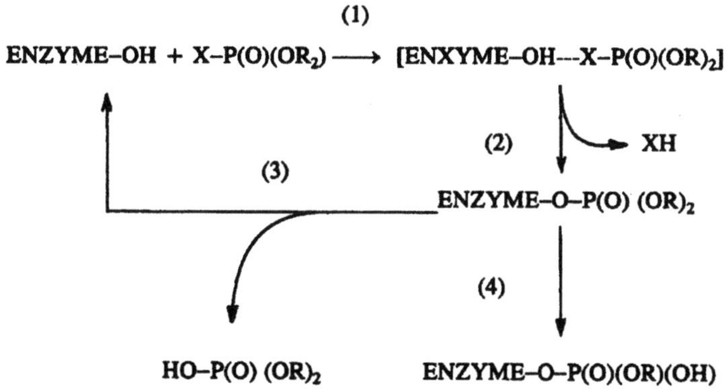

Fig. 11.4. Inhibition of an esterase enzyme by organophosphorus compounds. *(1)* Formation of Michaelis complex; *(2)* phosphorylation of the enzyme; *(3)* reactivation reaction; *(4)* "aging". (WHO 1986)

which is detached from the enzyme. The hydrolysis step is much slower in comparison with that of the normal substrate acetylcholine, so that the enzyme is no longer capable of dealing efficiently with bursts of acetylcholine entering the synapse as a result of nervous activity. Acetylcholine then builds up and proper nerve function is disrupted.

11.5 Metabolism

Most OP insecticides are converted in the insect or the animal body to compounds of greater activity (activation) or to those of lower or no biological activity (detoxification). Many of the OP insecticides containing the P=S bond (called "thion") are converted to P=O (called "oxon") by a system of oxidizing enzymes (called microsomal mixed-function oxidases or MFO) in which the enzyme cytochrome P-450 plays a major role. In most cases the P=O derivative is much more inhibitory to AChE than the parent P=S compound, for example, paraoxon is more potent than parathion, malaoxon is more potent than malathion, etc. On the other hand, the oxon derivatives are more easily broken down by hydrolytic reactions to yield less toxic or nontoxic metabolites which are quickly eliminated from the body. Other enzymes, such as glutathione transferase, carboxyesterase, carboxyamidase, etc., are also involved in the metabolism of OP compounds.

Because OP compounds are highly reactive, and as a result of this reactivity they disappear quickly, the amounts of unchanged insecticide found in food and in animal or human tissues are extremely low. This explains why it is difficult to detect OP residues in food and why determination of OP residues in human tissues or in urine is unlikely to indicate the degree of exposure of a person to OP insecticides (except in cases of attempted murder or suicide, or exposures to large doses of the poison). The individual or a combination of the following initial metabolic reactions occur on each of OP compounds in mammals, insects, and plants, or even in soils and by light. The metabolic rate is an important factor to affect the mammalian toxicity, insecticidal efficacy, persistency, selective toxicity, and resistance.

1. Conversion of Thion to Oxon. MFO catalyzes the reaction, but not only the oxon (II), but also several other products are produced concomitantly (Nakatsugawa and Dahm 1967) via an unstable phosphorus oxythionate (I) (Fig. 11.5, McBain et al. 1971). The intermediate is converted to the oxon (II) by releasing sulfur radical which binds macromolecules in vitro and is eventually excreted as sulfate ion in vivo. At the same time, it is hydrolyzed to give the phosphoric acid (III) and the thiophosphoric acid (IV) together with the leaving group (V). The intermediate also rearranges to (VI), which is also produced by combination of (IV) and (V) (Wustner et al. 1972; Miyamoto and Yamamoto, 1977).

2. Activation of Phosphorothiolates. In general, phosphorothiolates of $(RO)_2P(O)SZ$ type strongly inhibit AChE. However, certain thiolates such as methamidophos, profenophos, and prothiophos oxon are weak AChE inhibitors, but strongly insecticidal. MFO converts them to the sulfoxides which either inhibit AChE or decompose depending on the alkyl substituents on the sulfur atom (Fig. 11.6; Eto et al. 1977; Wing et al. 1982; Kimura et al. 1984).

3. Dealkylation of O-alkyl and N-alkyl Groups by MFO. The O-alkyl group (except methyl) is hydroxylated by MFO to the alkylol (I), then dealkylated. Likewise, the N-alkyl group is hydroxylated by MFO to the alkylol (II), probably via N-oxide (III), then dealkylated as shown in Fig. 11.7.

Schradan (OMPA) is poor in cholinesterase inhibition in vitro, but in vivo it is oxidatively converted to the metabolite(s) of strong antiAChE activity. The N-oxide or N-methylol might be the unstable active intermediate to inhibit AChE.

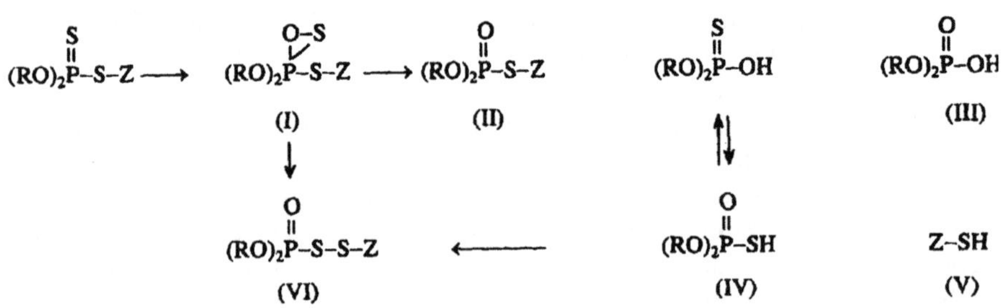

Fig. 11.5. Activation and degradation of thionophosphorus compounds. *I* Phosphorus oxythionate; *II* oxon; *III* Phosphoric acid; *IV* thiophosphoric acid; *V* leaving group. Formation of *VI* can be demonstrated by the detection of the stable reaction products with diazomethane (Miyamoto and Yamamoto 1980):

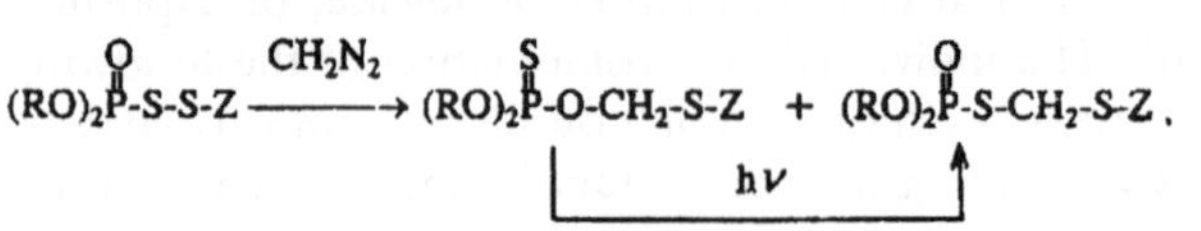

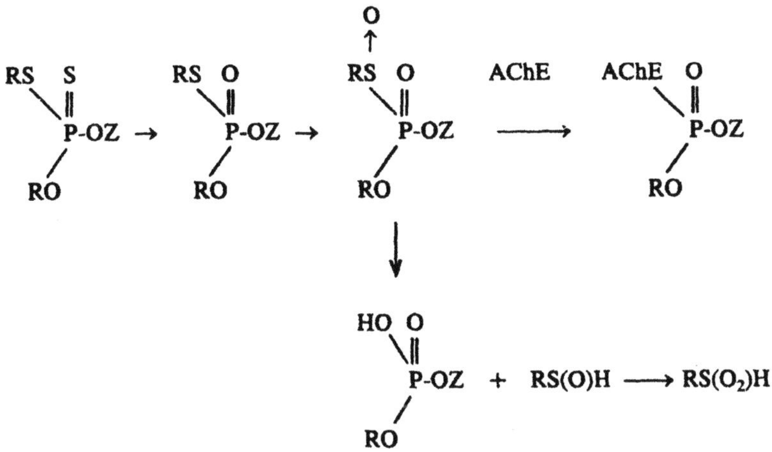

Fig. 11.6. Activation of S-alkyl phosphorothiolates

$$[(CH_3)_2N]_2P(O){-}O{-}P(O)[N(CH_3)_2]_2 \rightarrow {-}NO(CH_3)_2 \rightarrow {-}N(CH_3)\ (CH_2OH) \rightarrow {-}NH(CH_3)$$

Schradan — N-oxide — N-methylol — N-desmethyl Schradan.

4. Oxidation of Thioethers to Sulfoxides and Sulfones by MFO

$$=P(O){-}O{-}R{-}S{-}R \rightarrow =P(O){-}O{-}R{-}S(O){-}R \rightarrow =P(O){-}O{-}R{-}S(O_2){-}R$$

Thioether — Sulfoxide — Sulfone

These products are often stronger AChE inhibitors.

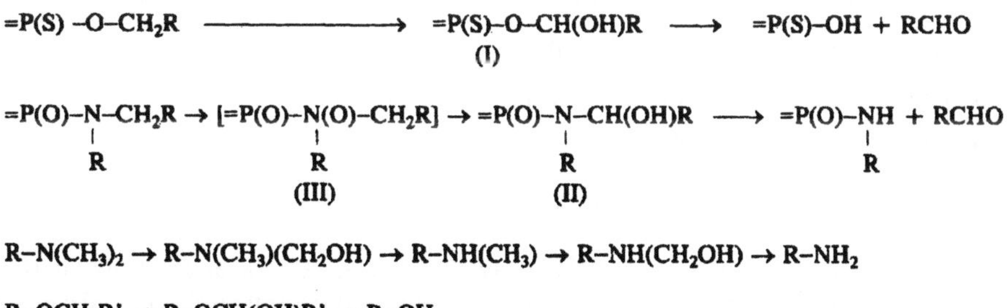

Fig. 11.7. Dealkylation of *O*-alkyl and *N*-alkyl op compounds. *I* *O*-alkyl group (except methyl) hydroxylated by MFO to alkylol *I*; *II* N-alkyl group hydroxylated by MFO to alkylol *II*; *III* N-oxide

5. Hydroxylation of Side Groups by MFO

$-CH(CH_3)_2 \rightarrow -C(OH)(CH_3)_2$
$\qquad\qquad\quad -CH(CH_3)(CH_2OH)$

$-CH_3 \rightarrow -CH_2OH \rightarrow -CHO \rightarrow -COOH$

$-OCH_2CH_3 \rightarrow -OCH(OH)CH_3 \rightarrow -OH.$

Conversion of tri-*o*-cresylphosphate to cyclic phosphate.

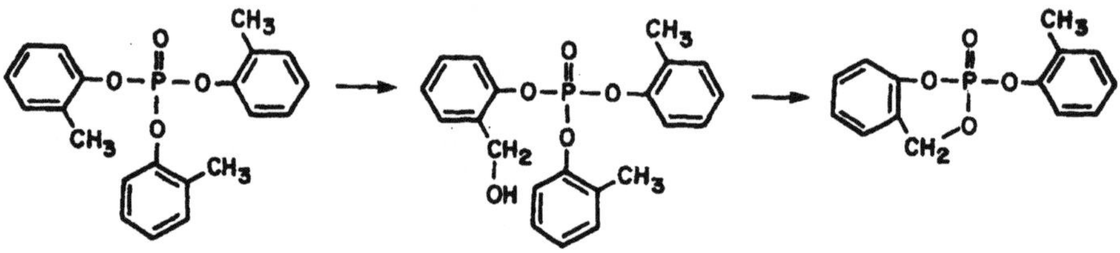

This metabolic reaction leads to a cyclic phosphate metabolite of remarkable anticholinesterase activity, which is of delayed neurotoxicity but not insecticidal. Modification to aryl ester to methyl ester results in the invention of salithion insecticide with no delayed neurotoxicity (Eto et al. 1962, 1963).

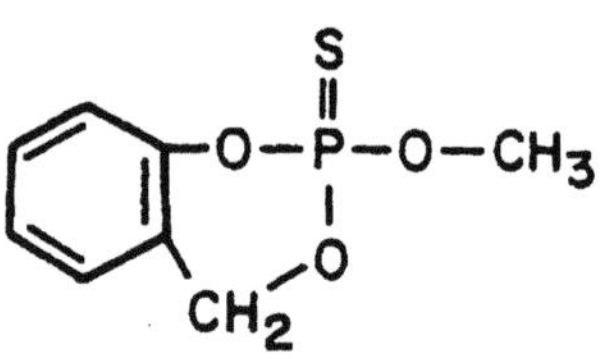

Salithion

6. Hydrolysis of Phosphates and Phosphorohalidic Acid Esters by Esterases. As compared with phosphorothionates, phosphates are preferred substrates.

$(RO)_2P(O){-}O{-}Z \rightarrow (RO)(HO)P(O){-}O{-}Z + ROH$

$(RO)_2P(O){-}O{-}Z \rightarrow (RO)_2P(O){-}OH + ZOH$

$(RO)_2P(O){-}X \rightarrow (RO)_2P(O){-}OH + X,$

where Z denotes the leaving group and X halogen, CN.

7. Hydrolysis of Ester Groups in the Leaving Groups. A typical example of the hydrolysis of ester groups in the leaving groups is the hydrolysis of malathion by carboxyesterase.

$$(CH_3O)_2P(S){-}\underset{\substack{|\\ CH_2-COOC_2H_5}}{CH}{-}COOC_2H_5 \rightarrow (CH_3O)_2P(S){-}\underset{\substack{|\\ CH_2-COOC_2H_5}}{CH}{-}COOH \;+\; -\underset{\substack{|\\ CH_2-COOH}}{CH}{-}COOC_2H_5$$

8. Degradation by Glutathion S-Transferases. The methoxyl group on the phosphorus atom is demethylated by glutathion-glutathion S-transferase system (GSH-GST), while higher alkoxyl groups are dealkylated by oxidation. Also, leaving groups (Z) may well be conjugated with GSH.

$$(CH_3O)_2P(S)\text{–}OZ \xrightarrow{\text{GSH–GST}} (CH_3O)\,(HO)P(S)\text{–}OZ + CH_3SG$$
$$(CH_3O)_2P(S)\text{–}OH + ZSG.$$

9. Isomerization of Phosphorothionates to the Thiolates. Isomerization of phosphorothionates to the thiolates occurs by light, heating, and metabolism, leading often to highly toxic products.

$$(RO)_2P(S)\text{–}O\text{–}Z \rightarrow (RS)\,(RO)P(O)\text{–}O\text{–}Z$$
$$(RO)_2P(S)\text{–}O\text{–}Z \rightarrow (RO)_2P(O)\text{–}S\text{–}Z.$$

10. Miscellaneous

(a) Conversion of trichlorfon and naled to dichlorvos:

$$\underset{\text{Trichlorfon}}{(CH_3O)_2P(O)\text{–}CH(OH)\text{–}CCl_3} \longrightarrow \underset{\text{Dichlorvos}}{(CH_3O)_2P(O)\text{–}OCH{=}CCl_2}.$$
$$\underset{\text{Naled}}{(CH_3O)_2P(O)\text{–}O\text{–}CH(Br)\text{–}CCl_2Br} \nearrow$$

They are all insecticidal with different physical properties: trichlorfon, rather water-soluble; dichlorvos, very volatile; naled, nonvolatile.

(b) Reduction of the nitro group by nitroreductase:

Parathion (R: C_2H_5)

The in vivo reduction of parathion, paraoxon, EPN, and fenitrothion to the amino derivatives is an important detoxication step in ruminants, but not so in other animals.

11.6 Aging of the Inhibited Cholinesterase

One of the characteristics of the inhibition of cholinesterase by OP compounds is that the rate of recovery of the enzyme becomes less and less as the time of enzyme-inhibitor contact becomes longer in the absence of treatment with oximes (Hobbiger 1955). This is known as "aging." The reason for aging is that an alkyl group is removed from the OP portion of the molecule, thus making the inhibited enzyme more stable and not subject to reactivation by oximes. Therefore, in cases of poisoning, first aid and therapy should be given as soon as possible in order to prevent the aging phenomenon.

11.7 Avian Teratogenicity

Certain organophosphorus (OP) and carbamate (MC) compounds cause teratogenesis in birds, when they are injected into developmental fertilized eggs. There are two types of teratogenicity. The symptoms of type 1 are shortened legs and no feathering, while the symptom of type 2 is vertebral abnormality. Carbaryl causes only type 1 and parathion causes mostly type 2 symptoms. Dicrotophos, diazinon, and eserine show both symptoms, but type 1 symptoms at lower doses. These compounds could inhibit kynurenine formamidase (Kfase) in yolk sac membrane and catalyze the convertion of tryptophan to nicotinamide adenine dinucleotide (NAD), decreasing NAD level in the embryo, and resulting in type 1 teratogenicity (Fig. 11.8.). Nicotinamide mitigates the action. On the other hand, type 2 is the result of disturbing the cholinergic system by inhibition of acetylcholinesterase, which is mitigated by pralidoxime. Type 2 is also caused by cholinergic ligands like carbamylcholine, decamethonium, and tetramethylammonium.

Avian teratology does not necessarily infer human because the avian embryo is nonplacental and therefore does not have the protection offered by maternal absorption, detoxication, and excretion of foreign compounds. Therefore, avian teratology is not considered in pesticide regulation.

11.8 Therapy

The treatment of OP poisoning by drugs is a case of antagonism between the OP compound and the drug. Two principles have been employed in therapy: (1) attempts to compensate for the excess acetylcholine in the synapse due to

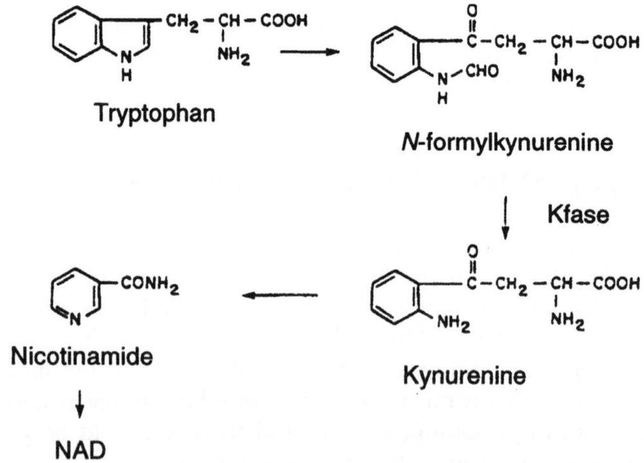

Fig. 11.8. Organophosphorus and methylcarbamate type 1 teratogens inhibit kynurenine formamidase (Kfase). (Eto 1981)

the inhibition of cholinesterase by using acetylcholine antagonists; (2) attempts to reactivate the inhibited cholinesterase. Atropine, an acetylcholine antagonist, is the oldest antidote used in poisoning by OP compounds. Atropine is a complex alkaloid derived from the plant *Atropa belladonna* which competitively blocks the action of acetylcholine but only at cholinergic receptors in the parasympathetic nervous system (see diagram of nervous system in Fig. 8.1). Effects at the neuromuscular junctions, such as restlessness, tremors, muscular spasms, fibrillation of facial muscles, etc., are counteracted by treatment with oximes in conjunction with atropine. The oxime combines with the inhibited cholinesterase and displaces the OP moiety from it, thus liberating the enzyme. The phosphorylated oxime produced should be further converted to nitrile as shown in Fig. 11.9. If not, the phosphorylated oxime may under some circumstances be a more potent inhibitor than the original inhibitor by re-phosphorylating the enzyme as is found for mixtures of sarin and 2-PAM. One of the best oximes for this purpose is pralidoxime (pyridine aldoxime methiodide or 2-PAM). Pralidoxime chloride is among the more useful, but it cannot be transported to the central nervous system.

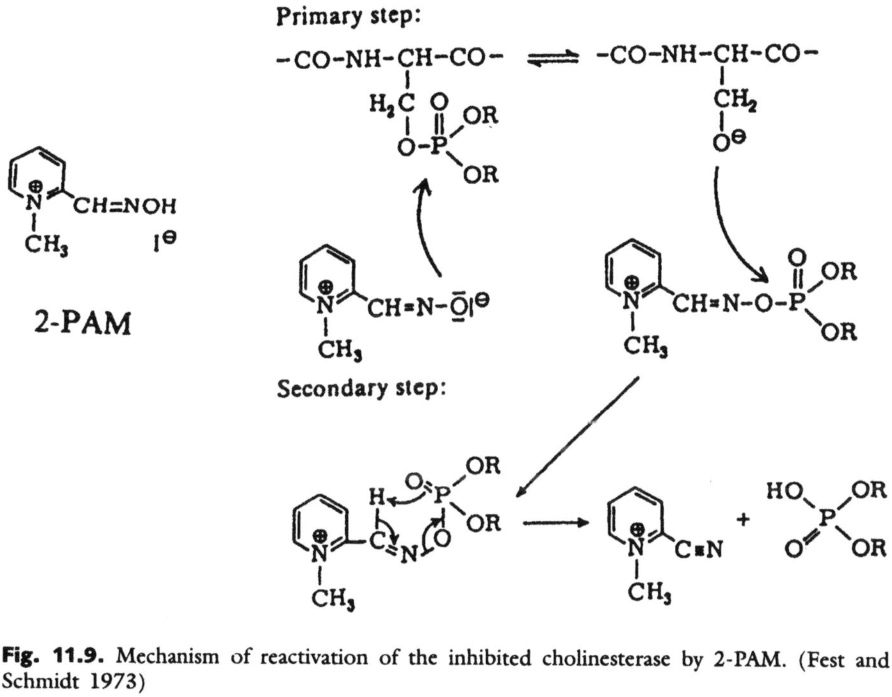

Fig. 11.9. Mechanism of reactivation of the inhibited cholinesterase by 2-PAM. (Fest and Schmidt 1973)

In serious cases of poisoning by OP compounds the order of treatment should be as follows:

1. Artificial respiration preferably by mechanical means.
2. Injection of 2–4 mg atropine sulfate intravenously and repeated at 5–10-min. intervals until signs of atropinization appear.

3. Injection of pralidoxime chloride (1 g) slowly, intravenously.
4. Decontamination of the skin, stomach, and eyes as indicated.
5. Symptomatic treatment (e.g., inhalation of oxygen).

In the more usual cases, the procedure should be as follows:

1. Injection of 1–2 mg atropine sulfate if symptoms appear, and repeated doses as necessary.
2. Decontamination of the skin, stomach, and eyes.
3. Injection of pralidoxime chloride (1 g) slowly, intravenously, if the patient fails to respond satisfactorily to atropine.
4. Symptomatic treatment (see Hayes, 1982 for more details).

CHAPTER 12

Carbamate Insecticides

12.1 Structure

Biologically active carbamates have been used as far back as the 17th century in the old Calabar region of southeast Nigeria. The Effiks who inhabited the area used to collect the coffee-colored Calabar beans of a plant later named *Physostigma venenosum*. The crushed bean called esere by the natives was used as a poison ordeal to establish the guilt of a prisoner. The accused to be judged was forced to swallow the milky potion of crushed Calabar bean seeds in a small quantity of water. If the prisoner was fortunate enough to regurgitate the "milk" and survive he was declared innocent, but if he developed shakes, froth at the mouth, and quickly died he was declared guilty.

Tentative identification of the bean extract as an alkaloid was made in 1861 and was named eserine from the Effiks' "esere." A year later the alkaloid was isolated, purified and renamed physostigmine. The molecular structure was established in 1925 and the compound was synthesized in 1935 (see Kuhr and Dorough 1976 for details).

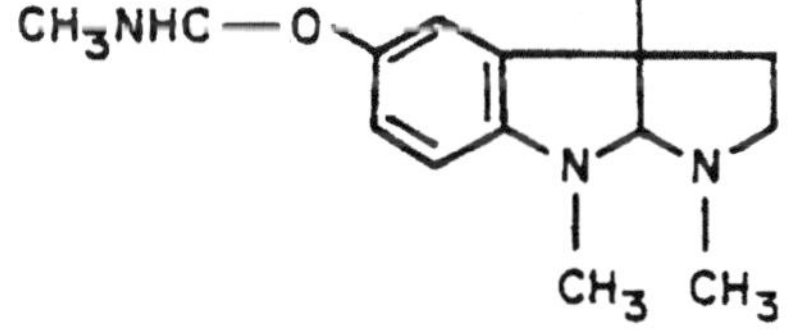

Physostigmine (eserine)

Following the synthesis of physostigmine a large number of methylcarbamates was synthesized and proven to be toxic to many insect species. The important carbamate insecticides in use worldwide are listed in Table 12.1.

12.2 Mode of Action

Carbamates react with acetylcholinesterase in the same manner as the OP insecticides. Thus, the reason that the carbamates are toxic lies in their ability

Table 12.1. Carbamate insecticides

Common name	Structure	Toxicity LD_{50} (mg/kg)	Comment
Aryl methylcarbamate			
Bendiocarb[a]	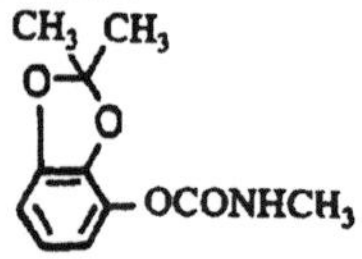	80–120 rats 45 mice	
BPMC (Fenobucarb)	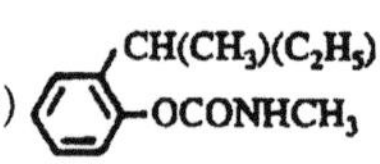	410 rats	Systemic
Carbaryl[a]	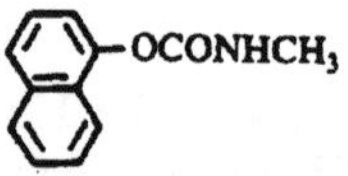	800 male rats 500 female rats	
Carbofuran	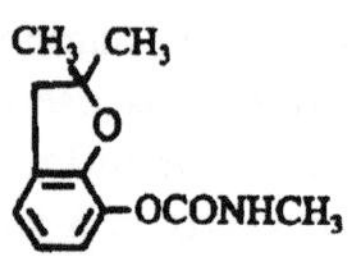	8–14 rats	Systemic Also nematicidal
Ethiofencarb	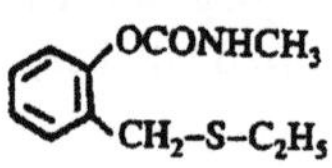	200 rats 240 mice	Systemic
Formetanate	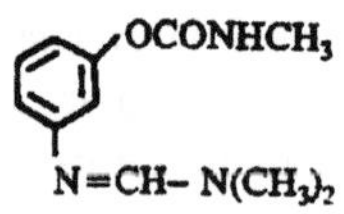	24 rats 13–25 mice	Also miticidal and molluscicidal
Methiocarb	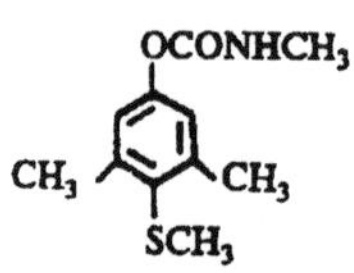	20 rats 52–58 mice	Also miticidal and molluscicidal
MIPC (Isoprocarb)	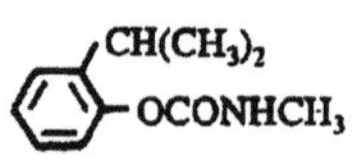	193 male mice 128 female mice 485 male rats	
MPMC (Xylylcarb)	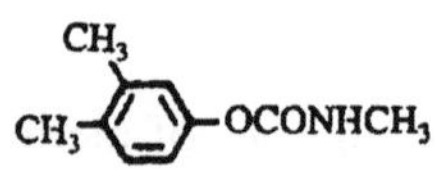	375 male rats 325 female rats	
MTMC (Metolcarb)	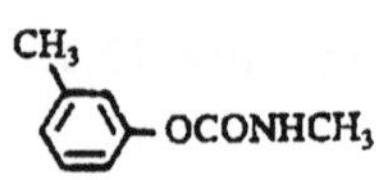	109 mice 580 male rats 498 female rats	Systemic
Promecarb	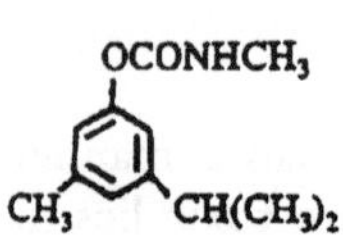	81–90 rats	

(Contd)

Common name	Structure	Toxicity LD_{50} (mg/kg)	Comment
Propoxur[a]	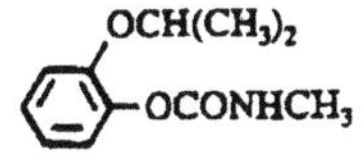	90–128 male rats 104 female rats 100–109 male mice	
XMC	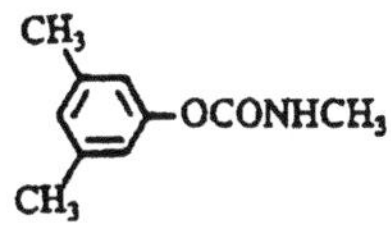	245 mice 542 rats	
Heterocyclic dimethylcarbamate			
Pirimicarb	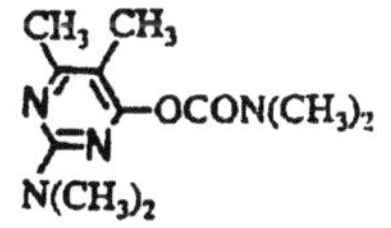	147 rats 107 mice	Systemic
Oxime methylcarbamate			
Aldicarb	SCH_3 | CH_3–C–CH_3=N–$OCONHCH_3$ | CH_3	0.93 male rats	Systemic Also nematicidal and miticidal
Butocarboxim	$CH(CH_3)SCH_3$ | CH_3–C=N–$OCONHCH_3$	158–215 rats	Systemic Also miticidal
Methomyl	SCH_3 | CH_3–C=N–$OCONHCH_3$	17–24 rats	Systemic Also nematicidal
Oxamyl	SCH_3 | $(CH_3)_2NCO$–C=N–$OCONHCH_3$	5.4 rats	Systemic Also miticidal and nematicidal
Carbamate derivative			
Alanycarb	SCH_3 CH_3 | | CH_3–C=N–$OCON(CH_2CH_2COOC_2H_5)CH_2C_6H_5$	301 rats	Also nematicidal
Benfuracarb		138 rats 175 mice	Also nematicidal
Carbosulfan	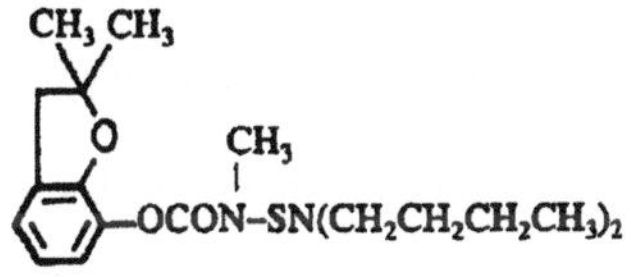	250 male rats 185 female rats	Systemic Also nematicidal

(Contd)

Common name	Structure	Toxicity LD_{50} (mg/kg)	Comment
Furathiocarb	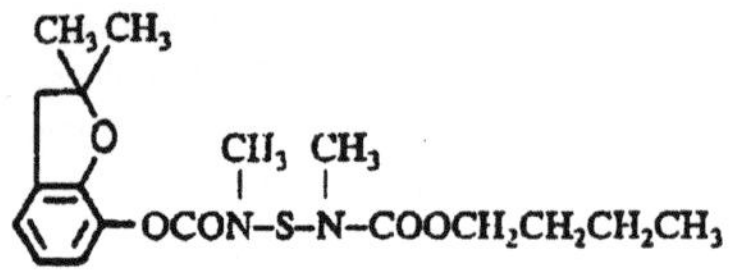	53 rats 327 mice	Systemic Also nematicidal
Thiodicarb	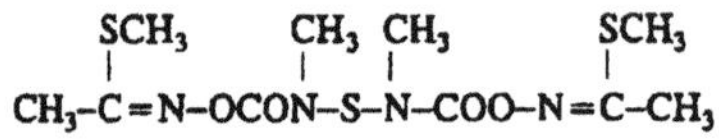	166 rats	Also molluscicidal
Cyclic carbamate			
Metoxadiazone	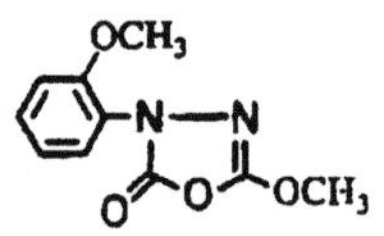		

[a]These compounds are also used in the control of insects in public health.

to inhibit the vital nerve enzyme acetylcholinesterase, resulting in the accumulation of acetylcholine in the nerve synapse and causing a disruption of nerve function (Fig. 12.1).

When a carbamate insecticide enters the synapse, which occurs during poisoning, it competes with acetylcholine for the active site on the enzyme. If successful, the insecticide binds to the enzyme and forms a complex with it (step 1). This complex could be reversible and the enzyme could be dissociated from the carbamate if pressure from the insecticide is relieved. However, most of the time the reaction will proceed to step 2 where the complex will decompose into a stable carbamylated (inhibited) enzyme and a leaving group. Finally, in step 3, the carbamylated enzyme is hydrolyzed to regenerate the free enzyme and methylcarbamic acid (Fig. 12.1). Steps 2 and 3 are thousands of times slower with carbamates than with acetylcholine but are faster than those with OP compounds. The half-life for decarbamylation, that is, liberation of the enzyme, has been calculated to be 30–40 min; thus, if all the acetylcholinesterase in a synapse were carbamylated (inhibited) and no further carbamate entered the synapse, the enzyme will begin to recover in only a few minutes and would be fully restored after several hours. This quick restoration of the enzyme explains

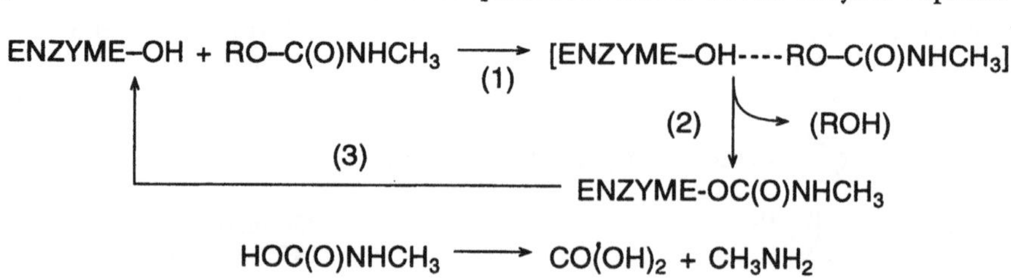

Fig. 12.1. Inhibition of an esterase enzyme by carbamate compounds. *1*, *2*, and *3* are steps in the mode of action

why animals showing symptoms of carbamate poisoning often recover within hours after exposure to carbamate insecticides. In contrast, the enzyme which is inhibited by OP insecticides (the phosphorylated enzyme) is considerably less susceptible to hydrolysis (step 3) so that regeneration of the free enzyme is slow and recovery from OP poisoning is a much slower process.

Heterocyclic carbamates (N-aryloxadiazolones) inhibit acetylcholinesterase in a similar way to conventional carbamates.

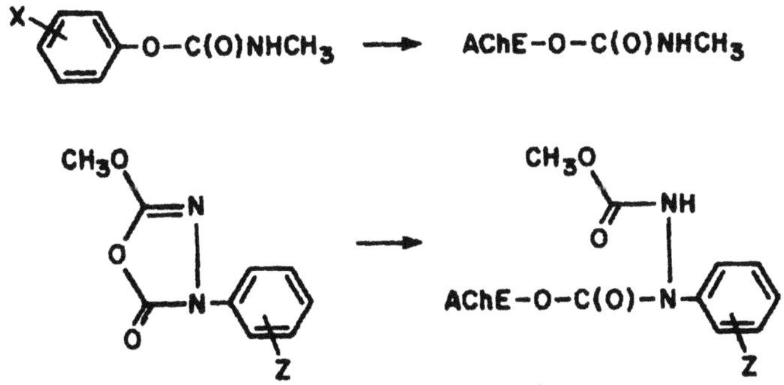

12.3
Toxicity to Mammals

In general, symptoms of poisoning by carbamate insecticides are similar to those produced by OP compounds and include miosis, blurred vision, weakness, nausea, and vomiting, sweating, salivation, tearing, urination, pulmonary edema, and convulsions. Death occurs from respiratory failure, but most likely it results from bronchial constriction, lowered blood pressure, neuromuscular block of respiratory muscles, and failure of the respiratory center of the brain. The rapidity with which these symptoms appear and their severity depend on the amount of chemical absorbed, the chemical structure of the toxicant, the mode of entry into the body, and the ability of the animal to metabolize the compound and eliminate the by-products. An acute dose of carbamate insecticide will produce most of the symptoms described above. Exposure of humans to sublethal doses usually causes headache, salivation, nausea, vomiting, muscular weakness, and diarrhea. Recovery from these symptoms is often complete in only a few hours after exposure, much faster than recovery from OP poisoning. Unlike OP compounds, carbamates do not affect the nerve myelin and do not produce the characteristic, delayed neurotoxic effect culminating in flaccid paralysis. Poisoning resulting from exposure to carbamates occurs less frequently than poisoning by OP compounds. Most carbamates are much less toxic by dermal absorption than by the oral route. However, there are exceptions, as shown in Table 12.2.

12.4
Toxicity to Insects

Many factors, both physiological and biochemical, influence the amount of insecticide required to kill a particular insect. The housefly is often employed as

Table 12.2. Acute toxicity of some carbamate insecticides to rats. (Gaines 1969)

	LD_{50} (mg (a.i.,/kg)			
	Oral		Dermal	
	Male	Female	Male	Female
Aldicarb	0.8	0.6	3.0	1.0
Isolan	23	13	5.6	6.2
Propoxur	83	86	> 2400	> 2400
Mexacarbate	37	25	> 1500	> 1500
Carbaryl	850	500	> 4000	> 4000

the standard laboratory insect to evaluate the toxicity of insecticides including carbamates. Several carbamates are broad-spectrum insecticides and are toxic to many insect species, while others are more selective. Carbaryl, for example, has a wide range of toxicity to insects with LD_{50}'s ranging from 0.1 to 1000 mg/kg. It is quite toxic to many insect species but is practically nontoxic to the housefly, the LD_{50} being 500 mg/kg. Carbamate insecticides possess both contact and stomach toxicity. In addition, some carbamates such as aldicarb (Temik) and carbofuran (Furadan) possess also systemic activity, that is, when applied to the soil the insecticide is absorbed by the plant roots and moves to other parts of the plant where it becomes toxic to sap-sucking insects. Systemic activity occurs also when the insecticide is sprayed on the leaves.

The symptoms produced by poisoning insects with carbamates include hyperactivity, incoordination (ataxia), convulsions, paralysis, and death. The remarkable feature of insect poisoning is the ability of the insect to withstand long periods of paralysis without dying. Complete recovery can occur in many instances after hours of prostration (knockdown). Presumably, the carbamate is metabolized quickly and the inhibited acetylcholinesterase is also regenerated rather rapidly, allowing recovery. The appearance, intensity, and duration of symptoms are dependent on dose, chemical structure of the compound, length of exposure, and mode of entry into the body.

12.5 Metabolism

Carbamate insecticides, like most other foreign compounds, are subject to biochemical transformations in vivo through many enzyme-catalyzed reactions. However, the principal mechanisms of detoxication are hydrolysis, oxidation, and conjugation.

Hydrolysis. Enzyme-catalyzed hydrolytic reactions convert the lipid-soluble compound into a more polar material for excretion or storage away from sensitive tissues (Williams 1959). In many cases, the insecticide is broken down leaving nontoxic products that can easily be deactivated by conjugation. Differences in enzymic hydrolysis rates depend on the structure of the insecticide and on the species contacting the chemical. Usually, mammals are more efficient in breaking the carbamate ester bond than are insects and plants. These differences are the basis for the selective toxicity of many carbamates in favor of mammals, but there are exceptions (Kuhr and Dorough 1976).

Although the carbamate ester linkage is broken with consequent liberation of the phenolic or oxime moiety directly by direct hydrolysis (1), the latter is liberated also by oxidation (2).

$$RO\text{-}C(O)NHCH_3 \rightarrow ROH + HOC(O)NH\ CH_3 \rightarrow CO(OH)_2 + CH_3\ NH_2 \quad (1)$$

$$\rightarrow RO\text{-}C(O)NHCH_2OH \rightarrow ROH + CO(OH)_2 + NH_2 + HCHO \quad (2)$$

Oxidation. The enzymic oxidation most intensively studied has been the microsomal mixed-function oxidase system through the mediation of the enzyme, cytochrome P-450, which is responsible for many types of reaction such as hydroxylation, dealkylation, deamination, desaturation, epoxidation, and N- or S- oxidation. The primary purpose of these biological reactions seems to be the enhancement of the solubility of the substrates in water, or providing them with an active functional group for conjugation and ultimately for excretion (for details see Matsumura 1985). Many carbamates are detoxified via this microsomal enzyme system both in mammals and in insects (Fig. 12.2). A similar mechanism may be operative in plants.

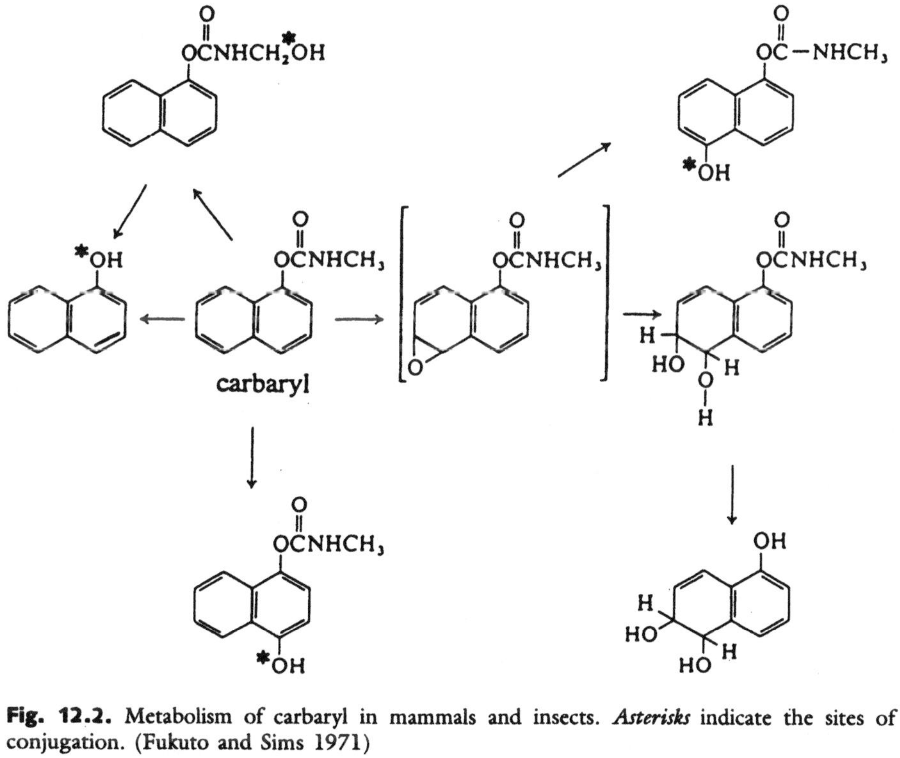

Fig. 12.2. Metabolism of carbaryl in mammals and insects. *Asterisks* indicate the sites of conjugation. (Fukuto and Sims 1971)

Conjugation. Conjugation is a process whereby endogenous substrates as well as foreign compounds react with natural body components such as glucuronic acid in mammals, glucosides in insects, glycosides in plants, as well as with other tissue components, principally, sulfate, phosphate, and amino acids (Fig. 12.3). The purpose of conjugation is to convert the metabolic products to water soluble compounds in order to facilitate their excretion or storage. In animals, excretion takes place via the urine and/or feces, whereas in plants conjugates tend to be stored as terminal metabolites in various tissues.

12.6 Therapy

Poisoning resulting from exposure to carbamate insecticides occurs less frequently than poisoning by OP insecticides and is easier to treat. Although the carbamate insecticides produce their toxic effects via the same mechanism as the OP compounds, that is, by inhibition of cholinesterase, the carbamylated (inhibited) enzyme recovers more rapidly than the phosphorylated enzyme. This quick recovery together with the more rapid metabolism of the carbamates results in the faster disappearance of symptoms of poisoning following exposure.

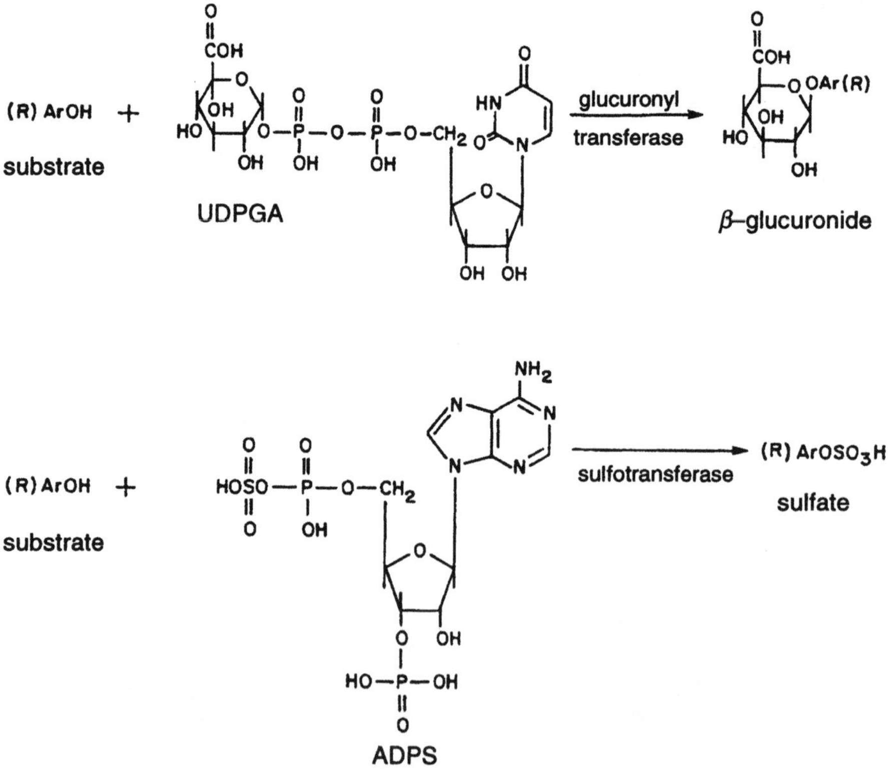

Fig. 12.3. Mechanism of glucuronide (*above*) and sulfate (*below*) conjugation of phenols (ArOH) and alcohols (ROH) in mammals. (Kuhr and Dorough 1976)

The antidote for carbamate poisoning is atropine (the same as that for OP insecticides). Atropine antagonizes the effect of excess acetylcholine which accumulated within the synapse by blocking the acetylcholine receptor sites, thus preventing continuous nerve stimulation. However, the antidote 2-PAM is not effective in accelerating the reactivation of cholinesterase (in some cases it might even be harmful); therefore, it should not be used as a therapeutic agent in human poisoning (Hayes 1982). The management with atropine is the same as that with OP insecticides.

CHAPTER 13

Botanical Insecticides

13.1 Nicotine

As early as 1690 water extracts of tobacco leaves were used to control sucking insects on garden plants (for the history of nicotine see Shepard 1951; for the reviews see Yamamoto 1965; Schmeltz 1971). The usefulness of this preparation was due to its content of the nicotine alkaloid which is present in tobacco leaves. In addition to nicotine, two other closely related compounds, i.e., nornicotine and anabasine, also have some insecticidal activity (Fig. 13.1). These compounds act as nerve poisons in both insects and mammals, but there is a great deal of variation in susceptibility among different species. Nicotine is obtained from the plants *Nicotiana tabacum* and *Nicotiana rustica* to the extent of 2–8%, but it also occurs in other plants. Nornicotine predominates in *Nicotiana sylvestris* and anabasine is extracted mainly from *Anabasis aphylla*.

13.1.1 Physical and Chemical Properties

The chemical name of the active nicotine compound is *l*-3-(1-methyl-2-pyrrolidinyl)-pyridine. The commercial name is Black Leaf 40 for the compound containing 40% nicotine sulfate. Free nicotine is a colorless or pale-yellow oily liquid with an acrid burning taste and the odor of pyridine. It is miscible with

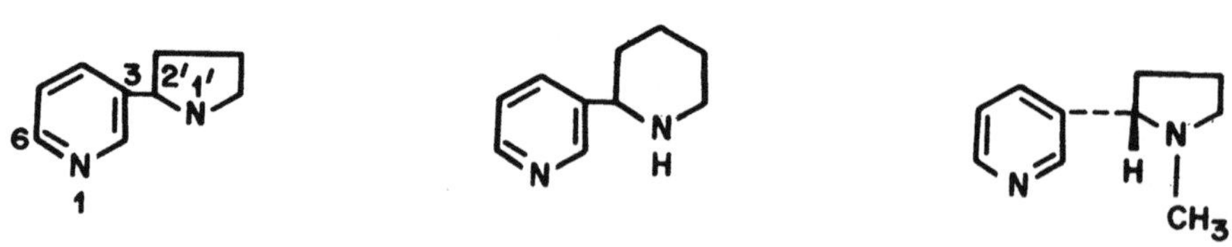

R = CH_3, Nicotine
R = H, Nornicotine

Anabasine

Stereostructure of nicotine

Fig. 13.1. Structures of insecticidal tobacco alkaloids. Natural nicotine and anabasine are laevorotatory ($[\alpha]_D^{20}$ = −169; −83.1°, respectively—and the absolute configuration is expressed by D- or S-. Although *l*-nornicotine ($[\alpha]_D^{20}$ = −88.8°) has the same configuration as *l*-nicotine, natural nornicotine occurs as the mixture of *d*- and *l*-forms

water below 60 °C, and is very soluble in alcohol, chloroform, ether, kerosene, and oils. It is very hygroscopic and turns brown and viscous, with a disagreeable odor, on exposure to air or light. The sulfate form is stable and freely soluble in water and alcohol. Free nicotine, because of its high volatility, is used mainly for fumigation in greenhouses. In agriculture nicotine is used as nicotine sulfate which acts as a stomach poison for leaf-eating insects. Also, the sulfate when sprayed gradually releases free nicotine and is useful to control insects over extended periods. Addition of alkaline compounds such as soap and calcium caseinate liberates the active free nicotine which then acts as a contact insecticide or as a fumigant against soft-bodied insects. Nicotine has largely been replaced by insecticides of greater effectiveness and greater safety to man.

13.1.2 Mode of Action

Nicotine mimics a part of the action of acetylcholine by interacting with the nicotinic acetylcholine receptor, that is, the receptor fails to distinguish between nicotine and acetylcholine. Hence, the effects produced by nicotine across the synapses and neuromuscular functions are called "nicotinic effects" in mammals. In mammals the primary effects of nicotine are at the neuromuscular junctions (voluntary muscles) and other cholinergic junctions such as smooth muscle and glands, whereas in insects the effects are on ganglia of the central nervous system, because the insect neuromuscular junction is not cholinergic. Both acetylcholine and nicotine produce stimulation at low concentrations and blockage at high concentrations. In actual poisoning one observes these two effects in sequence, first stimulation then blockage, so that early twitching and convulsions are followed later by paralysis.

Structure–toxicity studies of many nicotine-like compounds (Yamamoto et al. 1962) showed that essential structural moiety for higher toxicity is the 3-pyridylmethylamino group; the basicity of the nitrogen atom must be high enough to ionize at physiological pHs (pKa preferably 7.4–9.0), and the distance between the two nitrogen atoms should approximate to the distance between the two functional groups of acetylcholine (quaternary nitrogen and carbonyl) (for more details see Chap. 16). Nicotine has no inhibitory effect on cholinesterase activity (Richards and Cutkomp, 1945). However, at higher concentrations the enzyme is inhibited by nicotine and serves as a model of the nicotinic acetylcholine receptor (Yamamoto et al. 1968).

13.1.3 Toxicity to Mammals

For higher animals nicotine is one of the most rapid and toxic poisons known. The smallest single dose fatal to man appears to be about 60 mg and death can occur in 5–30 min. Nicotine is absorbed rapidly from the tongue or eyes, but more slowly from the stomach because of the acidity of the gastric juice. In the free form nicotine is rapidly absorbed through the unbroken skin. As little as 3 ml of 40% free nicotine rubbed on the skin of the abdomen of a cat can rapidly be fatal, whereas much larger amounts of nicotine sulfate have no effect.

Laboratory animals poisoned with nicotine show symptoms of salivation and vomiting, followed by muscular weakness, tremors, hyperexcitability, incoordination, and ultimately, convulsions and, cessation of respiration. The cause of death usually is the result of respiratory paralysis. If respiration is resumed after the convulsions stop, the animal usually survives. The LD_{50} of nicotine sulfate to the rat is oral 83 mg/kg and dermal 285 mg/kg (Gaines 1969). Since nicotine is strictly a nervous system poison interfering with nervous integration both in insects and mammals, it is much less toxic to lower organisms which are devoid of a well developed nervous system.

13.1.4 Toxicity to Insects

Although nicotine is usually classed as a contact insecticide, it probably acts largely as a fumigant even when applied as a dust or spray. One mode of entry of nicotine into the insect is thought to be as a gas passing through the spiracles into the tracheal system. Symptoms of poisoning by nicotine follow the general sequence of excitation, convulsions, paralysis, and death.

13.1.5 Metabolism

Nicotine is the active principle of tobacco smoke and as such its metabolism has been studied extensively in man and animals (Schmeltz 1971; Fig. 13.2). Oxidation occurs on 5'-carbon atoms, 1'-nitrogen atom, and on the N-methyl group of nicotine to form cotinine, nicotine-1'-oxide, and nornicotine, respectively. Methylation on the pyridyl nitrogen atom occurs also to give nicotine isomethonium. Different from animal metabolism, soil bacteria oxidized nicotine on 6-carbon to form 6-hydroxynicotine. In all animals tested the major metabolic product was found to be cotinine, which is further converted to several metabolites, and eventually to 3-pyridylacetic acid. Nornicotine follows a pathway similar to that of nicotine. Insects convert nicotine mainly to cotinine together with several unidentified products. Certain tobacco-feeding insects metabolize nicotine to avoid intoxication, while others (e.g., tobacco hornworm, tobacco budworm, and cabbage looper) do not metabolize nicotine, but efficiently excrete it into feces.

13.1.6 Therapy

Artificial respiration is always lifesaving provided it is given before there is a severe drop in blood pressure (Franke and Thomas, 1933, 1936). If the poison is swallowed give activated charcoal. Vomiting should be induced if the poison itself has not produced vomiting. Poisoning can also be treated with anticonvulsants.

13.2 Rotenone

There are a group of compounds in the roots of *Derris* and other plants of the family *Leguminosae* which are similar to rotenone in chemical structure, such as

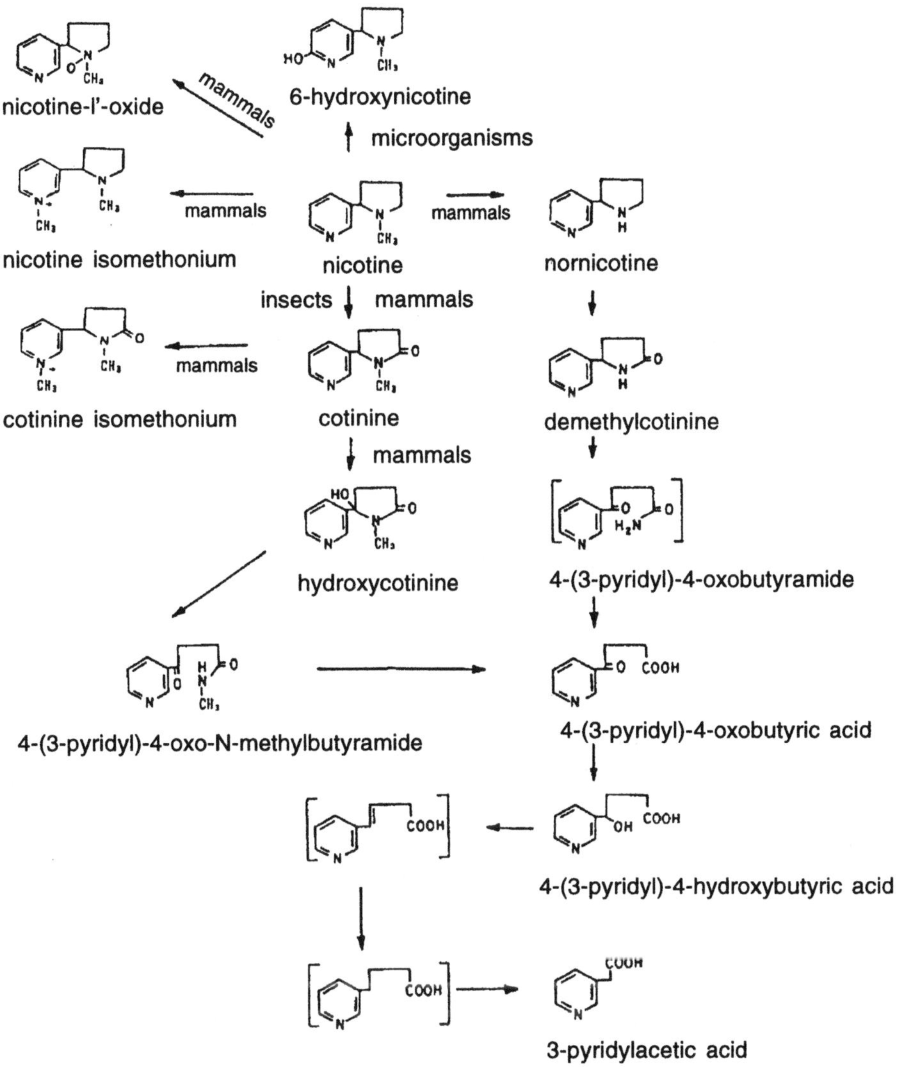

Fig. 13.2. Metabolism of nicotine in mammals, insects, and microorganisms. (Schmeltz 1971)

deguelin, toxicarol, sumatrol, elliptone, malaccol, dalpanol, pachyrrhizone, dolineone, amorphin, munduserone, and erosone, collectively called rotenoids. However, rotenone is the most important compound and insecticidal preparations of it have been used in crop protection since 1848 (for review see Fukami and Nakajima 1971). Rotenone occurs in certain leguminous plants (in as many as 68 species), but mainly in *Derris eliptica* grown chiefly in the Far East and in cube root from *Lonchocarpus* spp. which occur in the Amazon Valley of South America. The long slender roots are dried, ground to a powder, diluted with talc, pyrophyllite, clay, sulfur, or similar nonalkaline diluents, and used as insecticide dusts containing 0.75–1.5% rotenone. These dusts are commonly used for the control of beetles, aphids, weevils, and many other insect pests of truck crops, and against animal parasites such as fleas, lice, and ticks. Rotenone has been used as a lotion for chiggers, as an emulsion for scabies, and as a spray against cattle grubs and mange of dogs.

For use as liquid formulations the roots are extracted with ether or acetone. Such extractives known as derris and cube resins contain about 30–40% rotenone (see Shepard 1951 for details). Natives throughout the tropics have long used rotenone-containing plants as fish poisons. Rotenone is sometimes applied to lakes and ponds to kill undesirable fish before restocking with more desirable varieties. Derris and other rotenone-containing materials undergo decomposition in light and air. Colorless solutions of rotenone in organic solvents, when exposed to light and air, become yellow, orange, and finally red, due to oxidation.

13.2.1 Mode of Action

It has been determined that the mode of action of rotenone poisoning derives from the ability of rotenone to inhibit respiratory metabolism (Fukami 1956, 1961; Lindahl and Oberg 1961; Yamamoto 1970; Storey 1981). More precisely, rotenone interferes with the electrontransport system between NADH dehydrogenase and coenzyme Q at complex I as piericidin A, causing a severe reduction in oxygen consumption (see Chap. 17). Natural rotenone has *cis* or 6aβ, 12aβ, 5'β (6aS, 12aS, 5'R) configurations and is designated as 5'β-rotenone. Synthetically, 5'α-rotenone (6aα, 12aα, 5'α; the optical antipode of natural rotenone), 5'α-epirotenone (6aβ, 12aβ, 5'α), 5'β-epirotenone (6aα, 12aα, 5'β) (Unai and Yamamoto 1973), and two *trans*-rotenones (6aβ, 12aα, 5'β; 6aα, 12aβ, 5'β) (Begley et al. 1990) are produced. Natural rotenone is the highest in inhibiting mitochondrial respiration followed by *trans* (6aβ, 12aα, 5'β)-rotenone, the latter being epimerized to rotenone with acid; 5'α-epirotenone is less inhibitory and all others are not inhibitory.

13.2.2 Toxicity to Mammals

The mammalian toxicity of rotenone varies greatly with the animal species, method of administration, and type of formulation. The acute oral LD_{50} of crystalline rotenone to the rat is 132 mg/kg, to the guinea pig is 60 mg/kg, and to the rabbit the minimal lethal dose is about 1500 mg/kg (Haag 1931). Poisoning by rotenone in man is rare, but local effects include conjunctivitis, dermatitis, pharyngitis, and rhinitis. On ingestion, rotenone produces gastrointestinal irritation, nausea, and vomiting. The estimated fatal oral dose for a 70-kg man is 100–200 g. Inhalation of the dust is more hazardous and it can cause respiratory stimulation followed by respiratory depression, incoordination, clonic or tonic convulsions, muscle tremors, and death from respiratory failure (Haag 1931). Death is slow following a fatal oral dose and may take 2–10 days or more (Lehman 1951–1952).

13.2.3 Toxicity to Insects

Rotenone is both a contact and a stomach poison to insects. It kills insects slowly but causes them to stop feeding almost immediately. Like all other botanical insecticides its life in the sun is short (1–3 days). It is toxic to many insect species, but the poisoning effect is a slow process. In some cases the mouth parts of insects become paralyzed so that they stop feeding and die of starvation. The poisoning sequence is inactivity, locomotive instability, knockdown, paralysis, and slow death. The cause of death, most likely, is due to respiratory failure.

13.2.4 Metabolism

Initial metabolism of rotenone occurs mainly by oxidation at 8', 6'–7', 12 positions and on the methoxyl group as shown in Fig. 13.3. Combination of these reactions gives many more metabolic products (Fukami et al. 1967; Yamamoto et al., 1971b). All metabolites are less inhibitory to the mitochondrial respiration than rotenone, but 8'-hydroxyrotenone is as toxic as rotenone to mice (i.p. LD_{50} 2.8 and 2.6 mg/kg, respectively).

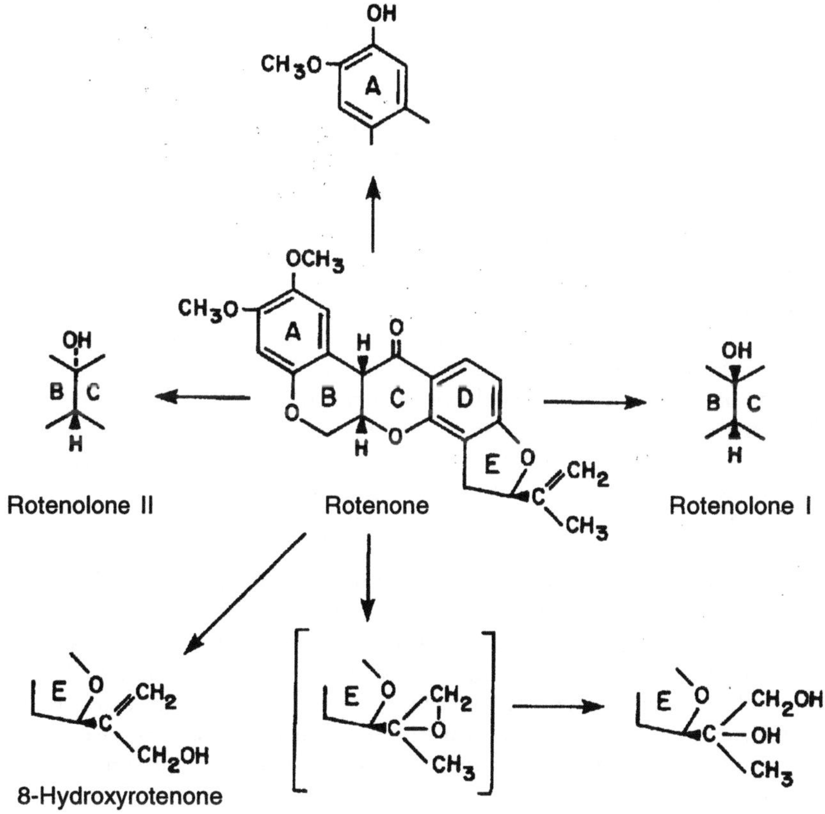

Fig. 13.3. Initial metabolic pathway of rotenone in insects and mammals

13.3 Pyrethrum

13.3.1 Composition and Biological Activity

Pyrethrum is one of the oldest organic insecticides known to man. Even today it is competitive with many of the new generation insecticides for domestic use. This is principally due to its low toxicity to mammals, nonpersistency, and high toxicity to insects coupled with its remarkable effectiveness in causing insect knockdown. The reader may consult excellent available reference books on pyrethrum (Casida 1973; Casida and Quistad 1995a).

"Pyrethrum" refers to the dried and powdered flower heads of a white-flowered, daisy-like plant belonging to the genus *Chrysanthemum*. The only species of commercial importance is *Chrysanthemum cinerariaefolium*. "Pyrethrins" are all the toxic constituents of the pyrethrum flowers and "pyrethroids" refer to the insecticides based on the pyrethrins as prototypes. "Rethrins" apply to compounds having similar alcohol moiety to pyrethrins. *Chrysanthemum cinerariaefolium* is native to the Dalmatian mountains, Croatia, but Japan was more important as a commercial source before World War II, with, principally Kenya, Uganda, and Rwanda holding the market today. Recently, Australia has become the world's second largest producer.

The pyrethrins are esters formed by a combination of two acids, i.e., chrysanthemic acid and pyrethric acid, with three alcohols (rethrolones), namely pyrethrolone, cinerolone, and jasmolone. The esters of chrysanthemic acid (chrysanthemates) are pyrethrin I, cinerin I, and jasmolin I, and are together known as pyrethrins I. The esters of pyrethric acid (pyrethrates) are pyrethrin II, cinerin II, and jasmolin II, and are together known as pyrethrins II (Fig. 13.4). These six components together account for the toxicity and knockdown properties of pyrethrum extract.

Pyrethrins are generally combined with a synergist in oil sprays and aerosols for use against a wide variety of flying insects. The action of the synergist is to prevent or minimize the breakdown of pyrethrins in the body of the organism. The usual household formulation contains 0.5% pyrethrins and 5% synergist. The most important commercial synergist is piperonyl butoxide. Other synergists, such as extracts of sesame (sesamin and sesamolin), sesamex, propyl isome, and MGK264, have also been used with pyrethrins (see Chap. 15).

Uses. Pyrethrum has been used on many agricultural crops for control of aphids, beetles, mealybugs, leafhoppers, thrips, cabbage worms, loopers, and many others. Formulations of pyrethrins and synergized pyrethrins have been found useful in the control of insect pests associated with households and food commodity storage and processing facilities. They are used extensively as protectants against the ravages of insect pests of stored products, and against insects attacking dried fruits, nuts, and cereal products. Stabilized pyrethrins have proved effective in suppressing certain forest insect defoliators.

In the field of public health, pyrethrins in combination with the synergist piperonyl butoxide have been used successfully in the control of the brown dog

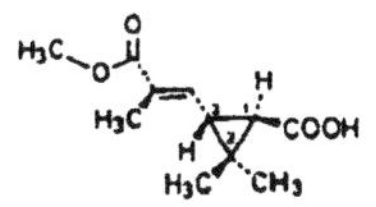

(+) - *trans*-Pyrethric acid

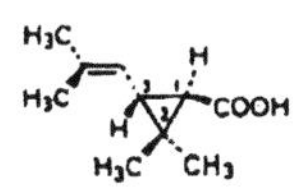

(+) - *trans*-Chrysanthemic acid

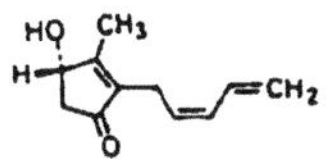

(+) - Pyrethrolone

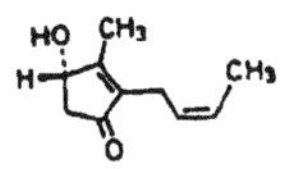

(+) - Cinerolone

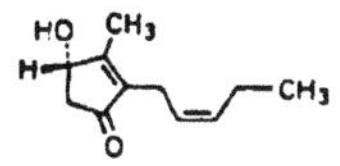

(+) - Jasmolone

	R	*R'*
"Pyrethrins I"—esters of chrysanthemic acid (chrysanthemates)		
Pyrethrin I	CH_3	$CH{=}CH_2$
Cinerin I	CH_3	CH_3
Jasmolin I	CH_3	CH_2CH_3
"Pyrethrins II"—esters of pyrethric acid (pyrethrates)		
Pyrethrin II	$CH_3OC(O)$	$CH{=}CH_2$
Cinerin II	$CH_3OC(O)$	CH_3
Jasmolin II	$CH_3OC(O)$	CH_2CH_3

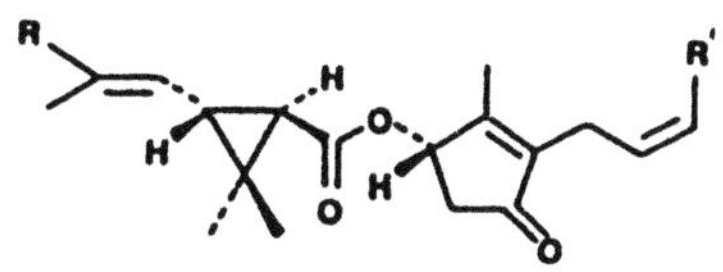

Fig. 13.4. Stereoformula of pyrethrins and constituent acids and alcohols

tick (*Rhipicephalus sanguineus*), for fly control in barns and outdoors, as larvicide for gnat (*Chironomus atrella*) control, as livestock sprays for protection against biting flies, such as stable flies (*Stomoxys calcitrans*), and horn flies (*Haematobia irritans*), for control of ear ticks (*Otobius megnini*), as louse powders, as mosquito repellents, as ultralow volume applications for control of tsetse flies (*Glossina morsitans* and *G. pallidipes*), against tabanid flies, and other insects of public health importance. A development of special significance in medical entomology was the use of pyrethrum aerosols in aircraft making international flights for control of insects, particularly, mosquitoes and other disease vectors, as a quarantine measure.

13.3.2 Mode of Action

Pyrethrins give rapid knockdown action, which is, in many cases, reversible, but could lead to paralysis depending on the dose. The mode of action of pyrethrins is similar to DDT, if not identical, in both insects and mammals. The depolarizing after-potential of a single spike is increased by the application of allethrin (a close analog of pyrethrins) to the giant axon of *Periplaneta americana*, leading to repetitive discharges and, at higher concentrations, conduction block. Knockdown may be caused by hyperexcitation of the nerve, and nerve block is responsible for paralysis. Pyrethrins, synthetic pyrethroids, and DDT are the only insecticides that possess a clear negative temperature coefficient of the insecticidal action. A housefly strain that is resistant to knockdown by DDT shows cross-resistance to pyrethrum. The gene responsible for kdr selectivity confers resistance to the entire category of both DDT and pyrethroid compounds. Unlike DDT there is a critical temperature below which no repetitive discharges are elicited (see Chap. 14 for further comments).

13.3.3 Toxicity to Insects

Pyrethrum is the insecticide well known for its rapid knockdown action causing temporary paralysis of the insect. The paralytic effect is, in many cases, reversible. This is especially evident in houseflies exposed to pyrethrins. Only minute amounts of the toxicant are needed to produce knockdown, but much larger amounts are required for producing mortality of the flies.

The symptoms of poisoning follow the typical pattern of nerve poisoning, that is, excitation, convulsions, paralysis, and death. The rapidity of action of pyrethrins upon contact with an insect is shown by the excitement with which cockroaches, affected by sublethal quantities of pyrethrum spray, emerge from their hiding places and run about frantically ("flushing effect"). Because they paralyze the insects so rapidly, the pyrethrins could reduce the annoyance from mosquito bites and reduce the opportunity for disease transmission by mosquitoes entering treated dwellings. In the case of DDT residues, the mosquitoes are able to bite before succumbing to the insecticide.

Pyrethrum does not have appreciable effect as a stomach poison and it contains no volatile agent of value as a fumigant, although pyrethrum smokes, produced by burning pyrethrum coils, have for many years been used against flies and mosquitoes. More recently, pyrethrum has been replaced by bioallethrin (a synthetic pyrethroid) for use in smoke coils and other forms of smoke generators.

13.3.4 Toxicity to Mammals

The pyrethrins have an acute oral LD_{50} to rats variously reported, 200–400 mg/kg probably being the range for the less pure preparations, and 400–1500 mg/kg for the more purified preparations (Barthel 1973). The acute toxic effects to laboratory animals proceed from excitation to convulsions to paralysis including muscular fibrillation. Death is caused by respiratory failure. In contrast with human experience, animal tests failed to show the sensitizing effect of pyrethrins.

Despite the long period of pyrethrum use in pest control, only a few documented cases of human poisoning have been reported (Casida, 1973; Hayes, 1982). The pyrethrins show little evidence of causing chronic poisoning; however, they cause some inhalation problems because of impurities in the formulations, resulting in asthma-like symptoms. Dermal toxicity of pyrethrins is negligible because they are poorly absorbed from the skin. The dermal LD_{50} ranges from 1350 to 5000 mg/kg. Many humans have had skin problems (dermatitis) from exposure to pyrethrins. This dermatitis is an allergic reaction (which may be associated with ragweed sensitivity) and disappears rapidly on removal from exposure. The dermatitis is caused by impurities in the extracts from the flowers which are not present in the current commercial preparations of pyrethrins. Pyrethrins do not pose a hazard of acute toxicity to humans. From the toxicity data available, there is no evidence that pyrethrins are carcinogenic, mutagenic, or teratogenic in test animals or, by extrapolation, in man, although the carcinogenic potential of pyrethrins has not been satisfactorily investigated by modern-day standards. Recent studies on typical manufacturing use extract (a.i. 57.6%) show 2.37 g/kg for males and 1.03 g/kg for females (Schoenig 1995). It can be said that pyrethrum extract has a low order of acute toxicity and very little potential to cause eye or skin irritation and skin sensitization. It is not a teratogen or a reproductive toxin and has minimal potential to induce tumors in the mammalian system, with little or no relevance to human health.

13.3.5 Metabolism

For many years it was considered that the major detoxification mechanism of pyrethrins in insects involved hydrolysis of the ester linkage, splitting the molecule into the acid and alcohol moieties. Later studies demonstrated that a nonhydrolytic mechanism was responsible for the metabolism of pyrethrins and related compounds in insects and mammals. Methylenedioxyphenyl type synergists are known to inhibit oxidative metabolism, by which they increased toxicity. The toxicity of allethrin to houseflies is enhanced by synergists. The major metabolic pathway involves oxidation by the mixed function oxidase system of the *trans*-methyl of the isobutenyl group to the hydroxymethyl metabolite. This initial oxidation step is sensitive to inhibition by the synergists. This metabolite is then conjugated and excreted, or is further oxidized to the aldehyde and then to the acid form (Yamamoto and Casida 1966; Yamamoto et al. 1969).

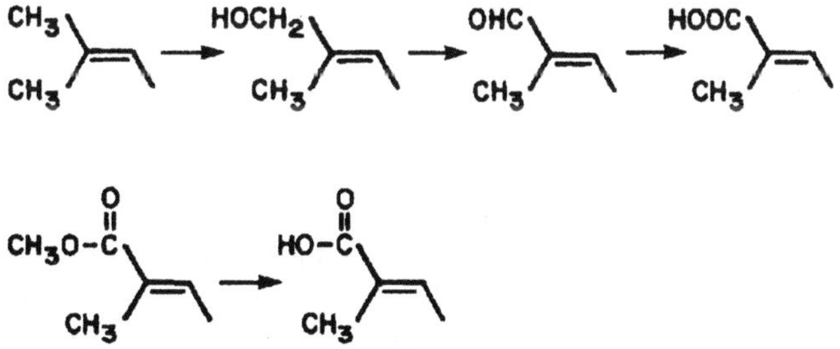

The terminal methoxycarbonyl of pyrethric acid moiety of pyrethrin II is hydrolyzed. However, oxidation of the alcohol moiety is also important in limiting the toxicity of pyrethrins I and II. The nature of the modification can be inferred from the result of pyrethrin I metabolism in mammals. In insects the initial stage of detoxication involves oxidation independently either on the acid or the alcohol moiety of pyrethrin I, but only on the acid moiety of allethrin, and either methoxycarbonyl cleavage of the acid moiety or oxidation of the alcohol moiety occurs with pyrethrin II, as shown in Fig. 13.5.

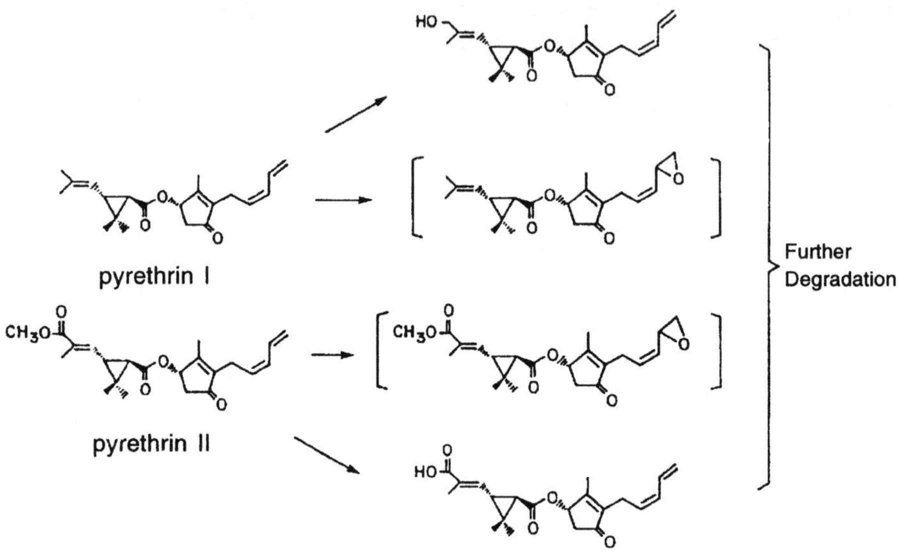

Fig. 13.5. Metabolism of pyrethrin I and pyrethrin II in insects. (Yamamoto 1973)

The scheme explains the facts that pyrethrin I and cinerin I are equal in toxicity to insects when synergized (without synergists cinerin I is inferior to pyrethrin I); pyrethrin II and cinerin II, when synergized, become equal in toxicity, but to a degree around half to one-third of the corresponding chrysanthemates; allethrin is not so synergized as compared with pyrethrins (Yamamoto 1973).

Pathways leading to the urinary metabolites in mammals are shown in Fig. 13.6 (Casida et al. 1971; Elliott et al. 1972).

13.3.6 Advantages and Disadvantages

The disadvantages of pyrethrum are few indeed. Pyrethrum is not a persistent chemical. It is destroyed by light and air and is biodegradable. This is a favorable point on the one hand, but, on the other hand, pyrethrum can only be used as a contact insecticide with little or no residual activity. The allergic responses in man seem to be of less concern when the purified preparations are used. Pyrethrum is toxic to fish and fish food (aquatic insects and crustaceans); however, the low insecticidal concentrations used in spraying operations appear to have a minimal direct impact on fish populations in an aquatic ecosystem (Pillmore 1973). Pyrethrum is also toxic to bees under laboratory testing.

Without doubt, the benefits of pyrethrum certainly outweigh its disadvantages. It is one of the oldest insecticides which has been tried for many years. Pyrethrum controls certain insects as effectively as and more safely than many of the currently used synthetic insecticides. There is no evidence of harm to humans, domestic animals, or wildlife when used as directed. From the toxicity data available, there is no evidence that pyrethrins are carcinogenic, mutagenic, or teratogenic in test animals or, by extrapolation, in man.

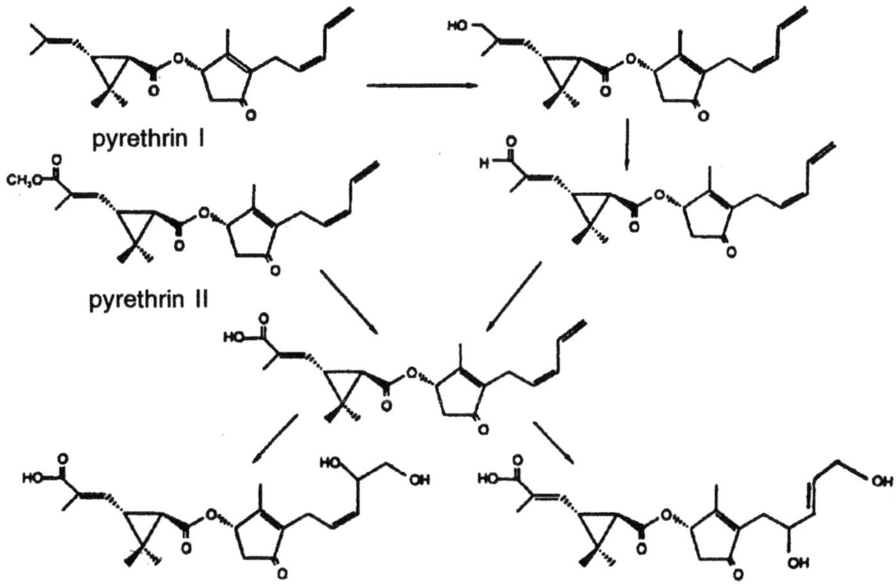

Fig. 13.6. Pathways leading to the urinary metabolites in rats. (Adapted from Casida 1973)

13.4
Neem

Native to India and Burma, the neem tree is a member of the mahogany family, Meliaceae, and is known as margosa tree or Indian lilac, ***Azadirachta indica*** A. Juss. It is perennial and requires little maintenance for growing. In India an estimated 18 million neem trees are grown, mostly for providing shade to the people. It was introduced to West Africa and other places, and has grown well there. A closely related species, the chinaberry tree or Persian lilac (***Melia azedarach*** L.), contains the similar active components as found in *A.* ***indica***.

Its insect control efficacy was first recognized by the withstanding of neem trees from the locust, which swarmed on the tree, but left without feeding. The extracts from seeds and leaves have pest insect control activities. Extraction is easy. The leaves or seeds are merely crushed and steeped in water, alcohol, or other solvents. The extract contains 4 major and perhaps 20 minor active compounds and can be used without further refinement, or the most effective ingredients can be isolated and formulated as commercial products. The affected insect can no longer feed, breed, or metamorphose, and can cause no further damage to crops. Neem oil from seeds is used for many purposes, such as heating, lighting, and lubricating, and in soaps, cosmetics, disinfectants, and others, and neem cake, left after the oil is pressed from the seeds, is effective for soil pests including nematodes (Schmutterer 1990. National Research Council 1992).

13.4.1 Composition and Biological Activity

The main neem chemical is a mixture of three or four related compounds. They belong to tetranortriterpenoid (limonoids) (Fig. 13.7).

1. *Azadirachtin:* the main agent for controlling insects, representing 90% of the effect on most pests; contents are usually 2–4 mg/g kernel. It is a feeding deterrent and growth regulator, repelling and disrupting the growth and reproduction of pests.
2. *Meliantriol:* a feeding inhibitor, causing insects to cease eating in extremely low concentrations.
3. *Salannin:* a powerful feeding inhibitor.
4. *Nimbin and nimbidin:* show antiviral activity.
5. *Others:* minor but also active in one way or another. Deacetylazadirachtin is as active as azadirachtin.

Some neem compounds act as systemics. When exposed to sunlight, neem products degrade and lose their pest-control activities.

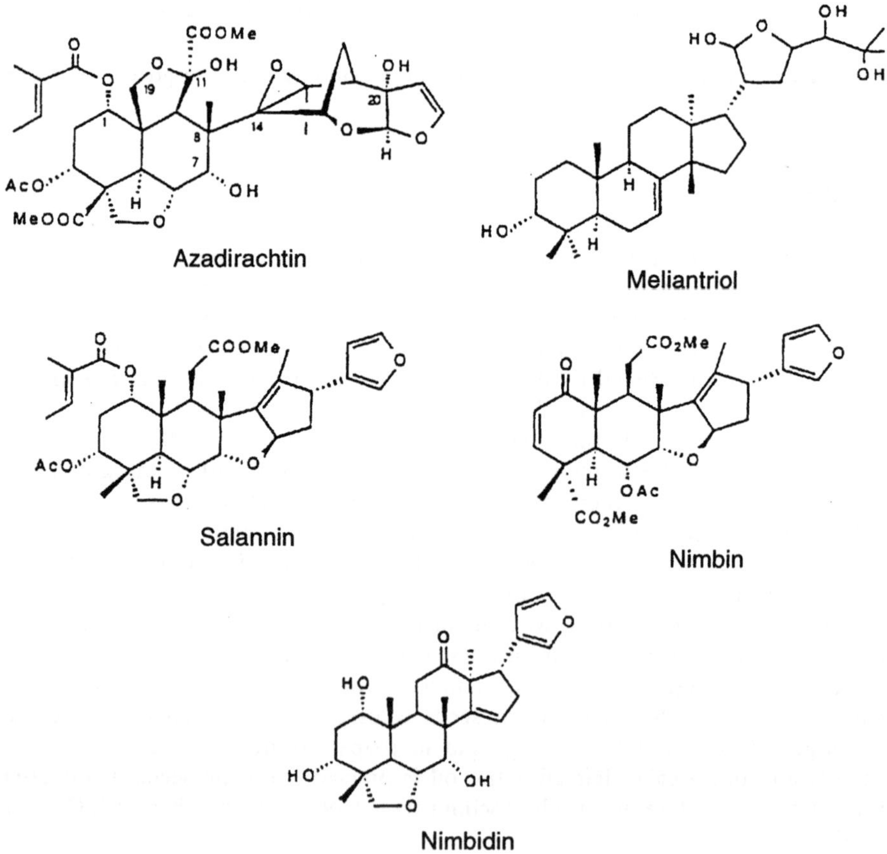

Fig. 13.7. Active neem components

13.4.2 Mode of Action

Azadirachtin is structurally somewhat similar to insect hormones, ecdysones, and seems to be an ecdysone blocker, so that insects will not molt. It also interferes with the function of taste, so that many leaf-eating insects are deterred by plants to which even small amounts of azadirachtin have been applied.

13.4.3 Effects on Insects

Neem preparations show the following effects:

1. Partial reduction or complete inhibition of fecundity and/or sometimes egg hatchability;
2. Reduction of the life span of adults;
3. Oviposition repellence against females;
4. Direct ovicidal effects;
5. Antifeedant effects against larvae, nymphs, or adults;
6. Formation of permanent larvae;
7. Growth regulating effects at molting between larval (or nymphal) instars and especially in the prepupal stage and giving rise to larval–pupal, nymphal–pupal, nymphal–adult, and pupal–adult intermediates, and to crippled adults (Ascher 1993).

13.4.4 Toxicity Problems

Neem has been utilized as a medicinal plant based on general antiseptic activity for generations in India and neem products appear to have little or no toxicity to warm-blooded animals, although no comprehensive toxicology data are available. Neem oil obtained from clean, fungus-free seed kernels showed no oral toxicity in rats at 5000 mg/kg. Neem extracts showed no mutagenicity in the Ames test. However, neem extracts should not be used for internal medicinal purposes. A problem only in the more humid neem-growing areas is that the fruits can carry *Aspergillus flavus*, which produces aflatoxins, the most potent carcinogen. The fungus can contaminate the seeds inside the fruits, and aflatoxins are extracted and concentrated along with pest control ingredients. Neem oil shows toxicity to certain fish species. The seed kernels, whole fruits, or leaves of *Melia azedarach* contain compounds highly toxic to mammals, like meliatoxins. For details about neem refer to Schmutterer (1995).

CHAPTER 14

Synthetic Pyrethroids

14.1 Insecticidal Activity and Photostability

Pyrethroids are synthetic compounds based on natural pyrethrins as models, generally arrived at by systematic variation of parts of the molecule for the purpose of improving photostability and insecticidal activity (Davies 1985). However, they differ markedly from the natural pyrethrins in their physical, chemical, and biological effects (Tables 14.1, 14.2). All pyrethroids are lipophilic compounds, almost insoluble in water; in these respects they resemble the organochlorine insecticides, but they differ from most organophosphorus and

Table 14.1. Relative toxicities of some important pyrethroids and other insecticides to four species of insect by topical application[a]

Compound	Musca domestica	Periplaneta americana	Glossina austeni	Boophilus microplus
Biorethmethrin (standard)	100	100	100	100
LC_{50} of standard (ng per insect)	5	2500	2.6	0.00014%[b]
Natural pyrethrins	2	(100)	(20)	170
Allethrin	3	–	–	2
Bioallethrin	6	150	–	12
S-Bioallethrin	10	–	–	–
Resmethrin	42	–	79	120
Cismethrin	42	500	260	–
Kadethrin	34	80	–	520
Phenothrin	30	–	26	–
Permethrin	60	290	87	200
Cypermethrin	210	–	350	330
Deltamethrin	1500	3000	3300	240
Fenvalerate	38	200	31	–
DDT	12	(15)	3	3
Dieldrin	20	(100)	26	21
Carbaryl	–	–	< 5	0.1
Malathion	1	–	<2	–
Dimethoate	45	60	5.5	–

[a]Results in this table are intended for general comparison only; figures in parentheses are from very approximate comparative data, not necessarily directly against bioresmethrin.
[b]Mean lethal concentration by Shaw Immersion Technique. (Elliot et al. 1978)

Table 14.2. Synthetic pyrethroid insecticides

Common name	Structure	Stereochemistry	Number of isomers	Content of active isomers (%)
Natural				
Pyrethrin I		1R *trans*; S	1	100
Pyrethrin II		1R *trans* E; S	1	100
Cinerin I		1R *trans*; S	1	100
Cinerin II		1R *trans* E; S	1	100
Jasmolin I		1R *trans*; S	1	100
Jasmolin II		1R *trans* E; S	1	100
Photolabile pyrethroids				
Allethrin	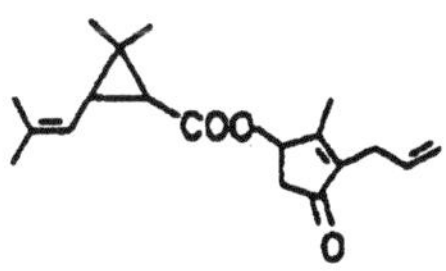	(±) *cis/trans*; RS	8	25
Bioallethrin		1R *trans*; RS	2	50
Esdepallethrin		1R *trans*; S	1	100
Barthrin	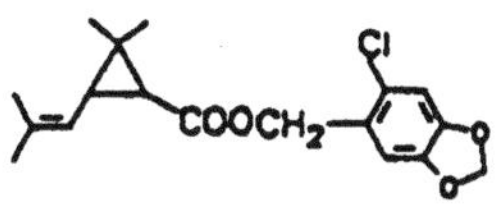	(±) *cis/trans*	4	50
Butethrin	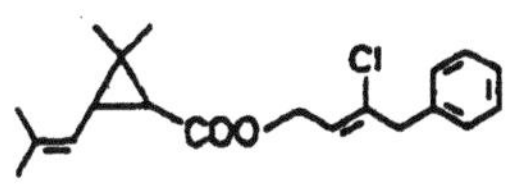	(±) *cis/trans*; Z	4	50
Cyphenothrin	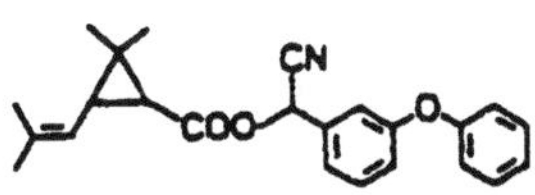	1R *cis/trans*; RS 20 : 80	4	50
Dimethrin	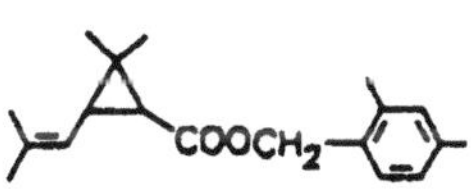	(±) *cis/trans*	4	50
Empenthrin (Vaporthrin)	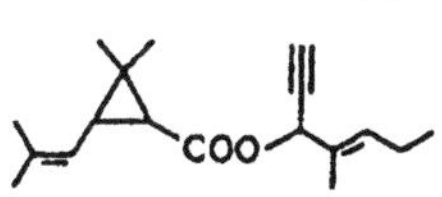	1R *cis/trans*; RS 20 : 80	4	50
Bioethanomethrin	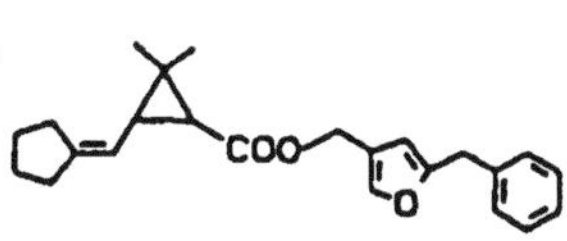	1R *trans*	1	100
Furamethrin	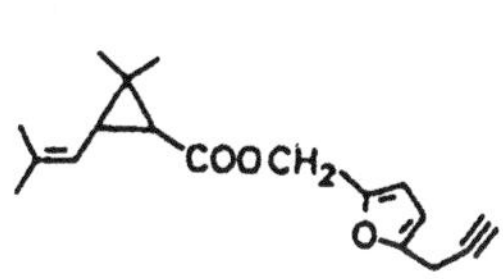	(±) *cis/trans*	4	50
d-Furamethrin		1R *cis/trans* 20 : 80	2	100

(Contd)

Common name	Structure	Stereochemistry	Number of isomers	Content of active isomers (%)
Kadethrin	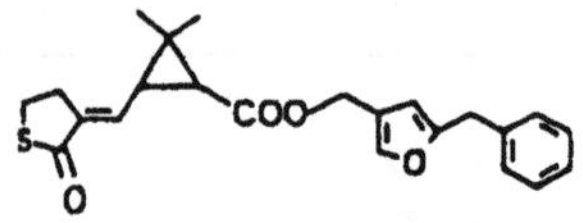	1R *cis* E	1	100
Phenothrin	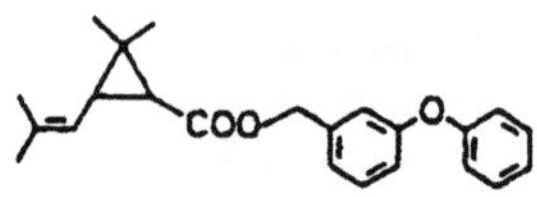	(±) *cis/trans*	4	50
d-Phenothrin		1R *cis/trans* 20 : 80	2	100
Prallethrin	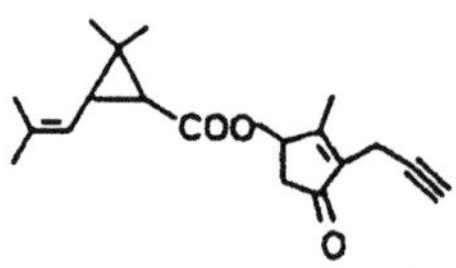	1R *cis/trans;* S	2	100
Proparthrin	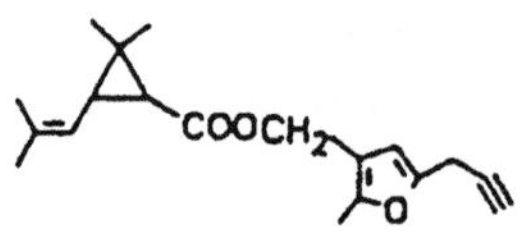	(±) *cis/trans*	4	50
Resmethrin	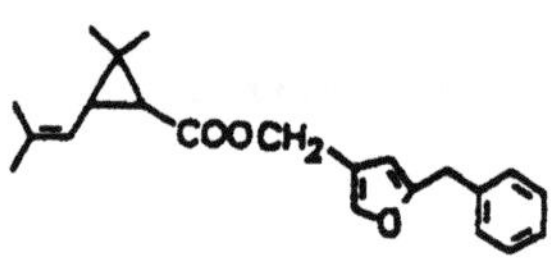	(±) *cis/trans*	4	50
d-Resmethrin		1R *cis/trans* 20 : 80	2	100
Bioresmethrin		1R *trans*	1	100
Terallethrin	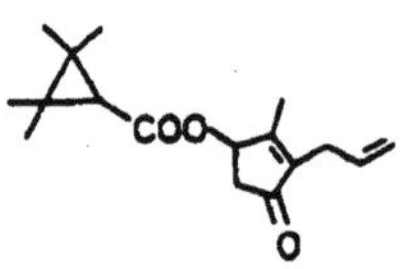	(±)	2	50
Tetramethrin	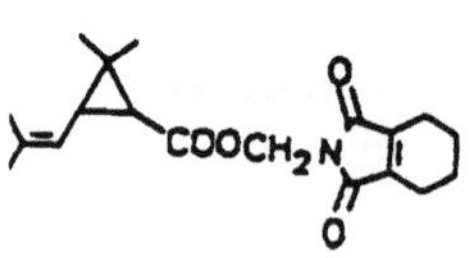	(±) *cis/trans* 20 : 80	4	50
d-Tetramethrin		1R *cis/trans* 20 : 80	2	100
Photostable or new pyrethroids				
Acrinathrin	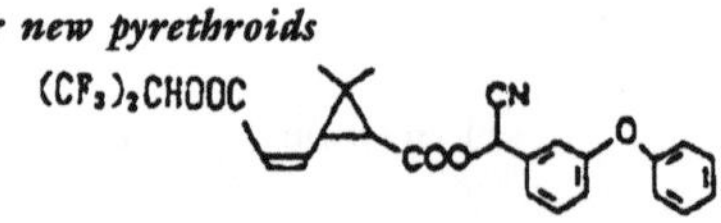	1R *cis* Z; αS	1	100
Bifenthrin	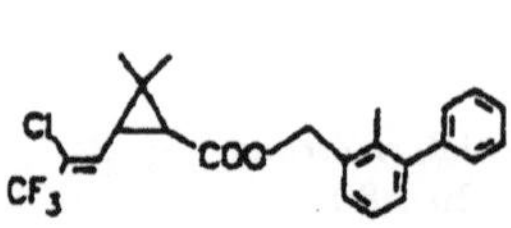	(±) *cis* Z	2	50

(Contd)

Common name	Structure	Stereochemistry	Number of isomers	Content of active isomers (%)
Cycloprothrin	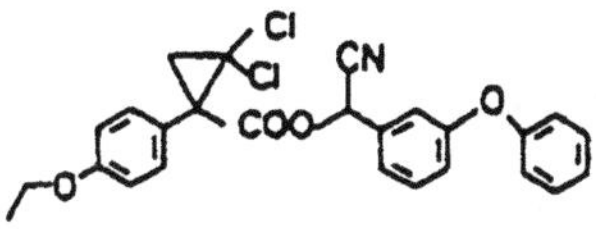	(±)	4	–25
Cyfluthrin	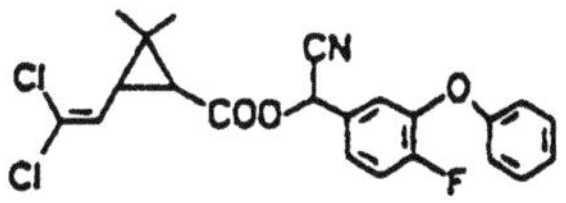	(±) *cis/trans* 40 : 60	8	20–25
Beta-Cyfluthrin		1R *cis/trans* αS 1S *cis/trans* αR *cis/trans* = 1 : 2	4	50
Cyhalothrin	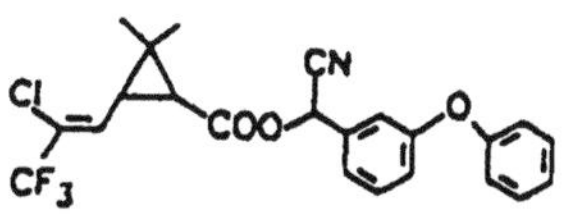	(±) *cis* Z	4	22
Lamda-Cyhalothrin		1R *cis* Z; αS 1R *cis* Z; αR	2	50
Cypermethrin	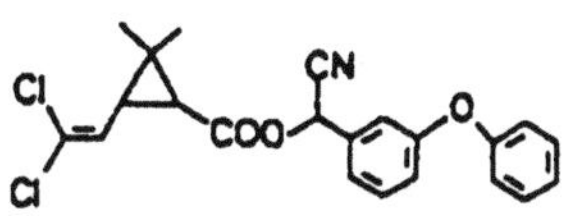	(±) *cis/trans* 40 : 60	8	18–25
Alpha-Cypermethrin		1R *cis* αR 1S *cis* αR	2	50
Beta-Cypermethrin		1R *cis/trans* αS 1S *cis/trans* αS *cis/trans*= 40 : 60	4	50
Zeta-Cypermethrin		(±) *cis/trans* αS	4	50
Deltamethrin	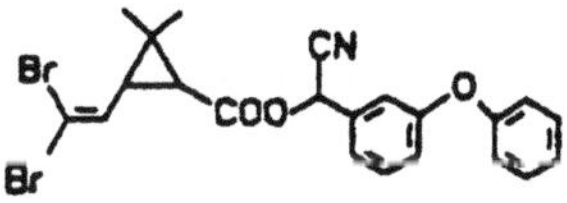	1R *cis*; αS	1	100
Ethofenprox	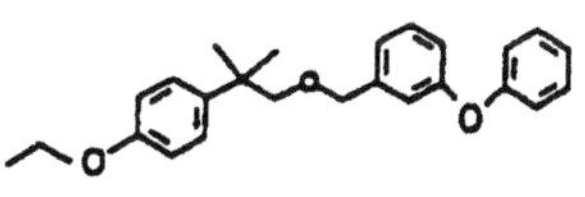	nonchiral	1	100
Fenpropathrin	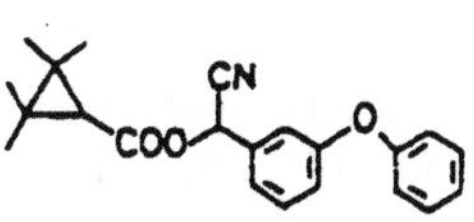	(±)	2	50
Fenvalerate	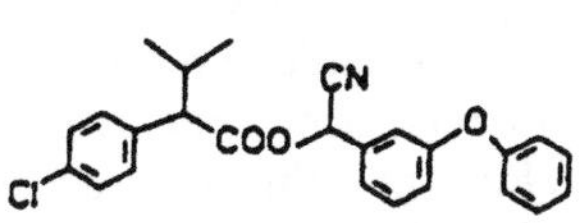	(±)	4	20–25
Esfenvalerate		(±) S; αS	1	100

(Contd)

Common name	Structure	Stereochemistry	Number of isomers	Content of active isomers (%)
Flucythrinate	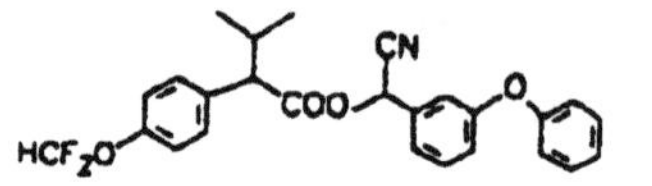	(±) S; α RS	2	50
Flumethrin	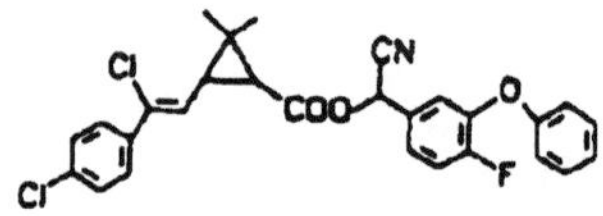	(±) *trans* Z	4	25
Fluvalinate	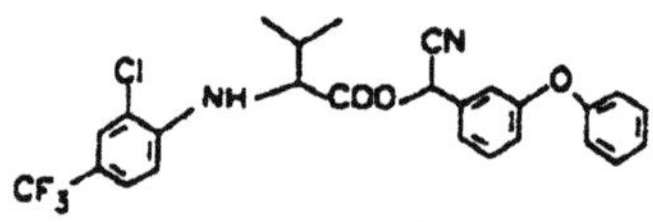	RS; α RS S; α RS	4 2	25 50
Halfenprox	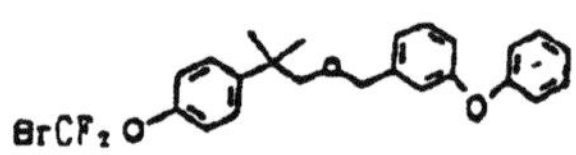	nonchiral	1	100
Permethrin	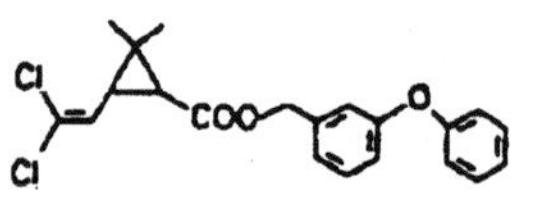	(±) *cis/trans* 40 : 60	4	50
Silafluofen		nonchiral	1	100
Tefluthrin	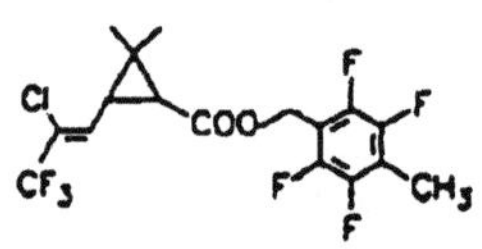	(±) *cis* Z	2	50
Tralomethrin	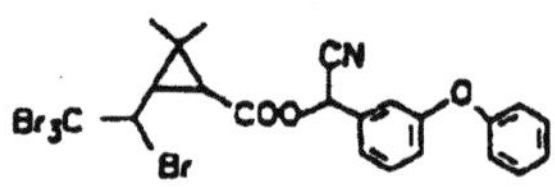	1R *cis*, R'S'; αS	2	100
Transfluthrin	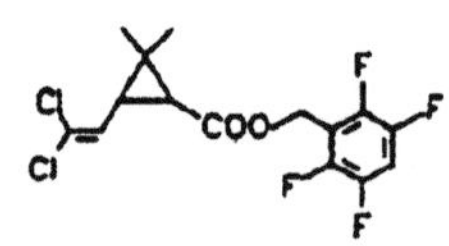	1R *trans*	1	100

carbamate insecticides. Most pyrethroids are relatively high-boiling, viscous liquids with low vapor pressures. Only a few (for example, allethrin, prothrin, and the natural pyrethrin I but not pyrethrin II) are sufficiently volatile to be useful constituents of mosquito coils. Vaporthrin is a volatile pyrethroid. These properties probably determine their fast action on insects, slow penetration into leaves, and low systemic movement in plants. Therefore, pyrethroids are effective as contact insecticides, but less effective as stomach poisons (Elliott 1977).

The more stable pyrethroids (such as permethrin, cypermethrin, deltamethrin, fenvalerate, ethofenprox, and others) were obtained by replacing the photolabile centers of the older compounds with other chemical groups that confer on them photostability (Fig. 14.1). Reactions by which the older and newer pyrethroids are photodegraded and metabolized by various organisms have been generally established (see summary by Holmstead et al. 1977). All decomposition products obtained are of lower toxicity than the parent compounds. Hence, there is little risk that toxic residues of decomposed pyrethroids will accumulate and contaminate the environment, especially since application rates are as low as one-tenth of those of other commonly used insecticides.

Although synthesis of analogs of the natural pyrethrins began as soon as the active constituents of pyrethrum were discovered (Staudinger and Ruzicka 1924), it was not until 1949 that the first commercially successful pyrethroid, allethrin, was introduced (Schechter et al. 1949). This constituted the first generation of the pyrethroids. Allethrin proved to be more stable and of longer residual activity than natural pyrethrins. It is very effective against flies and mosquitoes but less toxic to cockroaches and other insects. Its volatility, heat stability, and rapid knockdown makes it ideal for use in smoke coils and smoke mats for repellency, biting deterrence, and control of adult mosquitoes. It is used in aerosols for control of flying insects in households. It is used in agriculture for the control of aphids, beetles, thrips, mealybugs, loopers, leafhoppers, and other insects on many vegetable crops, on stored grain, and other commodities.

The second generation pyrethroids included the synthesis of dimethrin (1961), tetramethrin (phthalthrin, 1965), resmethrin and bioresmethrin (1969), and bioallethrin (1969). Tetramethrin was found to have greater knockdown activity than allethrin and it can be strongly synergized by pyrethrum synergists. Resmethrin

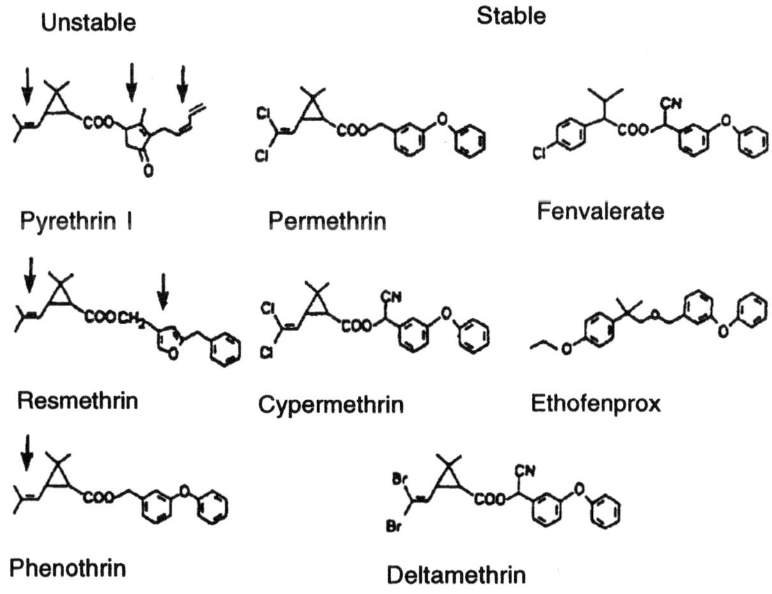

Fig. 14.1. Representative photostable and unstable pyrethroids. *Arrows* indicate photolabile moiety

and, especially, bioresmethrin possess greater toxicity and much greater knockdown activity than the natural pyrethrins against most insects studied. However, they cannot be synergized by the ordinary pyrethrum synergists. Resmethrin and bioresmethrin are more stable than the natural pyrethrins, but they, too, decompose rapidly on exposure to air and sunlight, and this is the reason why they have not been used as pest control agents in agriculture. Resmethrin found its greatest use, among others of the same group, in spray and aerosol applications for control of crawling and flying insects in glasshouses and dwellings. Bioallethrin has greater efficacy than allethrin, but it is not as active as resmethrin. A number of other compounds considered for commercial use during this period included prothrin, proparthrin, and butethrin. The last in this series was phenothrin (Sumithrin) which was introduced in 1973 and is used as a domestic insecticide.

The third generation pyrethroids comprise the most light-stable compounds which achieved wide application in agriculture. The first light-stable compound was fenpropathrin which was synthesized in 1971 but was commercialized as an acaricide only in 1980. During this period were introduced the most active and most photostable compounds permethrin, cypermethrin, deltamethrin, and fenvalerate.

The fourth generation pyrethroids were introduced during the period 1975–1983. Cypothrin has achieved commercial status in animal health for tick control. Flucythrinate was reported to be a broad-spectrum insecticide with activity against phytophagous mites. Fluvalinate, another compound of the same series, also proved to be effective against phytophagous mites. Cyfluthrin was introduced in 1981 against cotton insects. It has a level and spectrum of activity resembling those of cypermethrin. A later compound, flumethrin, has proved to be very effective against cattle ticks. A product of greater effectiveness and stability, cyhalothrin, has been introduced as an ectoparasiticide. Cycloprothrin and fenpyrithrin are being investigated as broad-spectrum insecticides and many other experimental compounds are in the stage of development.

Pyrethroid-induced death of insects is deemed to be the result of a cascade of events starting with various forms of hyperexcitation and leading to paralysis. There seems to be no single cause that triggers death in insects having no single respiratory center. However, the situation is different in mammals in which respiration is controlled by the respiratory center in the medulla. Hence, function including hyperexcitation and paralysis of the respiratory center as well as other regions of the nervous system that control cardiac function will cause drastic influence on the whole body function resulting in death.

Pyrethroids show negative temperature dependence of insect killing effect. In some cases, type II pyrethroids show positive temperature dependence; nerves respond to pyrethroids to produce repetitive discharges generally at low temperatures, but there is an optimal temperature range for this effect, and the negative temperature dependence is explainable in terms of more drastic changes in sodium currents at low temperature.

14.2 Uses of Photostable Pyrethroids

The advent of these new photostable pyrethroids opened a new chapter in the history of plant protection. In 1976, OP compounds represented a total of 40% of the world agricultural insecticide market, organochlorines 30%, carbamates 25%, and miscellaneous 5%. In 1983, OP compounds represented 35–40%, organochlorines 15%, carbamates 20%, pyrethroids 20–25%, and miscellaneous 5%. Of the pyrethroid insecticide market in 1983, permethrin occupied 10%, cypermethrin 22%, fenvalerate 30%, deltamethrin 35%, and others 3%. A graph showing global sales of pyrethroids and total insecticides (1980–1995) demonstrates the success achieved by this new family of insecticides (Fig. 14.2). A recent introduction, ethofenprox, unlike other pyrethroids, has a rather low fish toxicity which is a desirable property for control of insects in or near aquatic ecosystems.

The photostable pyrethroids have proven to be broad-spectrum insecticides, effective against a wide variety of insect pests, harmless to mammals and birds, and not phytotoxic. They combine high insecticidal activity with suitable persistence (they are as much as ten times more effective in the field than the most potent compounds of the other three principal groups of insecticides, namely, the organochlorines, organophosphates, and carbamates), they show high toxicity toward lepidopterous larvae on many crops, especially *Heliothis* and *Spodoptera* species on cotton, and against certain insect pests of forest trees (Elliott 1977; Elliott et al. 1978; Hervé 1985). They are also effective against eggs, larvae, and adults of many species of Coleoptera, Diptera and Heteroptera that cause damage to crops, such as fruits, oil seed rape, soybeans, and vegetables.

The photostable pyrethroids are also very effective against household pests,

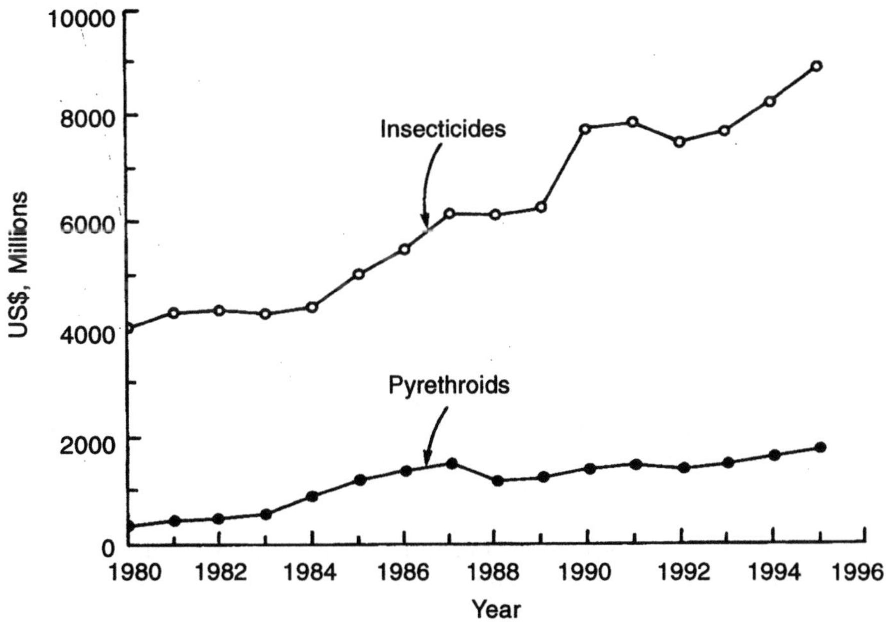

Fig. 14.2. Global sales of pyrethroids and total insecticides; 1980–1995 (J. McDougall and M. Phillips, Wood Mackenzie Agrochemical Service, pers. comm.)

fabric pests (for example, permethrin is excellent against clothes moths), against disease vectors and insect pests of public health (such as *Musca domestica, Glossina austeni, Anopheles stephensi, An. albimanus, An. quadrimaculatus, Aedes aegypti, Ae. nigromaculis, Culex pipiens, Stomoxys calcitrans*, and others). Permethrin has been found to be very effective against the body louse (*Pediculus humanus*) and the plague flea (*Xenopsylla cheopis*). Permethrin is more than 300 times as effective as DDT and 8 times as active as malathion against the body louse (Nassis and Kamel 1977). Permethrin also controls the dog flea (*Ctenocephalides canis*) and cat flea (*C. felis*). The stable pyrethroids have proved effective against cattle ticks, especially, OP- and carbamate-resistant strains of *Boophilus* species, as well as *Rhiphicephalus* and *Amblyomma* species.

The desirable properties of the pyrethroids make them useful for control of stored product insect pests. These properties are: 1. high toxicity to insects; 2. favorable toxicity to the consumer for the compound itself and any decomposition product that may be produced subsequently, such as during milling or cooking; and 3. sufficient persistence so as to minimize repeated applications.

14.3 Mode of Action

The signs of toxic action of the pyrethroids in insects and mammals are very similar. The early signs of hyperexcitation, tremoring, and convulsions are followed by paralysis and death at lethal levels of the insecticide. It appears, therefore, that the nervous system is the target in both insects and mammals. Although the site of action is similar in both insects and mammals, there is a large differential in the dose level at which the effects occur. The target species, the insects, are killed by an incredibly low concentration of a pyrethroid compound (for example, the LD_{50} of deltamethrin by topical application to the housefly is 0.01 mg/kg, whereas the rat oral LD_{50} is 130 mg/kg). The pyrethroids do not appear to act on any viable enzyme system; therefore most studies have concentrated on the site of action in the nervous system (see Litchfield 1985 for review). The mode of action of these compounds is similar to that of DDT.

14.3.1 Type I and II Pyrethroids

There are some differences in the mode of action of pyrethroids: knockdown vs. kill; peripheral vs. central action; type I vs. type II; insects vs. mammals; pyrethroids vs. DDT; negatively vs. positively correlated temperature effects. Pyrethroids are classified into two groups from poisoning syndromes in insects and mammals and actions on the nerve activity:

1. Type I pyrethroids: non-α-cyanopyrethroids including natural pyrethrins, allethrin, tetramethrin, resmethrin, kadethrin, phenothrin, permethrin, etc.
2. Type II pyrethroids: α-cyanopyrethroids including cypermethrin, deltamethrin, fenvalerate.

In mammals, the signs of pyrethroid poisoning are indicative of an action on the nervous system, and two distinct intoxication syndromes have been described. Type I pyrethroids are characterized by whole-body tremors similar to those produced by DDT. In contrast, type II pyrethroids like deltamethrin produced a distinctly different syndrome, characterized by sinus writhing convulsions (choreoathetosis) accompanied by profuse salivation. In insects exposed to these two groups, two syndromes of intoxication have also been noticed, but the differences are less clearly defined than those observed with mammals. Type I pyrethroids cause restlessness, incoordination, and hyperactivity, followed by prostration and paralysis. These actions generally resemble those of DDT. Type II pyrethroids produce a distinctly different syndrome from type I, and cause a characteristic pronounced convulsive phase; that is, within minutes of dosing, insects become uncoordinated.

Although we have an incomplete picture of the contribution that the various neurophysiological responses make to the poisoning symptoms in whole insects, the primary locus of pyrethroid action is the nervous system and pyrethroids have a capacity to affect all types of neurone.

14.3.2 Action at the Cellular and Molecular Levels

It is now well established that the sodium channel is the primary target site of pyrethroids in insects and other animals and that the pyrethroids bind to the closed and open sodium channels. Type I pyrethroids prolong the sodium current during excitation, causing depolarizing after–potential. When the after-potential exceeds the membrane threshold repetitive action potentials are generated, leading to hyperexcitation. The prolongation of sodium current by pyrethroids is caused by changes in channel opening and channel closing mechanisms. High concentrations reduce the amplitude of the action potential due to suppression of the sodium currents. These changes in sodium channel and sodium current are deemed responsible for the symptoms of poisoning leading to paralysis and death. Only a very small percentage (less than 1%) of sodium channel population needs to be modified by pyrethroids. Modification of the small fraction of sodium channel population is enough to elevate depolarizing after-potential to the level of threshold membrane potential to initiate repetitive discharges. This notion explains why pyrethroids are so potent (Tatebayashi and Narahashi 1994; Song and Narahashi 1996).

Type II pyrethroids also act on the sodium channel, prolonging sodium current to a greater extent than type I pyrethroids. However, the pattern of changes in excitability is somewhat different from that caused by type I pyrethroids. Type II pyrethroids depolarize the nerve membrane more strongly than type I pyrethroids. Because of membrane depolarization, nerve fibers do not initiate repetitive discharges, but sensory neurons discharge bursts of impulses and synaptic transmission is disturbed. The nerve conduction is eventually blocked due to membrane depolarization. The cockroach which develops a kdr type of resistance showed cross-resistance to pyrethrins and other type I pyrethroids, but not to type II decamethrin and cypermethrin.

These classifications of type I and II pyrethroids are not absolute, because there is a continuous transition from type I to type II structures and some pyrethroids such as cyphenothrin and fenfluthrin have an intermediate position in their effects on the axon. Two pyrethroids (fenpropathrin and an oxime of

0-α-cyanophenoxybenzyl ether) were classified as type I based upon electrophysiological criteria and as type II based upon in vivo symptomatology. It may not be necessary to invoke two different modes of action on sodium channels to explain the striking differences of the observed electrophysiological and toxicological effects.

Other than the neurophysiological effects, pyrethroids show several biochemical effects. Type II pyrethroids bind to the chloride ionophore component of the GABA receptor complex and inhibit GABA-dependent chloride flux, but the concentrations required to affect this system are quite apart from those capable of disrupting sodium channel function. Also, they have a direct effect on the vertebrate muscle, but the significance of this effect in terms of acute intoxication is unclear. Pyrethroids are good inhibitors of Ca–Mg and Ca–ATPases. The Ca–ATPase most likely represents the ATP-dependent phosphorylation and dephosphorylation system associated with the Na/Ca exchange, while Ca–Mg–ATPase is an enzyme responsible for calcium pumping and sequestration to maintain a proper intracellular calcium concentration and the gradient across the cellular membrane. However, functional significance of these effects of pyrethroids remains to be seen.

Widespread release of neurohormones resulting from a direct effect on neurosecretory cell or from increasing nervous activity would result in a wide variety of secondary disruptive effects.

14.3.3 Toxicity to Mammals

Acute toxicity to mammals depends to a large extent on the dosing vehicle used, the environmental conditions of testing, the strain and sex of the animal, and its dietary status. With the pyrethroids, it is also important to know the isomer content of the preparation since different isomer ratios have different toxicities. For the chrysanthemic or related acids, the esters of *cis* isomers are more toxic than those of the *trans* isomers. Also, it is important that comparisons between the potency of pyrethroids should all be done in the same laboratory. The LD_{50}'s shown in Table 14.3 indicate that the current synthetic pyrethroids have a wide range of acute toxicities. Acute dermal toxicities (Table 14.4) indicate that the pyrethroid molecule has a low order of dermal toxicity. Acute inhalation tests with aerosols containing either allethrin, phenothrin, permethrin, resmethrin, or tetramethrin (Miyamoto 1976) with particle sizes of 1–2 μ for periods of

Table 14.3. Comparative oral LD_{50} of pyrethroids. (Miyamoto 1976)

Pyrethroids[a]	LD_{50} in male mice (mg/kg)
Allethrin	500
Phenothrin	> 5000
Permethrin	490
Resmethrin	690
Tetramethrin	1920
Pyrethrins[b]	370

[a]Corn oil vehicle.
[b]Calculated as active ingredient.

Table 14.4. Acute dermal LD_{50} of pyrethroids. (FAO 1977, 1980, 1982; Kavlock et al. 1979)

Pyrethroid	Species (sex)	LD_{50} (mg/kg)
Permethrin	Rat (F)	> 4000
	Rat (M)	> 2500
	Rabbit (F)	> 250
Cypermethrin	Rat (F)	> 4800
	Rabbit (F)	> 2400
Bioresmethrin	Rat (F)	> 10 000
Fenvalerate	Rat	5000
	Rabbit	> 2500
Deltamethrin	Rat	> 800
	Rat (M)	> 2940
	Rabbit (M)	> 2000

F. female; M, male.

2–4 h showed no detrimental effects to the test animals. Other tests (Metker 1980) indicated severe tremoring in rats exposed to 500 mg/m^3 permethrin.

Several pyrethroids have been evaluated for carcinogenic potential by long-term studies with rodents, with very high doses of the chemicals, such as up to 3000 ppm fenvalerate, 5000 ppm permethrin, and 6000 ppm phenothrin. Some toxic effect occurred at the top dose for all compounds tested. Rats fed 3000 ppm permethrin in their diet for 6 months showed typical motor symptoms in the early stages of the study but not other changes, except for a slight increase in liver weight associated with an increase in smooth endoplasmic reticulum. Liver changes have been noted in rats exposed to high dietary levels of permethrin, but these were shown to be rapidly reversible with no evidence of toxic liver damage (Litchfield 1983). Rats fed 5000 ppm or more for 14 days developed acute poisoning and death (Hayes and Laws 1991). Rats fed 6000 ppm for 2 years showed only a small reduction in weight gain (FAO/WHO 1981). The conclusion by the US Environmental Protection Agency (Federal Register 1982) was that the oncogenic potential for humans was nonexistent or extremely low. The Joint Meeting on Pesticide Residues FAO/WHO in 1982 (FAO 1983) also concluded that the long-term rodent studies with permethrin did not indicate any oncogenic risk to humans. No evidence of teratogenicity (reproductive toxic potential) or mutagenicity (effect on the genetic material) was shown in a long-term study with several pyrethroids.

For a detailed account of the toxicology of pyrethroids see Litchfield (1985), and Hayes and Laws (1991).

14.4 Metabolism and Degradation

Sensitivity to photodegradation of the natural pyrethrins and the early pyrethroid insecticides is the reason why such compounds have been limited to control of pests of public health importance and had not achieved commercial uses in agriculture. However, an understanding of this photosensitivity was an important contribution which led to the synthesis of the photostable pyrethroids which assumed a major role in combating agricultural pests. The first such photostable compound to achieve commercial application was permethrin (Elliott et al. 1973a,b). The photostabilized permethrin was found to have a half-life in sunlight measurable in days rather than in hours as had been the case with the earlier compounds. Studies on the photodegradation of cypermethrin, deltamethrin, fenvalerate, tralomethrin, and tralocythrin in various organic solvents and in water, plant, soil, and on inert surfaces soon followed (Leahey 1985).

14.4.1 Mammals

The metabolism of the natural pyrethrins as well as the synthetic allethrin has been discussed in Section 13.3.5. It was pointed out that oxidation was the most important route of metabolism and that ester hydrolysis assumed minor roles (Yamamoto et al. 1969, 1971a; Yamamoto and Casida 1966, Yamamoto 1973). The metabolism of permethrin has been studied in great detail in a wide variety of animals. The in vivo and in vitro investigations established that permethrin is extensively metabolized by rats so that very little unchanged permethrin is excreted. The *trans* isomer is metabolized and eliminated much faster than the *cis* isomer. This difference is due to the greater susceptibility of *trans*-permethrin to esterase attack. Despite this difference, the major route of metabolism in vivo is via ester cleavage by esterase and oxidase attack, as well as hydroxylation of the terminal aromatic ring. These reactions yield conjugated products which can easily be eliminated from the body. Cypermethrin, deltamethrin, fenvalerate, and other more recent stable pyrethroids are similarly metabolized, with some modifications in the metabolic products (Fig. 14.3). Metabolism studies have also been made with cows, goats, chickens, and fish.

In summary: although the pyrethroids are highly lipophilic compounds, they are nevertheless not stored to a significant extent in fatty tissues or other tissues, in mammals. This is due to their rapid metabolism, with the production of metabolites of greater water solubility which can be conjugated and excreted. For most synthetic pyrethroids, the most important metabolic process in mammals is cleavage of the central ester linkage. This is in contrast to the natural pyrethrins and allethrin which are metabolized chiefly by oxidative attack with the intact ester linkage.

It is beyond the scope of this discussion to give detailed information on the metabolism of each compound separately. Such details can be found in a comprehensive review by Leahey (1985).

14.4.2 Insects

The metabolism of permethrin has been studied in several insect species, but in greater detail in the cockroach, adult housefly, and cabbage looper larvae. In all three species *cis*-permethrin was metabolized less readily than *trans*-permethrin (Ishaaya and Casida, 1980). Ester cleavage and hydroxylation of the alcohol moiety are the major metabolic routes in all three species. Hydroxylations at other positions in the molecule are also detected, being more important or less so depending on the species. Similar results have been obtained with isolated enzyme systems from the housefly and cabbage looper (Shono and Casida,

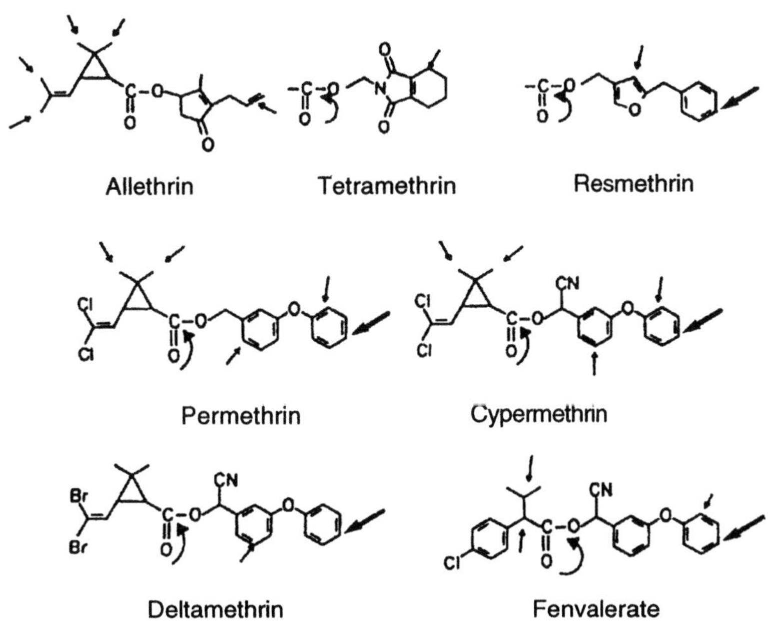

Fig. 14.3. Site of metabolism of synthetic pyrethroids in mammals. *Straight arrows* indicate the site of hydroxylation; *curved arrows* indicate the site of hydrolysis. (Modified from Nauman 1990)

1978; Shono et al. 1979; Ishaaya and Casida 1980), from the tobacco hornworm and bollworm (Bigley and Plapp, 1978), the porina moth (Chang and Jordan 1982), and the cattle tick (Schnitzerling et al. 1983). In general, it appears that metabolism of pyrethroids in insects is similar to that occurring in mammals, with slower rates of metabolism generated by insect enzymes (esterases and oxidases) than by those of mammalian systems (Jao and Casida 1974). This may explain, at least in part, why pyrethroids are much more toxic to insects than to mammals. Again, the *trans*-isomers are more readily metabolized by esterases than the *cis* isomers, but there are exceptions, such as with the lacewing (Ishaaya and Casida 1981) in which the reverse is true.

The importance of metabolism in the detoxification of pyrethroids by insects is evident from the increase in potency achieved by mixing synergists (specific esterase and oxidase inhibitors) with pyrethroids (Ishaaya and Casida 1980; Ishaaya et al. 1983, 1987). The mechanism and effects of such synergists with pyrethroids have been reviewed in detail by Yamamoto (1973), Soderlund et al. (1983), and Ishaaya (1993), and with other insecticides by Wilkinson (1971, 1976a,b).

14.4.3 Plants

The photostable pyrethroids have been shown to have half-lives of 1–6 weeks on plants under greenhouse conditions. Degradation under field conditions is faster. In both instances, the initial metabolic processes were found to be identical to those of mammals, that is, ester cleavage and hydroxylation reactions. However, oxidation of primary alcohols is a less important process in plants. The metabolites formed are usually conjugated with sugars or amino acids. With most compounds studied, it was shown that the pyrethroids are not translocated from the leaves or from the soil to other parts of the plant. In addition to metabolism, photoinduced reactions also occur on the surface of treated plants. In some cases, such as with phenothrin, phototransformation is in fact the predominant degradation process. It is possible that ester cleavage, which is a major degradation process in plants, is a photoinduced as well as a metabolic reaction. This may explain the faster degradation rates of pyrethroids on plants maintained in direct sunlight than under glasshouse conditions.

14.4.4 Soil and Water

Most of the stable pyrethroid insecticides undergo ready degradation in soils maintained under aerobic conditions, but degradation is slower under anaerobic conditions. The rate of degradation varies from compound to compound and also with the type of soil being investigated. Half-lives of 1–16 weeks have been reported. Under aerobic conditions, phenothrin degradation is very fast (half-life 1–2 days) with ester cleavage and hydroxylation and oxidation reactions yielding a number of metabolites which are further converted to carbon dioxide. Unextractable residues (bound residues) also form rapidly, reaching a maximum level of up to 55% within 30 days. The bound residue eventually decreases, probably by conversion to carbon dioxide. Under anaerobic conditions degradation is much slower, with a half-life of 2–8 weeks.

Permethrin under aerobic conditions in all soil types undergoes fairly rapid degradation (half-life 5–55 days) with conversion to CO_2. The unextractable residue appears to reach a maximum within 5–10 weeks, ultimately releasing CO_2. The degradation products due to ester cleavage, hydroxylation and oxidation are the same in all soils but appear to be at low levels. Under anaerobic conditions, degradation is even slower. In natural waters, permethrin is rapidly absorbed onto the sediment so that less than 2% remains in the aqueous phase after 7 days. Degradation also occurs in natural water/sediment mixture, with ester cleavage as the major degradation process.

Cypermethrin degradation under aerobic conditions, in soils takes place with half-lives of 1–10 weeks. The major degradation products are the same as those obtained with permethrin, with some minor additional products, and the eventual conversion to CO_2. As with permethrin, rapid absorption onto the sediment occurs, followed by rapid degradation (half-life approx. 5 days) when the pyrethroid is mixed with river water.

Fenvalerate degradation under aerobic conditions in different soils ranges from 2 to 14 weeks. The formation of bound residues and the evolution of CO_2 are major processes. In a model ecosystem study (Ohkawa et al. 1980) with radioactive (S)fenvalerate, maintained for 30 days, the distribution of the radioactive metabolites was as follows: soil 90–93%; water 0.94–1.1%; snails 0.35–0.4%; fish 0.10–0.24%; algae and daphnia 0.1%. The unchanged (S)fenvalerate accounted for 67–82% in soil, 33–40% in water, 61% in snails, 27–31% in fish, and 67–94% in algae and daphnia.

The potential for leaching of pyrethroids in soil has also been investigated. In all cases it was found that the stable pyrethroids are virtually immobile in soil, considering their high lipophilic nature. Such immobility, coupled with the formation of bound residues and their subsequent complete degradation suggests that the pyrethroids are unlikely to move from soil to any other part of the environment.

Extensive studies on the photochemistry and metabolic fate of the pyrethroids in insects, mammals, and plants have been made by Casida and coworkers, and by many others. The reader is referred to Leahey (1985) for references.

CHAPTER 15

Synergists

The term "synergism" is used for cases where two compounds together show a more concerted activity than that predicted from the sum of their individual activities. Often one component is not toxic or far less active than the counterpart component at the dosage employed, but when combined with the latter markedly increases the activity and is called a "synergist." As related to insecticides, a synergist is used at high doses (for example, 5–10 times more than the insecticide) in many cases and the cost justifies its use in limited cases like natural pyrethrum. The effectiveness of an insecticide synergist is commonly expressed by the ratio of the LD_{50} of an insecticide alone to the LD_{50} of the insecticide with the synergist (for reviews see Yamamoto 1973; Hodgson and Tate 1976; Wilkinson 1976a,b; Casida and Quistad 1995b).

In 1940 it was found that sesame oil remarkably enhanced insecticidal activity of pyrethrum and later the active components were identified as sesamin and sesamolin, both being methylenedioxyphenyl (MDP) compounds (1,3-benzodioxole). Since then, many MDP compounds and other types of synergists have been introduced, but because of their cost and efficacy only a few synergists such as piperonyl butoxide (PB) and MGK 264 are of a practical use. Synergists are found in practically all aerosol bombs containing pyrethrins and allethrin which are used against flying household insects, and in other type formulations containing bioallethrin and tetramethrin. Many of the synthetic pyrethroids (Chap. 14) are not synergized by the common MDP pyrethrum synergists. However, different types of synergists are still useful in insect toxicology to study mode of action, metabolism and resistance mechanism (Fig. 15.1).

Synergism can be obtained by simultaneous treatment, pretreatment with a synergist, or application at separate sites of the synergist and the insecticide. This implies the involvement of biochemical interactions in the insect body.

15.1 Types of Synergism

There are several types of synergism: (1) inhibition of mixed-function oxidase enzymes; (2) inhibition of hydrolyzing enzymes; (3) release of hydrogen cyanide from organothiocyanates by glutathion S-transferases; (4) probably serving as alternative substrates; (5) two compounds interact with different sites of the target.

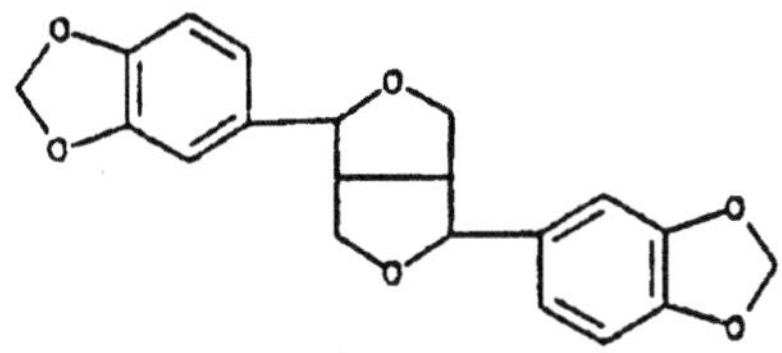

Sesamin

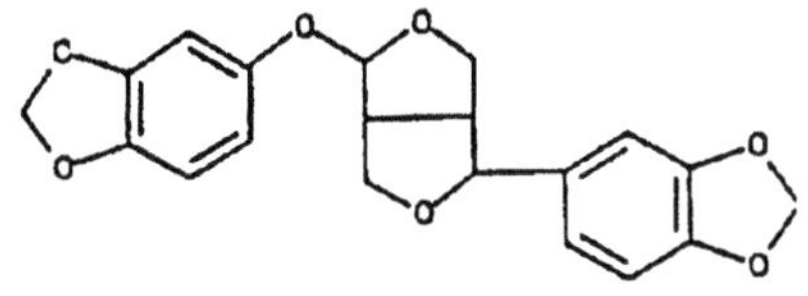

Sesamolin

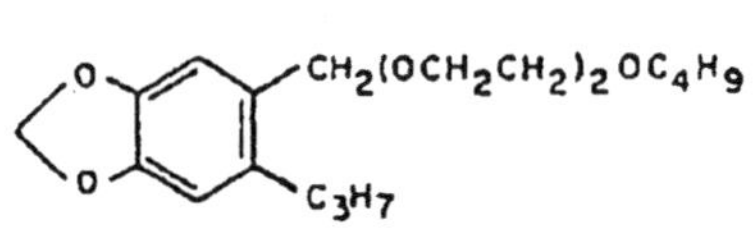

Piperonyl butoxide

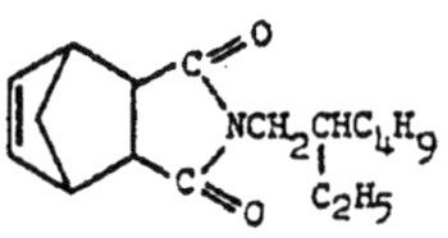

MGK 264

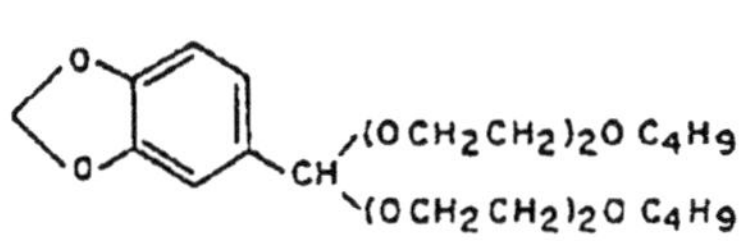

Tropital

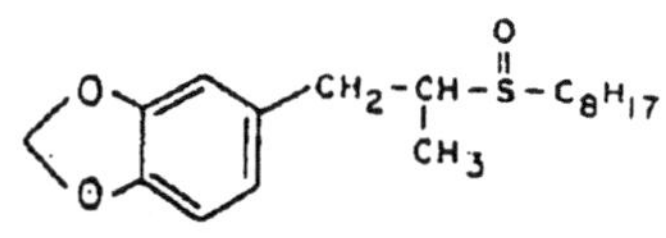

Sulfoxide

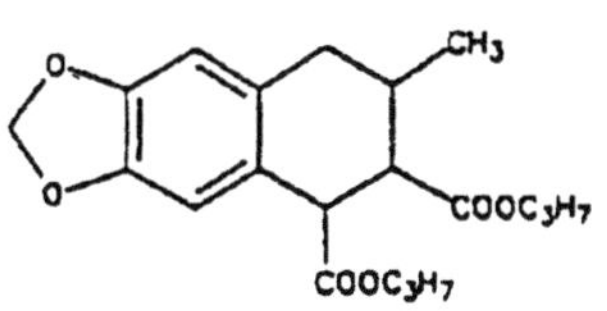

Propyl isome

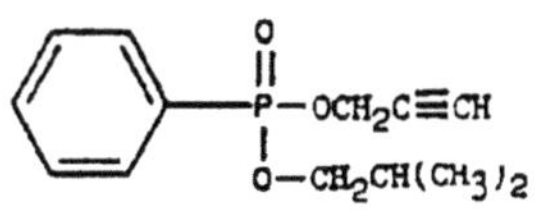

NIA 16, 824

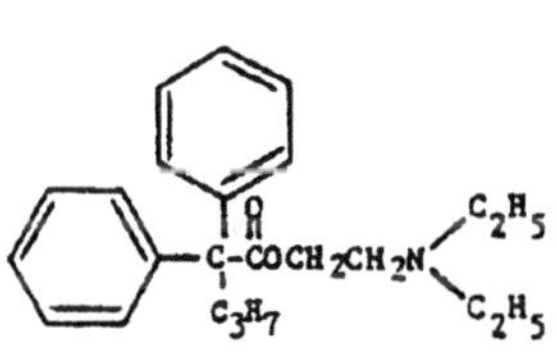

SKF 525-A

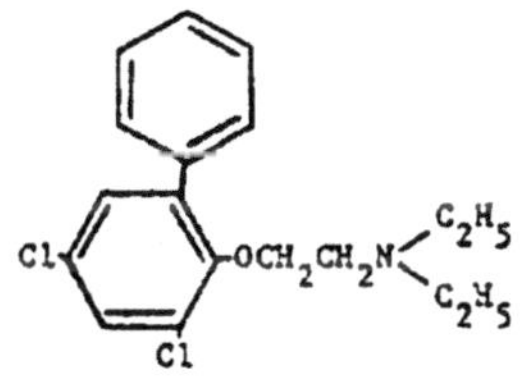

Lilly 18, 947

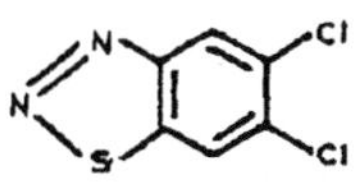

WL 19, 255

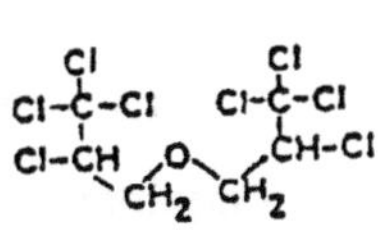

Octachlorodipropyl ether

Fig. 15.1. Insecticide synergists

15.1.1 Inhibition of Oxidative Metabolism

Inhibitors of oxidative metabolism of insecticides include diverse groups of compounds such as those containing the methylenedioxyphenyl (MDP) moiety, benzothidiazoles (e.g., WL 19,255), and imides (e.g., MGK 264). MDP compounds exert their synergism by inhibiting the mixed-function oxidase (MFO) system. Oxidation of methylenedioxy group by MFO first gives a hydroxylated intermediate, which takes two routes for further convertion. One is hydrolysis to give a catechol and formate as metabolites. Another is the production of a carbene, which complexes with P-450, resulting in the inhibition of MFO activity. The carbene may arise directly from MDP and degrades to the catechol (Fig. 15.2). These compounds can synergize not only pyrethrins, but also carbamates and other insecticides. When pest resistance is due to enhanced MFO metabolism, synergists often mask the resistance. However, in the case of organophosphorus insecticides, the MDP compounds inhibit both the oxidative detoxication and the conversion of P=S to P=O and the net result is often difficult to predict.

15.1.2 Inhibition of Hydrolysis

Inhibitors of esterases or hydrolyzing enzymes include EPN (O-ethyl O-p-nitrophenyl phenylphosphonothioate), TOCP (tri-o-cresyl phosphate), DEF (S,S,S-tributyl phosphorotrithioate), triphenyl phosphate, and others. They enhance the insecticidal activity of malathion to resistant strains of insects. The combination

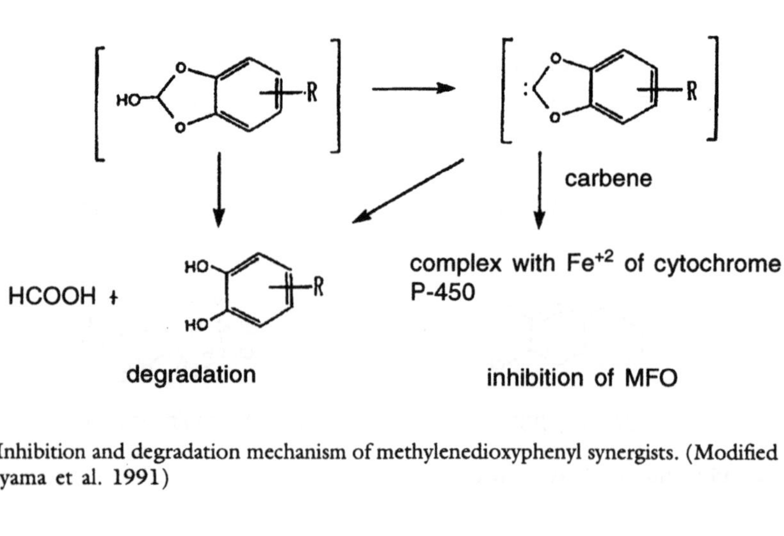

Fig. 15.2. Inhibition and degradation mechanism of methylenedioxyphenyl synergists. (Modified from Motoyama et al. 1991)

of malathion with iprobenfos (a fungicide) was practically used in Japan on the green rice leafhopper resistant to malathion by inhibiting carboxyesterase.

Many insects have esterases hydrolyzing pyrethroids named pyrethroid esterases. Monocrotophos, profenofos, acephate, methidathion, some propynyl phosphonates, and phenyl saligenin cyclic phosphate inhibit pyrethroid esterases, by which they increase the insecticidal activity (Ishaaya and Casida 1981; Ishaaya et al. 1987; Ishaaya 1993). In contrast, oxidase inhibitors such as PB, SV-1, and MPP synergize considerably the toxicity of pyrethroids in *Tribolium castaneum* and *Musca domestica* (Ishaaya et al. 1983, Ishaaya 1993). This implies that the predominant pathway for pyrethroid detoxication in insects, whether hydrolytic or oxidative, depends largely on the insect species.

15.1.3 Releasing Hydrogen Cyanide from Organothiocyanates

Certain organothiocyanates are effective insecticides and act as synergists for carbamates and pyrethrins. These compounds release hydrogen cyanide by metabolism with glutathion S-transferases-glutathion system (GST-GSH), which contributes to the insecticidal activity.

$$RCH_2SCN \xrightarrow[GSH]{GST} HCN + [RCH_2SSG] \xrightarrow[GSH]{} RCH_2SH + GSSG\ .$$

15.1.4 Analog Synergism

Synergists structurally related to the insecticides have shown their usefulness particularly with DDT. Compounds such as DMC [1,1-bis-(p-chlorophenyl) ethanol], a noninsecticidal structural analog of DDT, derivatives of diphenylamine and benzenesulfonamides, and several others, have been shown to increase the toxicity of DDT to DDT-resistant insects (Perry and Agosin 1974). The compound N,N-dibutyl(p-chlorobenzene)sulfonamide, designated as "WARF antiresistant" (Pramex), had been marketed as a DDT synergist but attained only limited success.

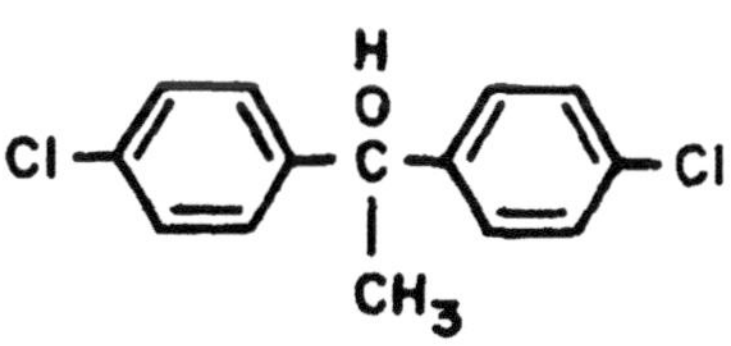

DMC

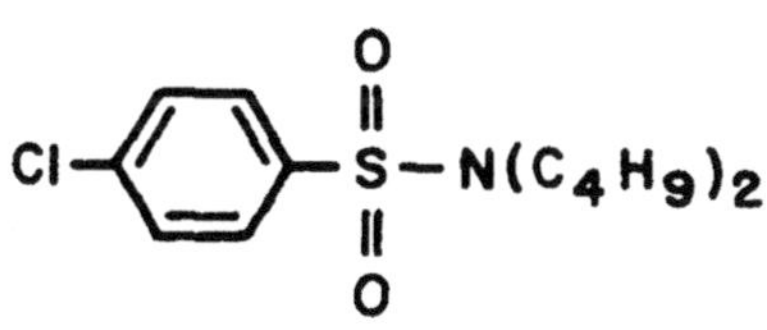

WARF anti-resistant

15.1.5 Target Site Synergism

Two compounds interact with different sites of the target. Aryl N-methylcarbamates are used to control the green rice leafhopper, which develops cross-resistance to them. Combination of an aryl N-methylcarbamate with an aryl N-propylcarbamate (Yamamoto et al. 1983) or with an oxadiazolone compound such as metoxadiazone (Yamamoto et al. 1993) overcomes the resistance. The resistance is due to the increased population having the modified acetylcholinesterase (AChE) insensitive to N-methylcarbamates. N-propylcarbamates or oxadiazolones are more inhibitory to the modified AChE. Both components of the combination interact with the preferred inhibition sites, thus achieving synergism (Table 15.1).

15.1.6 Alternative Substrates for Detoxification

Some compounds may function as alternative substrates for detoxification, thus protecting the insecticides. Insecticides having MDP moiety in the molecules may synergize themselves (self-synergism). However, such ideas had only marginal success.

Table 15.1. Synergism between an N-methylcarbamate and an N-propylcarbamate in green rice leafhoppers cross-resistant to N-methylcarbamates. (Yamamoto et al. 1983)

	AChE inhibition (I_{50} M)	Insecticidal activity (LD_{50} $\mu g/g$)
MTMC	1.2×10^{-4}	61[a]
N-Pr-MTMC	4.0×10^{-5}	50
MTMC + N-Pr-MTMC (1:1)	1.4×10^{-5}	19
Co-toxicity coefficient	4.3	2.9

MTMC, 3-tolyl N-methylcarbamate; N-Pr-MTMC, 3-tolyl N-propylcarbamate.
[a]2.0 to the susceptible strain.

CHAPTER 16

Neonicotinoids and Nitrogenous Insecticides

16.1 Neonicotinoids

Nicotine has never achieved the prominence that the synthetic insecticides have attained due to its expensiveness, lack of commercially applicable synthesis, extreme toxicity to mammals, and limited insecticidal spectrum. The mode of insecticidal action of nicotine was clarified by Yamamoto (Yamamoto et al. 1962; Yamamoto 1965) and many of the synthetic analogs of nicotine fitted with specific structural feature, required for nicotinoid action showed more or less insecticidal activity (see Chap. 13). This indicates the validity of the essential moiety, but none have shown any practical utility. Imidacloprid, a novel insecticide, derived synthetically from a nitromethylene insecticidal chemical, nithiazine, is closely related to nicotinoids in mode of action and structure–activity relationships. Later on several insecticides resembling imidacloprid appeared, which are collectively called neonicotinoids (Fig. 16.1):

Imidacloprid: 1-(6-chloro-3-pyridylmethyl)-*N*-nitroimidazolidin-2-ylideneamine
Acetamiprid: (E)-N^1-(6-chloro-3-pyridylmethyl)-N^2-cyano-N^1-methylacetamidine
Nitenpyram: (E)-*N*-(6-chloro-3-pyridylmethyl)-*N*-ethyl-*N*'-methyl-2-nitrovinylidene-diamine

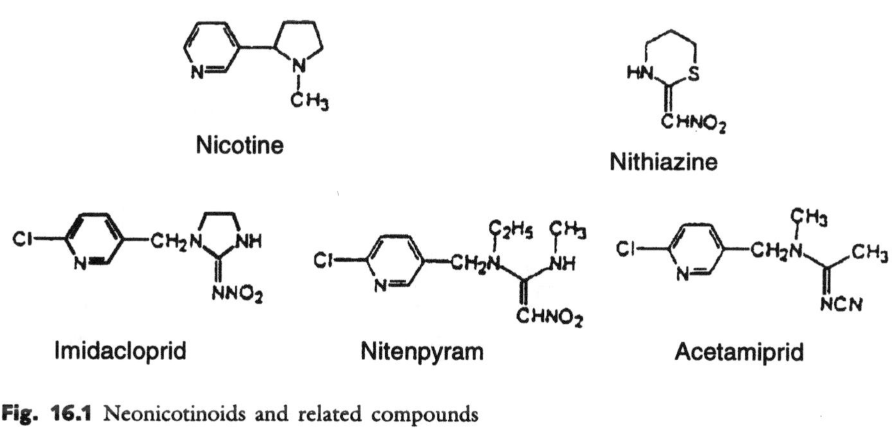

Fig. 16.1 Neonicotinoids and related compounds

16.1.1 Mode of Action

Imidacloprid and other neonicotinoids interact with the acetylcholine binding site of the nicotinic acetylcholine receptor as agonists, which cause excitation and eventually paralysis leading to death as nicotine does. The structure–activity relationships (Liu et al. 1993; Tomizawa and Yamamoto 1993) based on insecticidal activity and binding affinity to the receptor are summarized as follows:

1. 3-Pyridyl is preferred over 2-pyridyl, 4-pyridyl, or phenyl.
2. Between 3-pyridyl and a nitrogen atom which is a part of the imidazolidine ring, the presence of one methylene is preferred over no or two methylenes.
3. The presence of a strong electron withdrawing group such as nitromethylene, nitroimino, and cyanoimino is essential on the 2-position of the imidazolidine ring.
4. The imidazolidine ring can be replaced by other heterocyclics or even the ring can be cleaved as far as there is the presence of the above nitrogen, which is one carbon atom apart from the 3-pyridyl group, and the electron withdrawing group on the carbon atom neighboring to the nitrogen atom.

These features are similar to those for nicotinoids except that the above nitrogen is not ionized, but bears a partially positive charge induced by the neighboring electron withdrawing group.

It has been recognized that many nAChR ligands share the same feature: the presence of an electron donating group which makes hydrogen bonding to the receptor at 5.9 Å apart from the positively charged center which interacts with the anionic site of the receptor, while many mAChR ligands share this similar feature but the distance is 4.4 Å. For nicotine, the nitrogen atom of 3-pyridyl group is the electron donor and the protonated nitrogen atom is the positive center. From analogy, the partially positive center in neonicotinoids must be the center to interact with the anionic center. It is supposed that acetylcholine interacts with nAChR and mAChR by using the carbonyl oxygen and the ethereal oxygen, respectively, as the donor. The above situation is depicted in Figs. 16.2 and 16.3.

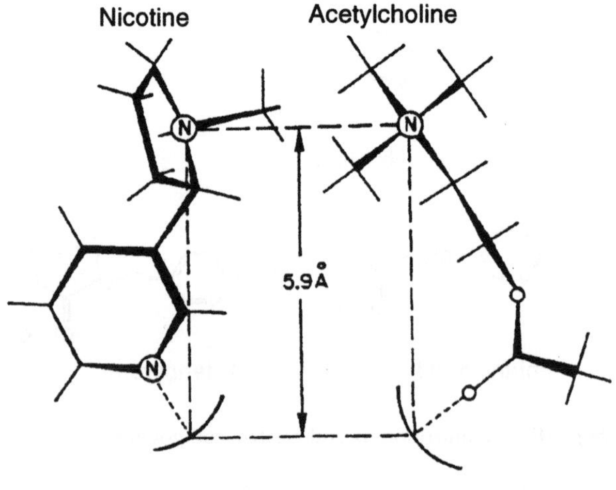

Fig. 16.2. *l*-Nicotine and acetylcholine. (Beers and Reich 1970)

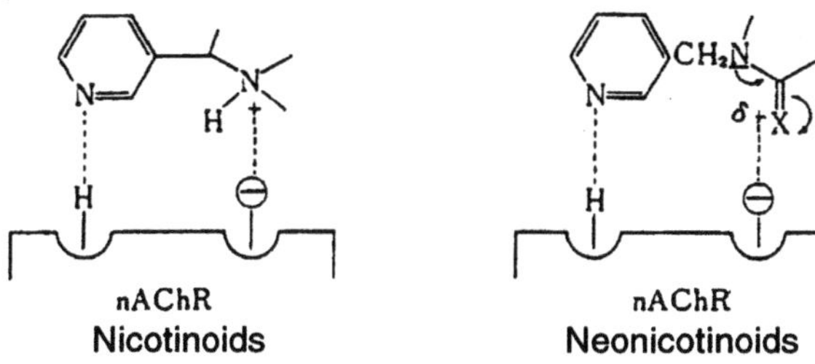

Fig. 16.3. Essential structural moieties of nicotinoids and neonicotinoids and their interaction with the nicotinic acetylcholine receptor. (Yamamoto et al. 1995)

The term "neonicotinoids" was derived from the recognition that imidacloprid and the related compounds share the same feature with nicotinoids, but with an additional feature: the partial positive charge on the nitrogen atom conferred by the neighboring electron withdrawing group. Based on the essentiality for higher insecticidal activity of the presence of a chlorine atom at the 6-position of the 3-pyridyl group, the term "chloro-nicotinyl" insecticides was also proposed. However, the introduction of a chlorine atom and methyl group somewhat increases and decreases the binding affinity as compared with the unsubstituted one, and only the unsubstituted one is inferior in insecticidal activity. The substitutions increase the hydrophobicity, which is good for penetration to the target site. These neonicotinoids share the above essential characteristics and systemicity, but differ in stability and insecticidal spectra, and thus can be complementarily utilized. Imidacloprid is mostly for sucking insects and persistent. Nitenpyram is less persistent and acetamiprid is effective to both hemipteran and lepidopteran insects.

16.1.2 Selective Toxicity

There is a distinct difference in the selective toxicity of neonicotinoids. Nicotine is highly toxic in mammals, while limited in insecticidal activity. Imidacloprid has the opposite and very favorable properties (Table 16.1). With regard to the binding affinity to the insect nAChR, nicotinoids including nicotine and anabasine and neonicotinoids including imidacloprid, nitenpyram, and acetamiprid, all having 3-pyridylmethyl moiety, bind well, while nithiazine and TMA (tetramethylammonium), lacking 3-pyridylmethyl moiety, are poor. Nicotyrine with almost no ionization is poor in binding. This implies that the presence of an electron donating atom, that is, the nitrogen atom in 3-pyridyl, which makes hydrogen bonding to the receptor at 5.9 Å apart from the charged center, either fully or partially positive, is required for potent interaction with insect nAChR. On the contrary, the vertebrate nAChR prefers the presence of one unit of positive charge for stronger binding.

Table 16.1. Toxicity of nicotine and imidacloprid

Insecticide	LD_{50}(mg/kg)	
	Rat (oral)	Housefly (injection)
l-Nicotine	53	272
Imidacloprid	450	22

Action in Insects

In insects, nAChR is present only in the central nervous system. Imidacloprid, once entering the insect body, is not ionized, is easily transferred into the central nervous system, and strongly interacts with the target site. Due to its lower hydrophobicity, imidacloprid is poor in penetration through cuticle as seen in lower insecticidal activity by topical application than by injection. This would be the cause of less activity to lepidopteran insects and more activity to sucking hemipteran insects. On the other hand, nicotine has a limited access to the target site due to its ionization, resulting in lower or limited insecticidal activity (Fig. 16.4).

Action in Vertebrates

In vertebrates nAChR is present both in the central and peripheral nervous systems. The interaction of imidacloprid with any nAChR is weak and also its penetration through the blood–brain barrier is poor. On the other hand, nicotine, although poor in penetration into the central nervous system due to its ionization, interacts strongly with peripheral targets like neuromuscular junction, resulting in higher toxicity.

16.2 Nereistoxin-Related Insecticides

Nereistoxin is a poison from a marine anelid, *Lumbrineris brevicirra* (= *Lumbriconereis heteropoda* Marenz), 4-dimethylamino-1,2-dithiolane. There are several 1,2-dithiolanes such as asparagusic acid, α-lipoic acid, brugine, brugierol, isobrugierol, gerrardine, ginesine, and caratoxin. Nereistoxin and caratoxin as

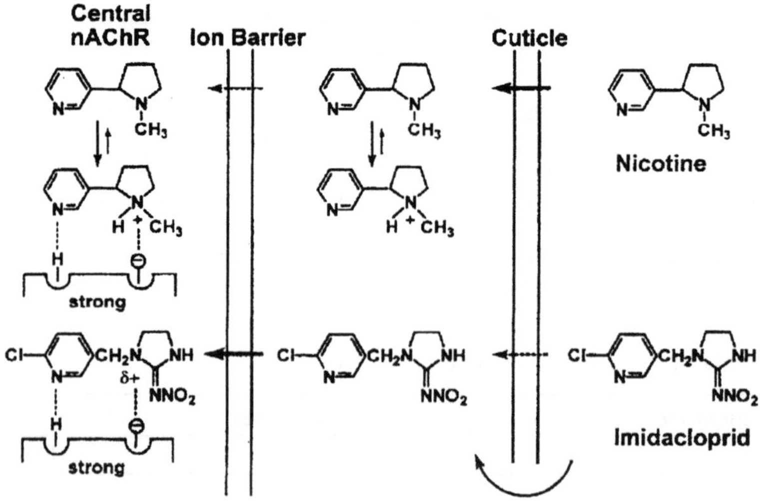

Fig. 16.4. Action of nicotine and imidacloprid in insects. (Yamamoto et al. 1995)

well as ginesine and gerrardine are insecticidal; gerrardine is also bactericidal. As the synthetic derivatives of nereistoxin, cartap, bensultap, and thiocyclam were commercially developed as insecticides:

1. Cartap: 1,3-bis(carbamoylthio)-2-(*N*, *N*-dimethylamino) propane; *S*, *S'*-2-dimethylaminotrimethylene bis(thiocarbamate).
2. Bensultap: 2-dimethylamino-1,3-bis(phenylsulfonylthio) propane; *S*, *S'*-2-dimethylaminotrimethylene di(benzenethiosulfonate).
3. Thiocyclam: *N*, *N*-dimethyl-1,2,3-trithian-5-ylamine.

They have broad spectra, particularly effective on Coleoptera and Lepidoptera, and are active as contact and stomach poisons with ovicidal, larvicidal, and adulticidal effects against sucking and leaf-biting insects.

Mode of Action. Cartap, bensultap, and thiocyclam are metabolically converted to nereistoxin, which interacts with nicotinic acetylcholine receptors (Fig. 16.5). Different from nicotine or imidacloprid, nereistoxin is an antagonist, blocking cholinergic transmission, and resulting in the paralysis and death of insects without convulsive symptoms. During the presence of nereistoxin at the receptor, ACh transmitter is not able to trigger excitation at the postsynaptic membrane.

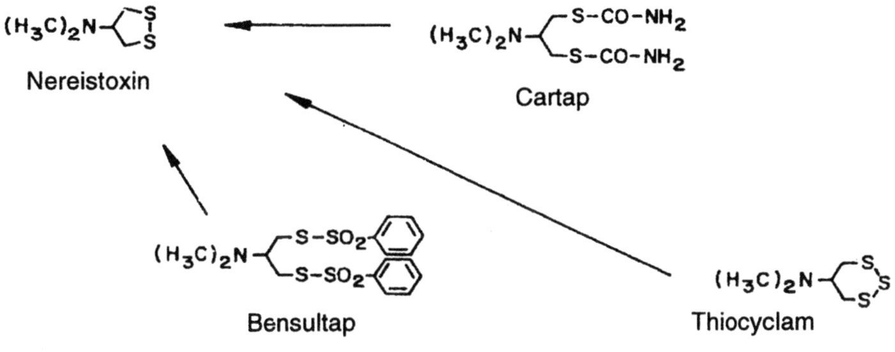

Fig. 16.5. Nereistoxin as the activated form of the related insecticides

16.3 Formamidines

Formamidines represented by chlordimeform and amitraz show, unlike conventional insecticides, very unique actions for the control of phytophagous mites, ticks, and certain insects (primarily Lepidoptera and Hemiptera) by acting as agonists of octopamine receptors. They are locomotor stimulants, anorexients, and eclosion inhibitors and because of the relative importance of sublethal and behavioral effects in pest control, they are termed "insectistatics" (Hollingsworth and Lund 1982). The expression of formamidine actions depends on the species, the life stage, chemicals, and test and use conditions, and one or some combination of these actions become crucial for pest control. With Lepidoptera and mites usually the eclosion from the eggs is the critical stage and the hatched larvae remain within the eggshell and die. Also, the insect larvae, mites, and ticks become restless, uncoordinated, and dispersed or detached from the treated host plants or animals. On lepidopteran larvae critical behavioral disruption occurs at much lower doses (0.1–10 μg/g) than the immediate lethal effects (LD_{50} 100–1000 μg/g). Anorexia, a loss of appetite or of the drive to feed, is also observed. Chlordimeform has a marked translaminar and systemic activity and a significant vapor pressure, so that the treated plant becomes as if a control releaser of the chemical. Amitraz is very effective in the control of cattle ticks and other animal ectoparasites and phytophagous mites.

The control of the Asiatic rice stem borer is described as an example. The rice plant treated with chlordimeform releases vapor which repels the female moths to oviposit and they fall down to paddy water by exhaustion. The egg mass pattern laid under disturbed behavior is disorganized and vulnerable to climatic conditions. The hatched larvae in the eggshell, which take chlordimeform absorbed onto the chorion by chewing a hole for emergence, cannot make ordinal eclosion, mostly remaining within the shell to die. The emerged larvae after biting the plant for penetration stop further feeding. Though penetrated, they die by starvation. All these factors combined together contribute to the field efficacy of chlordimeform. In another example, the tobacco cutworm larvae on vegetable leaves sprayed with a very low concentration of chlordimeform become restless, uncoordinated, and fall down from the leaf so that the feeding damage is drastically reduced without immediate lethal action. Cattle ticks feeding on cattle exposed to formamidines become hyperactive, detach from the host, wander, and eventually fall from the host without reattaching (tickfugal effects).

Mode of Action. The formamidines, chlordimeform and amitraz, are effective to all stages of mites. Both are converted to *N*-demethyl compounds (DCDM and U-40481, respectively) (Fig. 16.6), which act as the agonist to the octopamine receptor, overstimulating the receptor and elevating levels of the second messenger cAMP. At sublethal doses, they show detrimental effects on mite locomotion, feeding, and reproduction. Formamidines mimic octopamine and stimulate the octopamine receptors and this action is believed to be causative of many types of responses induced in insects or mites. Although there is a general lack of knowledge of octopamine functions in insects, octopamine is widespread in all insect nervous tissues and acts as a synaptic transmitter, neuromodulator and neurohormone.

Figure 16.6 shows several formamidines and the metabolic conversion. Chlordimeform itself is a poor agonist, but in vivo demethylated to give an active form, DCDM. U-46506 is also converted to DCDM. Amitraz is converted to a compound similar to DCDM. Chlormethiuron is activated by N-demethylation and probably converted to DCDM. Diafenthiuron is converted to a carbodiimide (DFCD), which shows octopaminergic action and, as the secondary action, inhibits mitochondrial F_1/F_0-ATPase. The structural relationship of clenpyrin with chlordimeform is similar to that of nicotine to 3-pyridylmethylamines.

In vitro studies indicate similar actions of octopamine and DCDM. For example, in the firefly light organ octopamine is the excitatory transmitter and the response

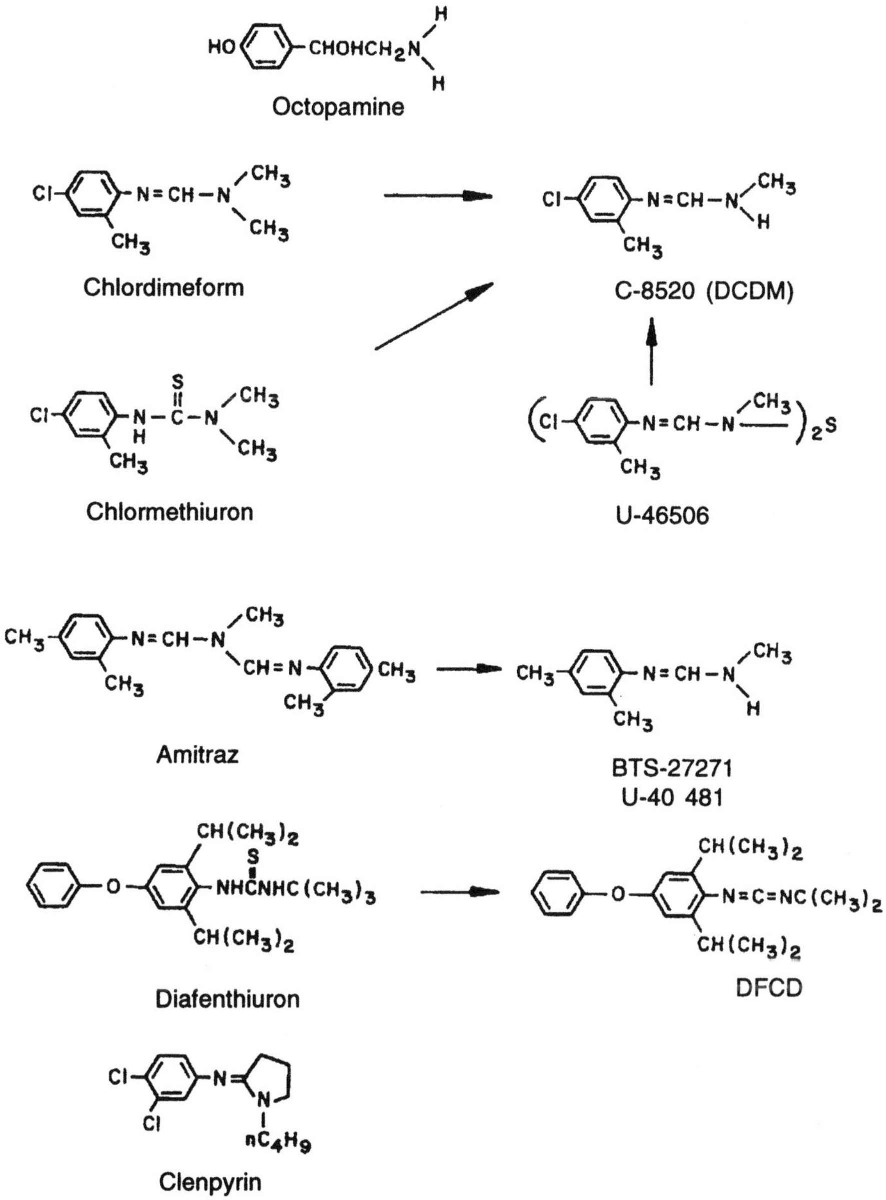

Fig. 16.6. Several formamidines and the metabolic conversion

to octopamine is mediated through an adenylate cyclase mechanism. DCDM, when applied to the firefly, lights up the lantern brightly. Both DCDM and octopamine stimulate the adenylate cyclase activity in the firefly light organ, resulting in an enhanced level of cAMP synthesis, and the activity is antagonized by both phentolamine and cyproheptadine (octopamine receptor antagonists).

Octopamine itself cannot be utilized as an insectistatic due to its high ionization and poor penetration into synaptic sites within the central nervous system through the ion barrier that surrounds it, while chlordimeform having a pKa of about 7 should be able to penetrate readily in its free base form, and after convertion to DCDM, it overstimulates octopamine receptors, thus disturbing normal physiology.

The close relationship between invertebrate octopaminergic and the α_2-adrenergic receptors of vertebrates may be related to the effects of formamidines in mammals.

CHAPTER 17

Compounds Interfering with ATP Synthesis

17.1
Respiratory Systems

In a respiratory system, the mitochondrial electron transport coupled with oxidative phosphorylation is the important step to produce ATP which is vital to any organism and vulnerable to pesticides. During various steps in the glycolysis and the tricarboxylic acid cycle, NAD is reduced to NADH. Also, flavin adenine dinucleotide (FAD) is reduced to $FADH_2$ when succinate is converted to fumarate during the tricarboxylic acid cycle. These two cofactors are reoxidized by passing their reducing power (H^+ + electron) through the mitochondrial electron transport system, eventually to oxygen, reducing it to water (Fig. 17.1). The electron transport system divided into four complexes (I, II, III, IV) and flavin mononucleotide (FMN), FAD, FeS, coenzyme Q, and cytochromes (b, c_1, c, a, a_3), constitutes a respiratory carrier chain. The reactions produce energy at the indicated sites, which is utilized for ATP formation from ADP and inorganic phosphate.

Sites 1 (in complex I), 2 (in complex III), and 3 (in complex IV) are the steps providing the driving force for ATP formation. The ATP formation coupled with the electron transport is called oxidative phosphorylation. The oxidation of NADH creates a proton gradient across the mitochondrial membrane (inside low), which is used by F_1/F_0-ATPase for ATP synthesis. There are four types of compounds which act on this system: (1) inhibitors of electron transport; (2) inhibitors of oxidative phosphorylation; (3) uncouplers of oxidative phosphorylation; and (4) inhibitors of F_1/F_0-ATPase. Type (1) inhibits the electron flow in the electron transport, resulting in the cessation of oxidative phosphorylation and ATP formation. Type (2) causes cessation of not only the oxidative phosphorylation but also the coupled electron transport. Type (3) does not impair the activity of electron transport with continuous consumption of oxygen, but without ATP formation. Type (4) inhibits F_1/F_0-ATPase with no ATP formation. There are many insecticides and miticides, old and new, which affect this system (Corbett et al. 1984; chemical structures of compounds, see Figs. 17.2, 18.1).

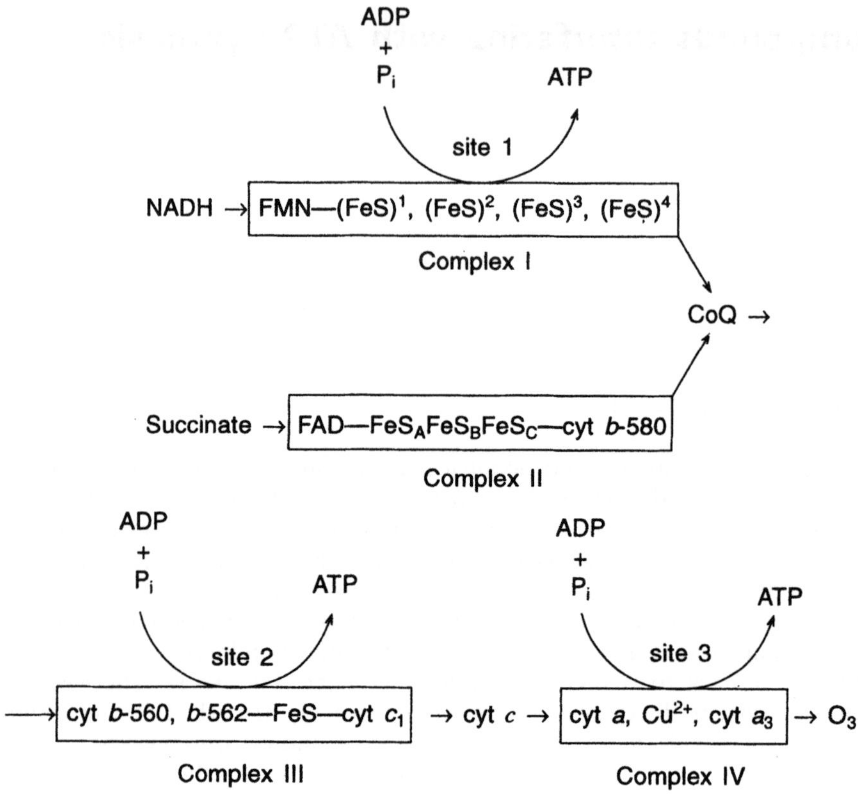

Fig. 17.1. Mitochondrial electron transport system. (Conn et al. 1987).

17.2 Inhibitors of Electron Transport (Complex I)

Only rotenone (an insecticide of limited use) (see Sect. 13.2) and piericidins (insecticidal but too toxic to use) were known to inhibit the mitochondrial electron transport of complex I. However, complex I has recently assumed new toxicological significance due to the appearance of many insecticides, particularly miticides, which inhibit this site (Hollingsworth et al. 1994). There is great variety in structures and origins of complex I inhibitors. Rotenone and *Annonaceous* acetogenins are of plant origin and piericidins are of microbial origin, and there are many synthetic nitrogenous heterocycles:

Fenazaquin. 4-*tert*-butylphenethyl quinazolin-4-yl ether, a miticide having activity also on fungal disease (Hackler et al. 1994)

Pyridaben. 2-*tert*-butyl-5-(4-tert-butylbenzylthio)-4-chloropyridazine-3(2H)-one. It shows excellent activity against a wide range of phytophagous mites including *Tetranychidae*, *Eriophyidae*, and *Tarsonemidae* (Hirata et al. 1995)

Fenpyroximate. *tert*-butyl(E)-α-(1,3-dimethyl-5-phenoxypyrazol-4-yl methylenamino-oxy)-p-toluate. It shows high activity against phytophagous mites such as *Tetranychidae, Eriophyidae, Tarsonemidae*, and *Tenuipalpidae*, but the activity on predacious mites is relatively low, and animal-parasitic and soil-living mites are almost unaffected (Hamaguchi et al. 1995)

Tebufenpyrad. *N*-(4-t-butylbenzyl)-4-chloro-3-ethyl-1-methylpyrazole-5-carboxamide. It is effective on eggs and adults of *Tetranychus* and *Panonychus* mites.

Pyrimidifen. 5-chloro-*N*-{2-[4-(2-ethoxyethyl)-2,3-dimethylphenoxy]ethyl}-6-ethylpyri-midin-4-amine. It is effective on various stages of mites as well as lepidopterans including the diamondback moth

Annonaceous acetogenins. Numerous members of the family *Annonaceae* produce many different kind of acetogenins. Their structures, insecticidal properties and other biological activities are reviewed (Rupprecht et al. 1990; Fang et al. 1993). The powder of the seeds is used locally in Southeast Asia, but the cytotoxicity restricted further development.

17.3 Uncouplers

Dinitrophenols. The dinitrophenols are a group of compounds with a wide range of biological activity and are useful as insecticides, acaricides, ovicides, herbicides, and fungicides. They were first used in the 1890s. They are yellow, relatively odorless solids, slightly soluble in water but readily soluble in organic solvents. Their acidic properties make them combine with bases and readily form water-soluble ammonium, sodium, potassium, and calcium salts. These water-soluble compounds are extremely toxic to plants and are used chiefly as herbicides. For use against insects and mites they are formulated with petroleum oils and are used as dormant sprays and ovicides. The major compounds include 4,6-dinitrocresol (DNOC), 4,6-dinitro-*o-sec*-butylphenol (dinoseb), 4,6-dinitro-*o*-cyclohexylphenol (dinex, no longer in use), binapacryl (Morocide), and dinocap (Karathane) (Fig. 17.2; see Metcalf 1971; Ware 1983).

The dinitrophenols are biologically active because of their ability to uncouple oxidative phosphorylation. Their rate of respiration increases markedly and the insects die from metabolic exhaustion. (Metcalf 1971). Some of the herbicidal effects may be due to interference with photosynthesis. The dinitrophenols are highly toxic to all forms of life. The oral LD_{50}s to rats are: DNOC 25 mg/kg; dinoseb 58 mg/kg; dinex 80 mg/kg; binapacryl 421 mg/kg; and dinocap 980 mg/kg. Signs and symptoms of acute poisoning in man include nausea, gastric upset, restlessness, sensation of heat, sweating, rapid respiration, tachycardia, fever, cyanosis, and finally collapse and coma. Blood levels of greater than 20 ppm produce toxic manifestations, and above 50 ppm are extremely dangerous. Treatment consists of ice baths to reduce fever, administration of oxygen, and electrolyte therapy (infusion of large quantities of isotonic saline) to replace loss by sweating (Murphy 1975).

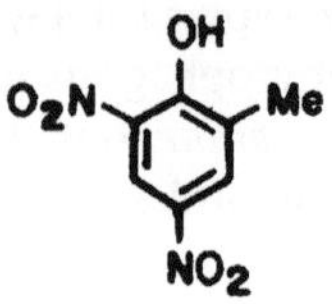

DNOC
(miticide,
fungicide)

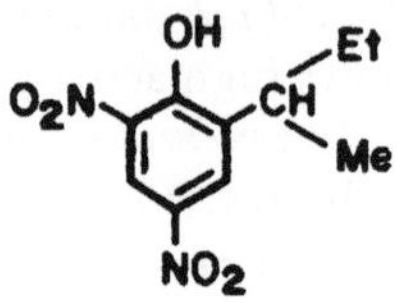

Dinoseb
(miticide,
fungicide,
herbicide)

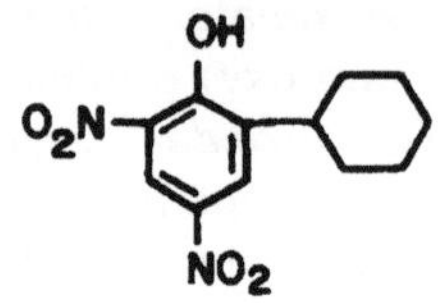

Dinex
(miticide)

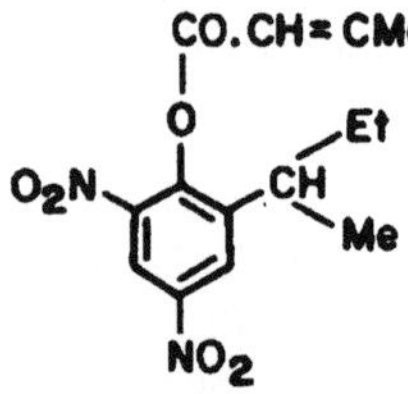

Binapacryl
(miticide,
fungicide)

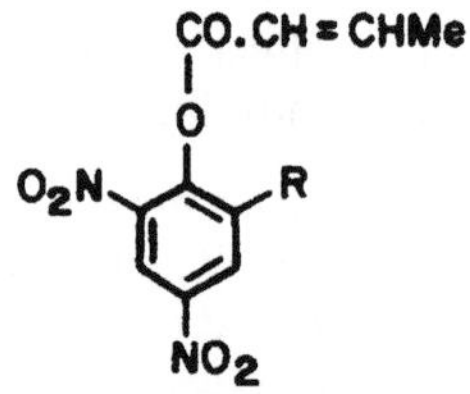

Dinocap
(miticide,
fungicide)

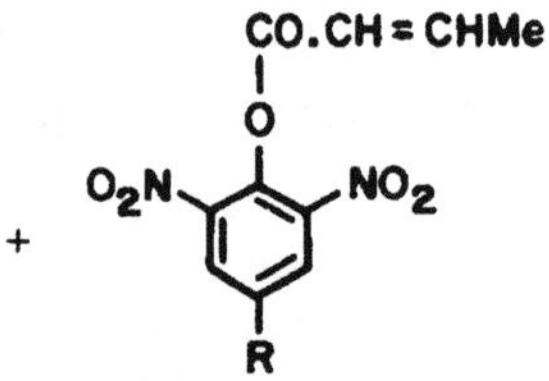

Mixture of compounds,
where R is C_8H_{17}
(1-methylheptyl,
1-ethylhexyl,and
1-propylpentyl)

Fig. 17.2. Dinitrophenol pesticides

Chlorfenapyr (AC 303, 630). The chemical structure of chlorfenapyr is 4-bromo-2-(4-chlorophenyl)-1-(ethoxymethyl)-5-trifluoro methyl) pyrrole 3-carbonitrile. Dioxapyrrolomycin from *Streptomyces fumanus* showed antibacterial, antifungal, insecticidal, and miticidal properties. Further modification of the structure lead to a wide range of highly active insecticidal, acaricidal, and molluscicidal 2-arylpyrroles, among which chlorfenapyr was commercialized. Actually, the compound is a propesticide, and its N-dealkylated product is a potent uncoupler (Black et al. 1994) (see Fig. 18.1).

17.4 Inhibitors of Electron Transport (Complex III)

Hydramethylnon. The chemical structure of hydramethylnon is tetrahydro-5, 5-dimethyl-2(1H)-pyrimidinone-[3-(4-trifluoromethyl) phenyl]-1-[(2-[4-

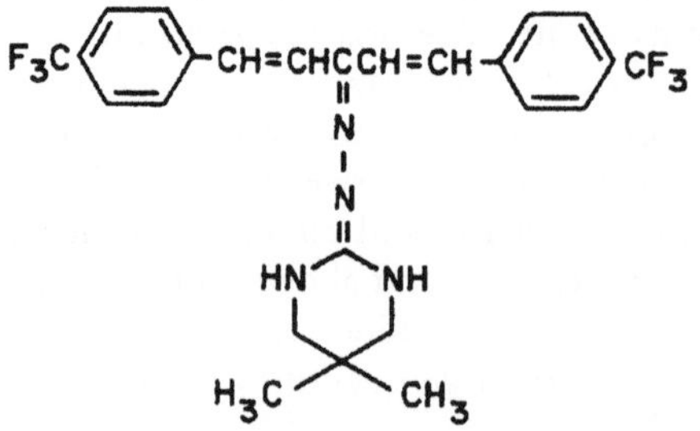

Hydramethylnon

(trifluoromethyl] phenyl]ethenyl)-2-propenylidene)hydrazone. A series of amidinohydrazones are selective to insects with chewing or sponging mouthparts. One of the members, hydramethylnon, is a slow-acting stomach poison effective on ants and cockroaches. It is an inhibitor of the electron transport chain in the region of site 2 in complex III (Hollingshaus 1987).

17.5 Inhibitors of F_1/F_0-ATPase

Diafenthiuron. Diafenthiuron is converted to a carbodiimide (DFCD). Like a simple dicyclohexylcarbodiimide, DFCD inhibits ATP synthesis by reacting with a proteolipid subunit of F_0 complex. However, it also functions as an agonist of the octopamine receptor (Ruder and Kayser 1992).

CHAPTER 18

Miticides

Phytophagous mites are different from insects, closer to spiders. There is no distinction between head, thorax, and abdomen, constituting a bag style. Mites develop to adults from eggs through larvae, the first and the second nymph during 2 weeks. The phytophagous mites are serious pests of over 150 agricultural crops. When nonspecific insecticides are used, natural mite predators are often eliminated without impairing mites. Their short development cycle tends to rapidly develop the resistant populations by removing unfavorable genes for their existence. The history of miticide development indicates the treadmill between chemicals and mites and many types of compounds should have been developed. The targets of miticide action were reviewed by Dekeyser and Downer (1994). The chemical structures of miticides are shown in Fig. 18.1.

CPCBS and Tetradifon. CPCBS is a selective miticide effective on eggs and larvae. The female adults treated with tetradifon bear infertile eggs. Eggs treated with CPCBS develop until hatching and then die, while eggs treated with tetradifon die at an early stage. There is no cross-resistance between the two compounds. CPAS and diphenylsulfide are converted to tetradifon. CPCBS inhibits olygomycin-sensitive Mg^{2+}-ATPase, but its significance is not clear.

Sulfite Esters. BPPS acts on mites, but is not ovicidal. The mode of action is not clear, though it inhibits olygomycin-sensitive Mg^{2+}-ATPase.

Organotin Compounds. Fenbutatin oxide and cyhexatin are effective on all stages except eggs. They inhibit olygomycin-sensitive Mg^{2+}-ATPase as well as Na^+, K^+-ATPase.

Quinoxalines. Quinomethionate is effective on all stages of mites. Hydrolysis gives a dithiol compound (QDSH), which inhibits SH enzymes.

Organophosphorus Compounds. Many organophosphorus insecticides are also miticidal. CMP (phenkapton) is specialized for mite control. Cross-resistance is often developed among them by increasing detoxification or decreasing sensitivity of acetylcholinesterase.

Pyrethroids. Many pyrethroids cause resurgence of mites, but some pyrethroids

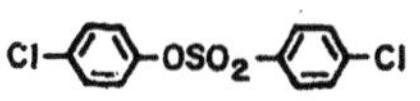

CPCBS

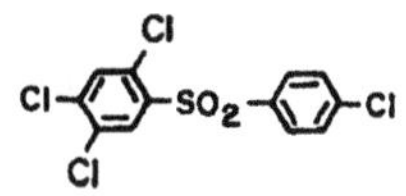

Tetradifon

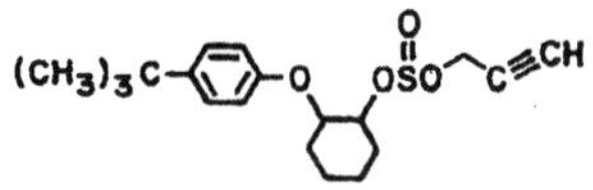

BPPS

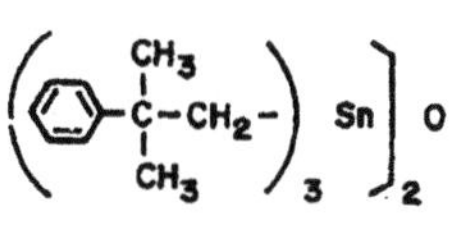

Fenbutatin oxide

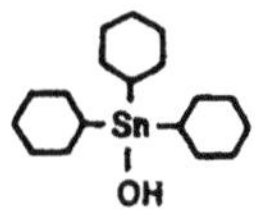

Cyhexatin

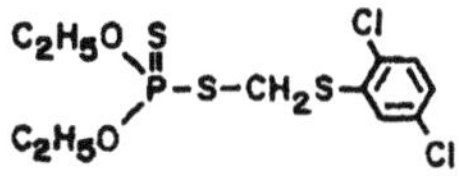

CMP

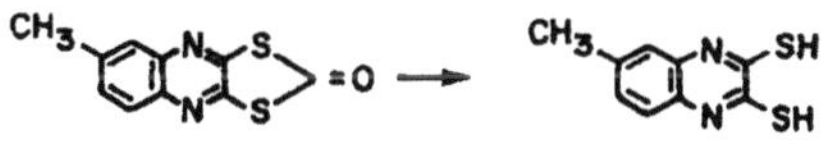

Quinomethionate QDSH

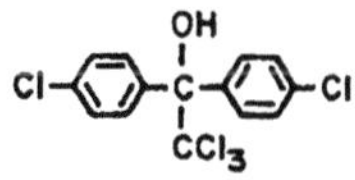

Dicofol

Chlordimeform N-Demethylchlordimeform (DCDM)

Amitraz U-40481

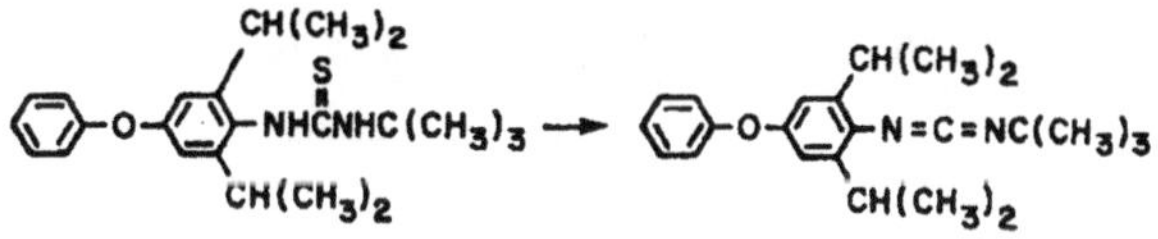

Diafenthiuron DFCD

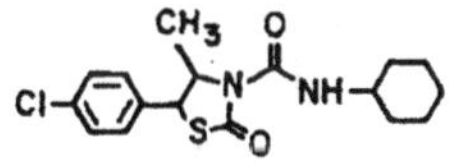

Hexythiazox

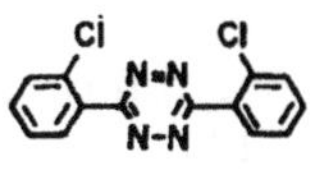

Chlofentezine

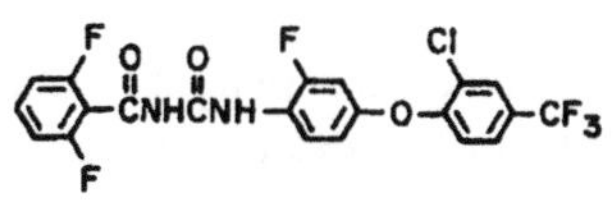

Flufenoxuron

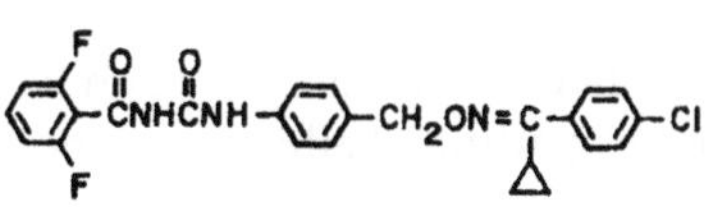

Flucycloxuron

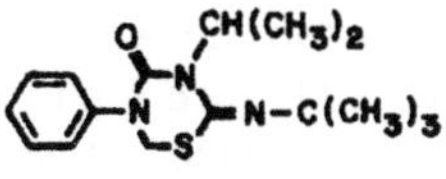

Buprofezin

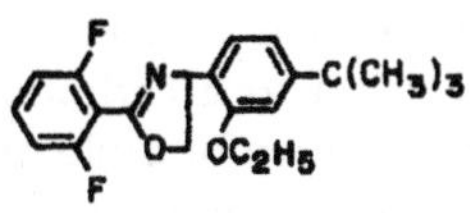

Etoxazole

Fig. 18.1. Miticides *(Contd)*

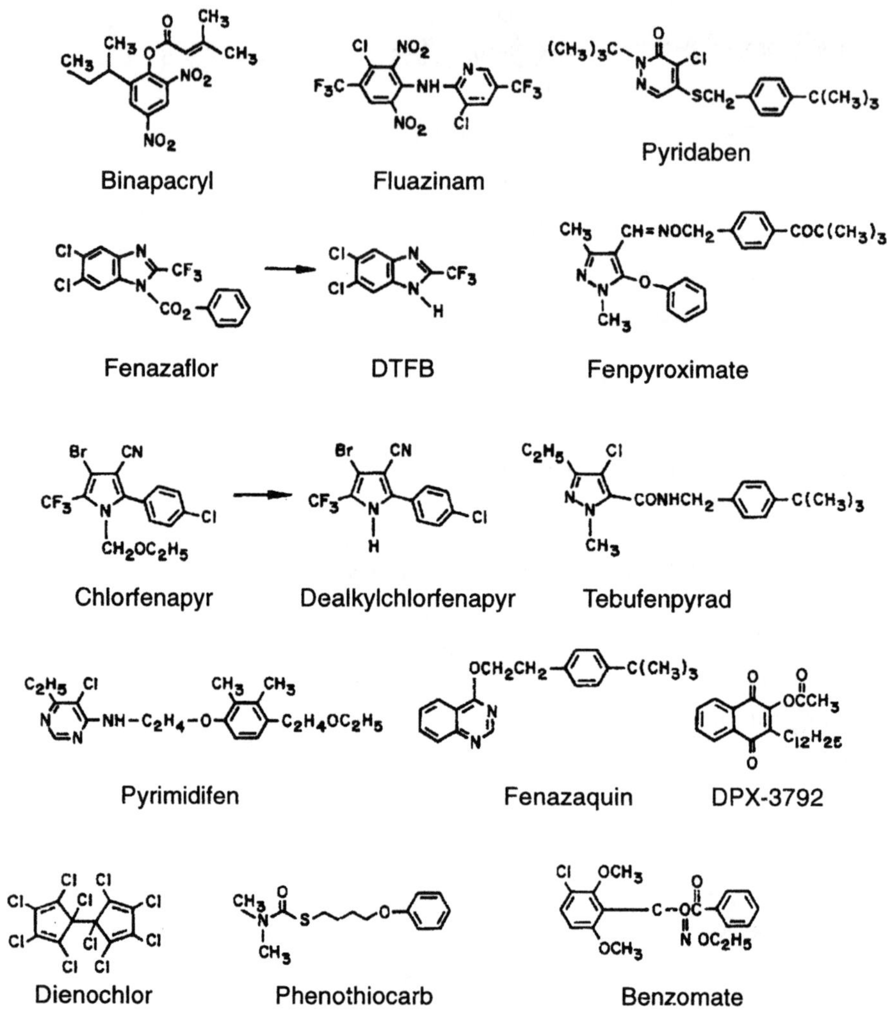

Fig. 18.1. Miticides

are effective on mites, though having no ovicidal activity. Fluvalinate, flucythrinate, fenpropathrin, halfenprox, and bifenthrin are examples (see Chap. 14).

Formamidines, Octopaminergic Compounds, and Dicofol. The formamidines, chlordimeform and amitraz, are effective on all stages of mites by convertion to N-demethylchlordimeform (DCDM) and U-40481. Diafenthiuron shows the octopaminergic action after conversion to a carbodiimide (DFCD) (see Chap. 16). Dicofol (Kelthane) is a representative of diphenyl carbinols, a detoxication product of DDT in insects, but effective on all stages of mites, causing gradual paralysis and eventual death. Dicofol resistance is not so high and the adaptation of resistant mites to the environment is often too low to survive. DDT first causes excitation of insects, but the dicofol-treated mites are slowly paralyzed.

Dicofol may act on the octopamine receptor or inhibit cAMP synthesis by acting on the GTP-binding protein.

Hexythiazox, Clofentezine, and Chitin Synthesis Inhibitors. Hexythiazox and clofentezine affect all stages of mites except adults. Both are different in chemical structure, but often cross-resistance occurs. The mites die without molting by abnormality of cuticle formation. Mode of action is not clear, but cross-resistance occurs between clofentezine and benzoylphenyl urea-type miticides such as flufenoxuron and flucycloxuron. Benzoylphenyl ureas and buprofezin are both insecticides and miticides (see Chap. 20). Etoxazole, a 2,4-diphenyloxazoline derivative, has excellent activities on aphids as well as eggs, larvae, and nymphs of mites, but not adult mites. Inhibition of a molting process similar to hexythiazox occurs, but it is effective on the hexythiazox-resistant mites (Ishida et al. 1994).

Uncouplers. Binapacryl is converted to the phenol to affect all stages of mites. Fluazinam acts on eggs and larvae, but not on adults. Fenazaflor is converted to DTFB to affect all stages of mites. Chlorfenapyr is a broad-spectrum insecticide, but also effective on *Tetranychus* mites (see Sect. 17.3).

Inhibitors of Electron Transport System (Complex I). Pyridaben, fenpyroximate, tebufenpyrad, pyrimidifen, and fenazaquin are effective on all stages of mites. Cross-resistance among them must be considered seriously (see Sect. 17.2). Though the exact site is not known, DPX-3792 inhibits the electron transport system, which is active on all stages of mites.

GABA-ergic Compounds. Dienochlor is a kind of chlorinated cyclodiene, utilized on ornamentals. It is supposed to inhibit GABA receptor function. There are several miticidal antibiotics. Abamectin, a mixture of avermectin-B_{1a} and -B_{1b}, is a macrolide disaccharide from *Streptomyces avermitilis*. Milbemectin, a mixture of milbemycin A_3 and A_4, is from *Streptomyces hygroscopicus* subsp. *aureolacrimosus* and the structure is closely related to abamectin but without disaccharide moiety. Both are effective on all stages of various mites by acting as GABA agonists (see Chaps. 10 and 23). Polynactin is also a miticidal antibiotic, produced by *Streptomyces aureus*, which consists of tetranactin, trinactin, and dinactin. The major component is tetranactin.

Phenothiocarb. Phenothiocarb is particularly effective on all stages of *Panonychus* mites, particularly on eggs just before hatching.

Benzomate. Benzomate is effective on all stages of *Panonychus* mites.

CHAPTER 19

Fumigants and Nematicides

It has been estimated that plant-parasitic nematodes are as important economically as fungi, bacteria, or viruses. Nematodes distribute about 0.3 m below the soil surface. *Meloidogyne, Heterodera, Tylenchulus, Rotylenchulus,* and *Protylenchulus* are important agricultural pest nematodes. To be effective, chemicals must have high volatility at room temperature. Hence, nematicides are generally gases or volatile liquids of comparatively low molecular weight enabling them to penetrate through soils. Nonvolatile nematicides diffuse in soil by watering or mixing in soil. Fumigation is an important method of soil sterilization for the control of soil insects, nematodes, fungi, and weed seeds. It is also very important in the control of stored product insects and fungi provided that the fumigants do not leave toxic residues on the fumigated commodities. Chemical fumigants, because of their high volatility, must be used in enclosed spaces such as greenhouses, food stores, warehouses, and the like. For soil treatment, the surface of the soil must first be covered with a plastic sheet to prevent loss of the fumigant. One of the earliest fumigants used commercially probably was hydrogen cyanide which was used for the control of scale insects on citrus trees in California.

19.1 Hydrogen Cyanide (HCN)

Hydrogen cyanide (HCN) has a boiling point of 26 °C and forms a gas slightly lighter than air (specific gravity 0.948 g/l). It is an effective fumigant for empty buildings or containers, but has a poor penetrating power, and is, therefore, not used for insect control in stored grain or for soil sterilization. It is used by quarantine personnel for treatment of plant material to avoid the introduction of insect pests and plant diseases.

HCN is a very dangerous acute poison. An air concentration of 300–400 mg/m^3 or a single oral dose of 50 mg can quickly cause human death. The oral LD_{50} to rats administered as a solution is 4.17 mg/kg. The gas can be detected by its slightly bitter-almond odor, but the common practice of detection is to add chloropicrin (tear gas) or cyanogen chloride to the concentrate to warn people of the presence of the gas by the effects of the added chemicals on the eyes, nose, or throat. Being highly poisonous to man and other warm blooded animals, HCN should be applied only by personnel thoroughly trained in its safe application with the availability of a suitable antidote. HCN has been largely replaced by safer gases.

19.2 Methyl Bromide (CH_3Br)

Methyl bromide is a colorless gas with a burning taste and a sweet, chloroform-like odor at high concentrations. It has a melting point of –93.6 °C, a boiling point of 4.5 °C, and a solubility in water of 17.5 g/l at 20 °C. It is freely soluble in alcohol, chloroform, ether, carbon tetrachloride, and benzene. It is stable and noncorrosive. It is a highly volatile fumigant used for soil sterilization in glasshouses, nematode control in soil, control of insect pests of stored products, and control of dry wood termites in buildings. Fumigation of fruits and vegetables in storage and transit, although practised extensively, may leave undesirable residues of methyl bromide or bromide ion (Hargreaves et al. 1978; Love et al. 1979). Methyl bromide is toxic to insects in all stages of development including eggs. It is less toxic than HCN but is a cumulative poison. It is the most widely used fumigant. However, it is too phytotoxic and not self-warning; hence, it is extremely dangerous to applicators.

High concentrations of methyl bromide can cause rapid unconsciousness in man during exposure. The most common symptoms in fatal and nonfatal cases are malaise, headache, visual disturbances, nausea, and vomiting. Later, a great variety of central nervous system manifestations can appear, including numbness, ataxia, tremor, myoclonus, exaggerated (or absent) deep reflexes, positive Romberg's sign, paroxysmal abnormalities of the EEG (electroencephalograph), great agitation, change of personality, coma, and clonic, tonic (often Jacksonian) convulsions. The cause of death usually is the result of pulmonary edema leading to respiratory failure or cardiovascular collapse (Hayes and Laws 1991). Treatment of poisoning is symptomatic and the use of diazepam or clonazepam is effective.

Lately, methyl bromide has come under scrutiny of health and environmental protection agencies because of the potential adverse effect it might have on human health and the environment, groundwater, and ozone layer, and the possibility of its being carcinogenic (Danse et al. 1984). However, after the banning of many fumigant nematicides there is no substitute for methyl bromide.

19.3 Phosphine (H_3P)

Phosphine is a colorless gas with a garlic-like odor. It has a density of 1.214 and a boiling point of –87.7 °C. It is spontaneously flammable and reacts violently with oxygen and the halogens. Phosphine is a fumigant for grain and other stored products. It is liberated by the interaction of aluminum phosphide with water. Most good commercial formulations of aluminum phosphide include about 40% ammonium carbonate, which liberates ammonia and carbon dioxide in the presence of water, resulting in a gas mixture that is not flammable or explosive. Formulations come in the form of tablets, pellets, or crepe paper bags. Phosphine also may be used for rodent control by placing the tablets in outdoor burrows, the openings of which are closed. It is an electron transport inhibitor like carbon monoxide.

Inhaled phosphine is freely absorbed by the lungs and is distributed in many tissues. Exposure to high concentrations leads to a profound fall in blood pressure and death. Even low concentrations, such as 7.5 mg/m^3, can lead to pulmonary edema and respiratory failure. There is no specific antidote for phosphine. The use of oxygen is indicated as well as maintenance of normal blood pressure and control of pulmonary edema. Methylphosphine (CH_3PH_2) is in the process of development and may prove to be an important fumigant for controlling stored product pests.

19.4 Carbon Disulfide (CS_2)

Carbon disulfide is a colorless or yellowish liquid with a foul smell due to impurities. In the pure form it has a sweet odor. It has a liquid density of 1.232 and a vapor density 2.67 times that of air. It has a boiling point of 46.5 °C and a vapor pressure of 357.1 mmHg at 25 °C. The vapor is extremely flammable and is explosive when mixed with air or when heated between 125 and 135 °C; hence, great caution must be exercised in its use. Carbon disulfide first was used as an insecticide in 1854. It is highly toxic to all forms of life including plants and some seeds, and is therefore not used for greenhouse fumigation. It is used for fumigation of stored grain and for soil treatment against nematodes. Usually, it is mixed with carbon tetrachloride to reduce fire hazard.

Mild chronic poisoning by carbon disulfide produces giddiness, weak and irregular pulse, dilated pupils, and shortness of breath. Acute poisoning acts on the nervous system causing mental disturbances, affected vision, nausea, vomiting, and paralysis. At high oral dosages, it causes biochemical changes in the liver of rats, particularly effects on microsomal enzymes and cytochrome P-450 (Bond and de Matteis 1969; Bond et al. 1969). Treatment of poisoning is symptomatic. Summon a physician at once. Remove patient to fresh air, give a stimulant (hot coffee or tea), and keep patient warm.

19.5 Chloropicrin (Trichloronitromethane; CCl_3NO_2)

Chloropicrin a fumigant, known during World War I as tear gas, is a colorless liquid with a boiling point of 112.4 °C, specific gravity of 1.692 and vapor pressure of 23.9 mmHg at 25 °C. The liquid volatilizes slowly and at low concentrations causes intense irritation of the eyes and induces vomiting. It injures the lungs. Symptoms may be delayed for several hours after exposure. It was first used as an insecticide in 1907. Later it was used as a soil fumigant and disinfectant because it possesses some fungicidal and bactericidal activity. It is nonflammable and nonexplosive, nonreactive with metals or fabrics, and has ability to penetrate bulk commodities. It is also added to other gases as a warning agent. It is highly toxic to plants.

19.6 Carbon Tetrachloride (CCl_4)

Carbon tetrachloride is a colorless liquid with a sweetish odor. It has a density of 1.589, a boiling point of 76.7 °C, and a vapor pressure of 114.5 mmHg at 25 °C. Its solubility in water is 0.28 g/l and it is miscible in alcohol and many organic solvents. It is nonflammable and noncorrosive. Carbon tetrachloride is used primarily for the fumigation of stored grain. Its toxicity to insects is rather low, but due to its safety from fire hazard it is often used in admixture with more potent insecticide fumigants such as ethylene dichloride, ethylene dibromide, carbon disulfide, and methyl bromide. Since the compound has good power of penetration it is useful for fumigation of deep grain silos.

The oral LD_{50} in the rat was reported as 2800 mg/kg (Smyth et al. 1970). Carbon tetrachloride has been used as a therapeutic agent in the control of hookworm (*Ascaris lumbricoides*) in humans but has been replaced with other drugs having less objectionable side effects. It was found to be carcinogenic in mice and rats, producing liver tumors after administration by various routes. It also produced liver tumors in trout and hamsters following oral administration. Several case reports described tumors associated with cirrhosis of the liver in humans (International Agency for Research on Cancer 1979). Treatment of poisoning is symptomatic. The use of mannitol to minimize kidney damage has been recommended (Stewart et al. 1963).

19.7 Ethylene Dibromide (EDB; $C_2H_4Br_2$)

Ethylene dibromide is a heavy, colorless liquid with a chloroform-like odor. It has a density of 2.172 at 25 °C, a boiling point of 131–132 °C, a vapor pressure of 11 mmHg at 25 °C, and a solubility in water of 4.3 g/l at 30 °C. It is soluble in ethanol, ether, and most organic solvents. It is nonflammable and turns brown on exposure to light. Ethylene dibromide has been used as a grain and soil fumigant in addition to its many other industrial uses.

The acute oral LD_{50} to male and female rats is 146 and 117 mg/kg, respectively, and 420 mg/kg for female mice (International Agency for Research on Cancer 1977). It is an alkylating agent. Following oral administration of radioactive ethylene dibromide to rats, radioactivity was incorporated into liver DNA, RNA, and protein (Nachtomi and Sarma 1977) and glutathione was depleted in the liver (Nachtomi 1970). It was also found to be mutagenic in Ames tests (International Agency for Research on Cancer, 1977). Evaluation of all data indicated that ethylene dibromide is carcinogenic; therefore it is no longer used as a fumigant.

19.8 Ethylene Oxide [$(CH_2)_2O$]

Ethylene oxide is a colorless, flammable gas at room temperature. It has a boiling point of 10.7 °C, a density of 0.869 at 20 °C, and a vapor pressure of

1095 mmHg at 20 °C. It is soluble in water and in most organic solvents. It is often used with carbon dioxide to reduce fire hazard. Ethylene oxide is used as an insecticidal fumigant for stored products, as a fungicide for the fumigation of books, for sterilization of dental, pharmaceutical, and scientific equipment, bedding for experimental animals, clothing, furs, furniture, and transportation vehicles including aircraft. It has also been used to accelerate the maturing of tobacco leaves (International Agency for Research on Cancer 1976). It is strongly sorbed in plastics. When used for the sterilization of foodstuffs and spices, or plastic materials containing chlorine, this compound will react with inorganic chloride in the commodity to form ethylene chlorohydrin which is fairly toxic (oral LD_{50} to male rats 71.3 mg/kg), and the latter can be further metabolized to form chloroacetaldehyde which is even more toxic and more irritating than ethylene chlorohydrin. The threshold limit value (TLV) for ethylene oxide in the USA is 1.8 mg/m^3 (American Conference of Governmental Industrial Hygienists 1989–1990).

Symptoms of poisoning are rare, but long exposure can produce headache, vomiting, diarrhoea, lymphocytosis, and skin burns. Ethylene oxide is a strong alkylating agent and is suspected of being carcinogenic (Hayes 1982; American Conference of Governmental Industrial Hygienists 1989–1990). Treatment of poisoning is symptomatic.

19.9 Dibromochloropropane (DBCP; CH_2Br-$CHBr$-CH_2Cl)

Dibromochloropropane (DBCP) is a dark-amber to dark-brown liquid with a pungent odor. The boiling point is 196 °C, the density 2.08 at 20 °C and the vapor pressure 0.8 mmHg at 21 °C. It is slightly soluble in water (0.1% w/w) and is miscible in most organic solvents. It is corrosive to aluminum, magnesium, tin, and their alloys. It is stable in neutral and acidic media, but is hydrolized by alkali. DBCP has been used as a soil fumigant for nematode control on a variety of fruit, vegetable, nut, and ornamental crops, as well as for garden use. DBCP is a potent nematicide, effective at low application without affecting many living plants.

The LD_{50} to male rats is 170 mg/kg. DBCP is stored mainly in fat tissue. The most striking injury produced by repeated doses of DBCP was atrophy of the testes. Male rats, rabbits, guinea pigs, and monkeys showed severe atrophy and degeneration of testicular tissue when exposed 70 or more times to 12 ppm vapor concentration (Hayes 1982). In addition, rats showed injury to the liver, lung, and kidney. Reduced spermatogenesis and sterility in workers exposed to DBCP in a chemical plant in California was first reported by Whorton et al. (1977, 1979). Contamination of water by DBCP has resulted in the closure of about 1000 drinking water wells (Berteau and Spath 1986), and an additional 2500 wells in California (Cohen 1986). Evaluation of toxicity data on DBCP (International Agency for Research on Cancer 1977) indicated that the compound is carcinogenic in mice and rats, producing carcinomas of the forestomach in both species and mammary carcinomas in rats. DBCP was banned from use in 1977.

19.10
1,3-Dichloropropene ($C_3H_4Cl_2$)

1,3-Dichloropropene is a white or amber-colored liquid with a chloroform-like odor. The density is 1.217 at 25 °C, and boiling points of the *cis* and *trans* isomers are 104 and 112 °C, respectively. Solubility in water at 20 °C is 1 g/l and it is miscible with many organic solvents. This compound is a soil fumigant for the control of nematodes in vegetable and tobacco crops. A common formulation of 1,3-dichloropropene with 1,2-dichloropropane is known as D-D.

The oral LD_{50} of 1,3-dichloropropene in rats ranged from 130 to 713 mg/kg in males and 110 to 510 mg/kg in females, depending on the vehicle and strain used. Major target tissues of 1,3-dichloropropene following an acute, high-dose oral or inhalation exposure are the liver, kidneys, and lungs (inhalation) in rats and guinea pigs. Following a more prolonged lower-level exposure, the stomach, liver, and nasal mucosa of rats and the stomach, nasal mucosa, lungs, and urinary bladder of mice may be affected. In man, this compound causes edema, redness and necrosis of the skin (Torkelson and Oyen 1977). Inhalation exposure causes headache, irritation of the mucous membranes, and chest discomfort. Treatment of poisoning is symptomatic (Hayes and Laws 1991).

19.11
Naphthalene ($C_{10}H_8$)

Naphthalene forms colorless, flaky crystals with a characteristic odor of moth balls. The density is 1.162, melting point 80 °C, and vapor pressure 4.92×10^{-2} at 20 °C. The solubility in water at room temperature is 30 mg/l. It is slightly soluble in ethanol, but soluble in many other organic solvents. Naphthalene has long been used as a household fumigant against clothes moths.

The oral LD_{50} values in male and female rats are 2200 and 2400 mg/kg, respectively. Dermal values for both are greater than 2500 mg/kg. Naphthalene is readily absorbed if inhaled or ingested. Symptoms of poisoning are characterized by hemolytic anemia, but nonspecific symptoms such as headache, listlessness, anorexia, vomiting, and diarrhea appear before hemolysis. The ordinary use of naphthalene is safe. The danger is almost exclusively to children who eat them in the form of mothballs or to babies whose clothes have been stored in naphthalene for prolonged periods. The threshold limit value is 52 mg/m^3. Treatment of poisoning is symptomatic.

19.12
p-Dichlorobenzene (PDB; $C_6H_4Cl_2$)

p-Dichlorobenzene (forms colorless, volatile crystals with a strong odor. It has a density of 1.458, melting point 53–54 °C, and vapor pressure 1 mmHg at 25 °C. Solubility in water is 80 mg/l at room temperature and it is readily soluble in organic solvents. PDB is used as a fumigant for clothes moths and for mites in fungal cultures. The intraperitoneal LD_{50} in rats is 2562 mg/kg. An oral dose of 4000 mg/kg was fatal to rats, but all survived a dose of 1000 mg/kg. PDB is irritating to the eyes and nose of most people at concentrations between 300 and 962 mg/m^3 (50 and 160 ppm). There has been no case of poisoning of a person by PDB due to the low acute toxicity and the irritant nature of the vapor which discourages long exposure. The threshold limit value for occupational exposure is 450 mg/m^3. Treatment is symptomatic.

In addition to the fumigants mentioned in Sections 19.1–19.12, there are several similar compounds which have found little or no use as fumigants in agriculture or in public health, and are used mainly as industrial solvents. These include methylene dichloride, ethylene dichloride, ethylene trichloride, tetrachloroethylene, and a few other.

19.13 Nonvolatile Nematocides

Volatile nematocides are limited in flexibility of the structure and more recent attention has been paid to nonvolatile nematocides, mostly organophosphorus and carbamate compounds. These are insecticides and applied to the soil at nematicidal dosages, which are usually 2–4 times higher than those required for insect control. Organophosphorus compounds such as dichlorofenthion, thionazin, parathion, demeton, phorate, disulfoton, diazinon, nellite, fensulfothion, phenaminophos, ethoprop, terbufos, and TH-285N can control nematodes. They must be distributed through the upper soil profile by percolation of water or by mechanical incorporation. Carbamates such as carbofuran, aldicarb, and oxamyl are effective in the control of nematodes and soilborne insects. Oxamyl is unique in that it can be applied to the foliage of plants and is translocated downward to affect nematodes in the roots. Avermectins, a natural product from *Streptomyces avermitilis,* is nematicidal, but the field efficacy is limited due to its instability in soil. The use of organophosphorous compounds, carbamates, and avermectins is rather limited to certain crops and not widely used.

CHAPTER 20

Insect Growth Regulators (IGRs)

In recent years many of the conventional methods of insect control by broad-spectrum synthetic chemicals have come under assault and scrutiny because of their undesirable effects on human health and the environment. As a consequence, new approaches have been tested and implemented. One of these approaches which has captured worldwide attention is the use of analogs and antagonists of insect growth regulators (IGR) such as juvenile hormones (JH), ecdysones, chitin synthesis inhibitors, and related compounds. It must be understood that compounds of this type are also chemicals, but because of their low toxicity to mammals, their selective toxicity toward insect species, and their safety to the environment they can assume a prominent role in the "integrated pest management" program.

20.1 Insect Growth Hormones and Mimics

The growth and molting of immature insects is regulated by three main groups of brain hormones, α-ecdysone, β-ecdysone, and the JHs. The brain hormones are secreted by neurosecretory cells in the protocerebrum and they activate the prothoracic glands. The activated prothoracic glands, presumably, secrete several closely related steroids, the ecdysones, or some precursors of ecdysone. Soon after their appearance in the body, these ecdysones cause the insect to molt. After each molt, a new cuticle is deposited by the epidermal cells. This process is controlled by the JHs which are secreted by the corpora allata. As long as JHs are secreted, the larva will molt into another larva. However, at the end of the last larval instar, the corpora allata are turned off by some physiological mechanism and secretion of JHs ceases or their concentration falls rapidly. In the absence of a sufficient concentration of the JHs, the immature form (larva or nymph) will change into the adult (in hemimetabolous insects, such as locusts, cockroaches, etc.), or the larva will metamorphose into a pupa and then into the adult (in holometabolous insects, such as moths, beetles, flies, wasps, etc.). When larval epidermal cells are stimulated by ecdysone, and in the presence of high concentrations of JH, they secrete a larval cuticle, whereas in the absence of JH or its presence in low titer, they secrete a pupal or adult cuticle. In summary, the molting hormone ecdysone (ecdysterone) initiates the molting process and induces metamorphosis, whereas the JH modifies the outcome of the molt (Fig. 20.1).

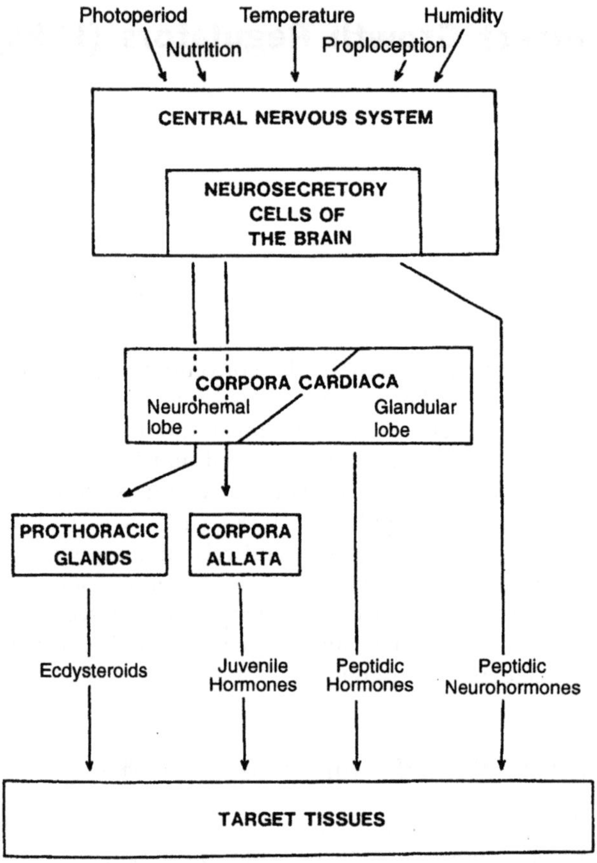

Fig. 20.1. Main neuroendocrine system of an insect. (Richards and Hoffman 1987)

In several orders of insects JH is also necessary for egg yolk formation in adult females and for development of the male sex glands. There are also other effects. For example, insect eggs are sensitive to JH, so the application of JH in the process of embryogenesis will prevent the development of the larva. Treatment of adult females with JH results in the deposition of eggs that fail to hatch. As stated above, in order for mature larvae to metamorphose into adults the concentration of JH has to decline; hence, application of JH at this stage of development can cause death of the insect (Henrick 1982). It is now known that JHs are also involved in the control of diapause, in the production of sex pheromones, and in behavior (Bowers 1982).

The possibility that JH analogs may have potential as insect control agents was first recognized by Williams (1956, 1967). Since then, the search for safer and more selective insecticides prompted extensive research in compounds showing JH activity (termed juvenoids or "third-generation insecticides".) Literally hundreds of compounds have been synthesized and screened for JH activity. The promise of practical field application was realized in the commercialization of the compounds methoprene and hydroprene (Fig. 20.2). Methoprene (Altosid) was introduced in 1975 by Zoecon Corporation as a mosquito growth regulator, for use against

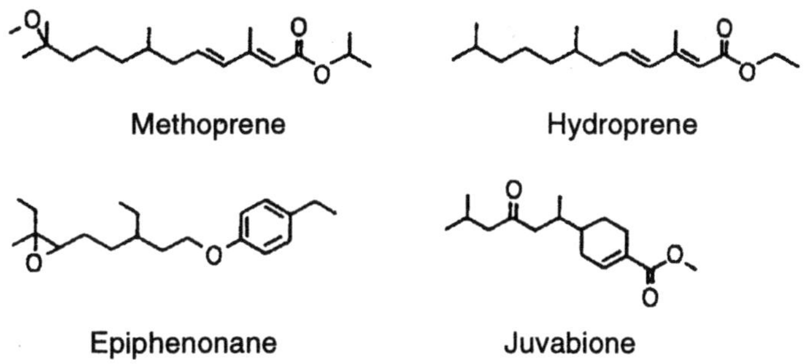

Fig. 20.2. Juvenile hormone mimics, known as insect growth regulators

second to fourth larval stages of floodwater mosquitoes to prevent adult emergence. Larvae exposed to methoprene continue their development to the pupal stage when they die. Methoprene has no effect when applied to pupae or adult mosquitoes. In addition, methoprene has been formulated for indoor control of dog and cat fleas. It acts in the same way by interrupting the fleas' life cycle at the larval stage, thus preventing the emergence of the adult. It is desirable to use methoprene in combination with an insecticide to control the adult fleas which are not affected by the growth regulator. Methoprene is also registered for control of cigarette beetles on tobacco, horn flies on cattle feed, and fungus gnats on mushroom-growing media. The World Health Organization has approved it for use in drinking water to control mosquito larvae. Kinoprene is another IGR developed by Zoecon for control of aphids, whiteflies, mealybugs, and scale insects. It acts by inhibiting development, reducing egg-laying, killing eggs already laid, and sterilizing adult insects.

For insect control purposes, JH activity depends on: (1) the particular insect species; (2) the critical phase of the insect's life cycle; (3) the mode of administration; and (4) the concentration of the active substance. The timing of the application of JH is very critical, since young larvae are not killed by JHs, analogs. In order for these agents to kill insects they must act during the last larval instar or during metamorphosis where they cause the production of deformed intermediates. Or, JH analogs must act upon adults where they can function as chemosterilants, or upon eggs to disrupt embryogenesis. Hence, JH analogs must persist for several days to be effective.

During the intensive research effort to isolate and identify the insect's JHs, Slama and Williams (1965) found that the bug *Pyrrhocoris apterus* could not complete its development in the laboratory, but molted into intermediary forms as if exposed to JHs. They were able to demonstrate that the paper used to line the rearing jars contained an active JH mimic. The paper was a product of the balsam fir tree *Abies balsamea*. They called the active compound "the paper factor." Later, the active compound was isolated and identified, and called "juvabione" (Bowers et al. 1966). This compound is especially active against insects of the family Pyrrhocoridae, which includes many economically important species known as "cotton stainers." Other plant products with high JH activity were soon discovered, including compounds with molting hormone activity (ecdysone-type compounds) which were called "phytohormones."

A further development was the discovery of compounds which would antagonize the insect hormone activity and derange the insect's development. These were called "precocenes" due to their ability to induce precocious metamorphosis of the immature insect (Bowers 1976; Bowers et al. 1976). Precocenes were also found to affect insect diapause, reproduction (causing female sterility in certain insects, and low reproduction potential in male Aedes mosquitoes), and behavior (such as sex pheromone production).

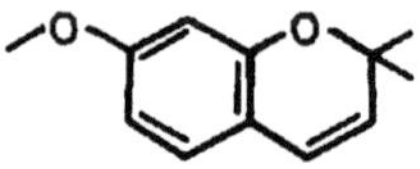

Precocene I

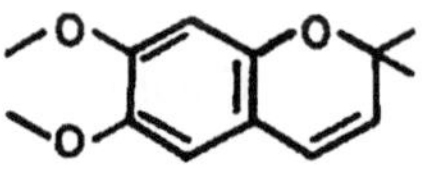

Precocene II

The mode of action of precocenes seems to be the prevention of JH biosynthesis in the corpora allata, since application of exogenous JH can reverse their action. It was demonstrated (Bowers and Martinez-Pardo 1977) that the corpora allata of young female milkweed bugs treated with precocene failed to develop. Application of precocene to older females caused sterility. Atrophy of corpora allata caused by precocene was also demonstrated in locusts and grasshoppers (Pener and Orshan 1977; Pener et al. 1977, 1981), and in many other insects (Staal 1986). The mode of action of precocenes, from a biochemical standpoint, involves the epoxidation of the 3–4 double bond in the precocene molecule by cytochrome P-450-linked enzymes in the corpora allata. The epoxides formed are highly reactive and unstable, causing cellular disruption. However, hydrolytic epoxide hydrase enzymes present in the corpora allata and in other tissues can detoxify the epoxides by converting them to polar nontoxic metabolites that are readily excreted. Not all insect taxa are affected by precocenes. The reason for their specificity is not known. Several reviews on JHs and anti-JHs have appeared in the literature from time to time. A recent comprehensive review of the anti-JH agents, their mode of action, and the insect taxa affected by them has been published by Staal (1986). It has also been found that precocenes are hepatotoxic (Hsia et al. 1981; Grossman and Hsia 1982) and nephrotoxic (Schrankel et al. 1982) in vertebrates. For compounds to be used as insect control agents, these are undesirable effects. The reason for this toxicity, probably, is the same as that in insects, that is, the reactive epoxides of precocenes that are formed are also injurious to mammalian cells. The precocenes share a common type of enzymatic bioactivation with many other carcinogenic compounds (Ellis-Pratt 1983). Newer type precocenes may offer some advantages.

Another type of IGR is cyromazine (Neporex) introduced by Ciba-Geigy.

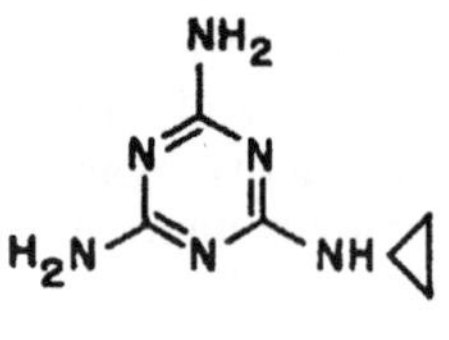

Cyromazine

This compound belongs to the chemical class of triazines and is useful, at low dosages, in the control of housefly larvae, and larvae of mosquitoes *Culex pipiens.* Young larval stages are more susceptible to the chemical which interferes with their development and with pupation; the adult insects are not affected. The rat oral LD_{50} is 3300 mg/kg.

20.1.1 Juvenile Hormone Mimics of Agricultural Importance

Juvenoids of agricultural importance are fenoxycarb and its derivative pyriproxyfen (Fig. 20.3). Pyriproxyfen which proved to be more potent than fenoxycarb on various insect species (Itaya 1987; Kawada 1988; Langley 1990; Koehler and Patterson 1991) was introduced recently for controlling whitefly in cotton (Ishaaya and Horowitz 1992; Horowitz et al. 1994) and scale insects in citrus (Peleg 1988). Pyriproxyfen is a potent suppresser of embryogenesis and adult formation of the sweetpotato whitefly, ***Bemisia tabaci*** (Gennadius), and the greenhouse whitefly, ***Trialeurodes vaporariorum*** (Westwood) (Ishaaya and Horowitz 1995). Dipping of cotton or tomato seedlings infested with 0–1-day-old eggs in 0.1 mg a.i. l^{-1} resulted in over 90% suppression of egg hatch of both *B. tabaci* and *T. vaporariorum* (Ishaaya and Horowitz 1992; Ishaaya et al. 1994). Exposure of whitefly females to cotton or tomato seedlings treated with pyriproxyfen resulted in oviposition of nonviable eggs. The LC_{90} values for egg viability of *B. tabaci* and *T. vaporariorum* exposed to treated plants were 0.05 and 0.22 mg a.i. l^{-1}, respectively (Table 20.1). Treatment of second instar *B. tabaci* with 0.04–5 mg a.i. l^{-1} resulted in normal development until the pupae stage; however, adult emergence was totally suppressed (Table 20.2). Inhibition of egg-hatch on the lower surface of cotton leaves was observed when their upper surface was treated with 1–25 mg a.i. l^{-1}, indicating a pronounced translaminar effect. Thus, the compound suppresses egg hatch and adult formation upon exposure of adult females and larvae, respectively, and exhibits, a strong translaminar activity. The compound is considered today as a major component for controlling whiteflies in cotton and ornamentals (Horowitz and Ishaaya 1994). The compound has a mild or no effect on parasitoids but a considerable effect on ladybird beetles (Mendel et al. 1994). Hence, the compound should be evaluated carefully before its introduction in integrated pest management programs.

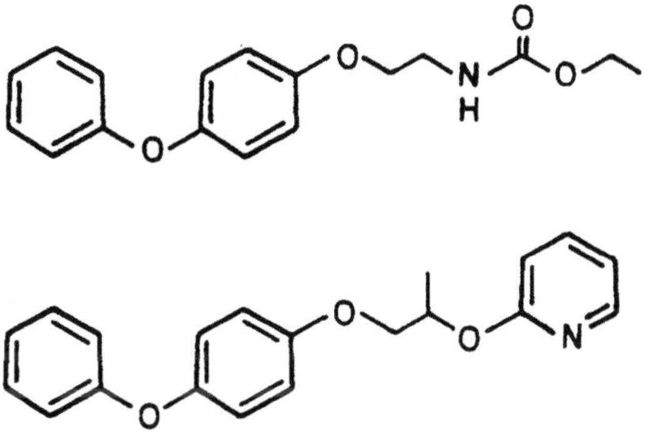

Fig. 20.3. Juvenile hormone mimics of agricultural importance. *From top* Fenoxycarb and pyriproxyfen

Table 20.1. Transovarial activity of pyriproxyfen on *Bemisia tabaci* and *Trialeurodes vaporariorum*. (After Ishaaya and Horowitz 1995)

Whitefly species	Slope (± SEM)	LC_{50} (mg l^{-1})	LC_{90} (mg l^{-1})
B. tabaci	4.6	0.03	0.05
	(0.2)	(0.02–0.03)[a]	(0.05–0.05)
T. vaporariorum	4.8	0.12	0.22
	(0.4)	(0.08–0.16)	(0.16–0.53)

[a]95% F.L.

Table 20.2. Effect of pyriproxyfen on pupation and emergence of *Bemisia tabaci* treated at the second instars. (After Ishaaya and Horowitz 1992)

Concentration, (mg l^{-1})	No.	Percent pupation	Percent emergence
Untreated	1014	96	95
0.04	455	95	0
0.2	1191	92	0
1.0	1153	95	0
5.0	489	94	0

Data are averages of 5–10 replicates of 80–140 larvae each.

20.1.2 Ecdysone Agonists of Agricultural Importance

Physiological and biochemical processes occurring during metamorphosis in insects depend largely on the presence of juvenile and ecdysone hormones, the titer and activity of which are due to biosynthesis, enzymatic degradation, and compartmentalization of the hormonally active compounds (Gilbert et al. 1980; Koolman and Karlson 1985). During the last decade many investigations have been directed to elucidate the possible use of ecdysteroid receptors as target site for developing insecticides with novel modes of action (Robbins et al. 1970; Bergamsco and Horn 1980; Horn et al. 1981). Several nonsteroid ecdysteroid agonists have been developed recently by Rohm and Haas Co., Spring House, Pennsylvania, USA (Fig. 20.4). The first compound of this series, RH-5849, binds to the ecdysteroid receptors, thereby initiating the molting process (Wing 1988; Wing et al. 1988). The compound acts on lepidopteran pests such as *Manduca sexta* (Wing et al. 1988), *Plodia interpunctella* (Silhacek et al. 1990), *Spodoptera frugiperda* (Monthean and Potter 1992), *S. littoralis* (Smagghe and Degheele 1992a,b), and *S. exempta* (Smagghe and Degheele 1993), and on certain dipterans (mosquitoes and flies) and coleopterans such as *Leptinotarsa decemlineata* and *Popillia japonica* species (Aller and Ramsay 1988; Darvas et al. 1992; Monthean and Potter 1992; Smagghe and Degheele 1993). Two novel derivatives in this series are the RH-5992 (tebufenozide) and RH-2485; both act on lepidopteran pests (Ishaaya et al. 1995) and are much more potent than RH-5849 (Smagghe and Degheele 1992a). Toxicity assays demonstrated that larvae of various lepidopteran species treated with either RH-5849 or

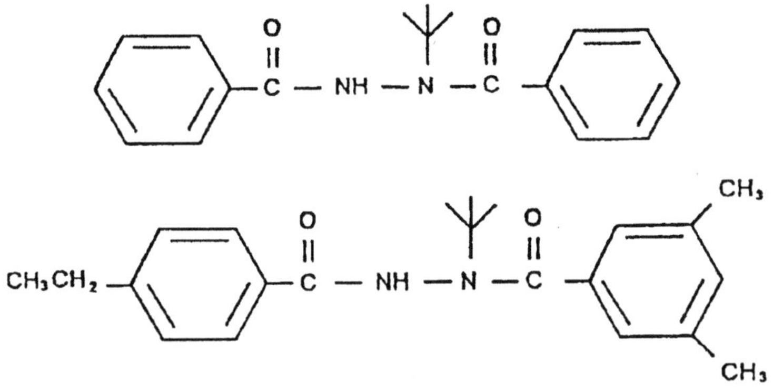

Fig. 20.4. Structures of RH-5849 and RH-5992 (tebufenozide)

tebufenozide undergo premature and lethal larval molting within 12–24 h after treatment (Smagghe and Degheele 1994a,b). According to LC_{50} values, RH-2485 is three- to seven-fold more potent than tebufenozide against a susceptible laboratory strain of *S. littoralis* and 7–14-fold more against the field strain. The LC_{50} values of RH-2485 for the first and sixth instars of the susceptible strain are 0.32 and 0.57 mg a.i. l^{-1}, respectively and those for tebufenozide are 2.1 and 1.9, respectively (Ishaaya et al. 1995). The high selectivity of both RH-2485 and RH-5992 to lepidopteran pests along with their low toxicity to mammals, fish, and ectoparasitoids (Wing and Aller 1990; Brown 1994) render these compounds potential agents to be used in IPM programs for controlling *S. littoralis* and other lepidopteran pests in various agricultural crops.

20.2 Chitin Synthesis Inhibitors

Chitin is a linear amino-sugar polysaccharide, chemically known as β-(1–4)-2-acetamido-2-deoxy-D-glucose polymer (or N-acetyl-D-glucoseamine polymer), which is insoluble in most solvents. Together with the deacetylated polymeric glucoseamine (known as chitosan), they form the basic structural components of the cell wall of fungi (Muzzarelli 1977). Chitin and chitosan can represent as much as 50% of the total fungal cell wall mass. The cell wall of fungi contains other polysaccharides, as well as proteins, lipids, inorganic salts, etc. Chitin and protein are also crucial elements of arthropods and other invertebrates, forming the main constituents of the body wall. In insects, they form the framework of the cuticle and are also constituents of the peritrophic membrane of the gut (Richards 1951, 1978). Synthesis of chitin and deposition of the cuticle in insects are regulated by the molting hormones (ecdysones), and are mediated by enzymes which catalyze a series of complex biotransformations starting with glucose or trehalose and ending in chitin formation. The enzyme chitin synthetase is the key enzyme in chitin formation. Chitin is degraded by the hydrolytic enzymes, chitinase and chitobiase, which are also vital in the molting process of arthropods and cell growth in fungi (Marks et al. 1982; Cohen 1987, 1990).

Obviously, any interference with chitin synthesis and degradation can lead to derangement of metamorphosis and growth of the organism. These considerations led to extensive research into chitin synthesis inhibitors as agents for insect control. Compounds which interfere with chitin biosynthesis exert their toxic effects at the time of molting. Therefore, such inhibitors elicit symptoms of poisoning a few days after treatment, unlike the conventional neurotoxic insecticides which are fast-acting.

Several test systems have been employed for the evaluation of candidate chitin synthesis inhibitors, the most common being the fungus Phycomycetes test, the cockroach leg test (Marks et al. 1982), and the in vitro enzyme preparation of chitin synthase (Cohen and Casida 1980a, 1983; Mayer et al. 1980). Compounds which show chitin synthesis inhibitory activity are polyoxin-D, nikkomycin-Z (Fig. 20.5), and the fungicide Captan. In addition, the chlorinated hydrocarbon insecticides DDT, chlordane, dieldrin, and endosulfan, the pyrethroid permethrin, the herbicide Barban; several benzimidazoles, and other compounds have also shown some inhibition of chitin synthesis in vitro (Marks et al. 1982; Cohen 1987).

The polyoxins are antibiotics derived from *Streptomyces cacaoi* (Suzuki et al. 1965; Isono and Suzuki 1968). Polyoxin-D has been used, principally in Japan, to control soil-borne fungal phytopathogens that cause diseases in rice, vegetables, and fruit trees. Its mode of action has been elucidated by Endo et al. (1970). It inhibits chitin synthesis in vitro in both fungal and insect test systems by competing with the natural substrate (UDP-N-acetyl-D-glucoseamine). Nikkomycin is also a peptide nucleoside antibiotic derived from cultures of *Streptomyces tendai* (Dähn et al. 1976; Kobinata et al. 1980). It has a similar mode of action to polyoxin-D (Cohen and Casida 1980a,b). These antibiotics are effective fungicides but they are poor insecticides, due perhaps to their polar nature, their high molecular weight, and their inability to penetrate the insect cuticle. However, they are toxic when fed to larvae of several insect species (Cohen 1987). The use of nikkomycins as fungicides has been discontinued.

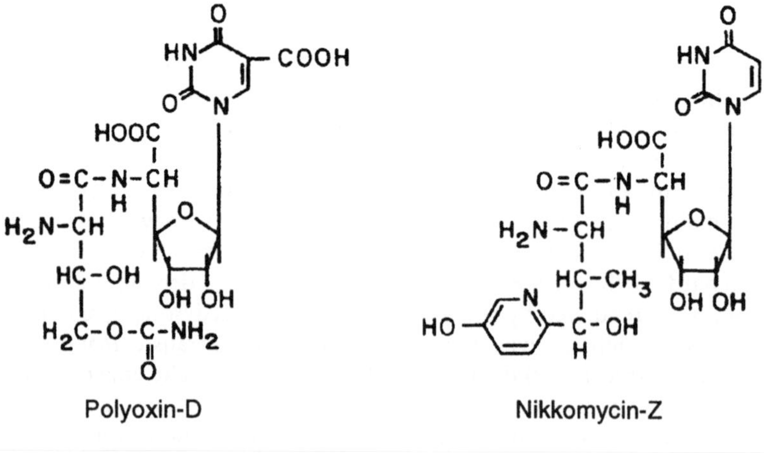

Fig. 20.5. Structures of polyoxin-D and nikkomycin-Z

The first potent chitin synthesis inhibitor of the benzoylphenyl urea type was discovered by scientists at Philips-Duphar in Holland (van Daalen et al. 1972; Verloop and Ferrell 1977). This compound named diflubenzuron (registered trade name Dimilin; Fig. 20.6) was found to affect mainly the larval stage (Grosscurt 1978). Because diflubenzuron does not inhibit the enzyme chitin synthetase in insect or fungi cell-free systems, several modes of action have been suggested, the most recent being that it inhibits DNA synthesis in larval epidermal cells during pupation, thus preventing the formation of an imaginal abdominal

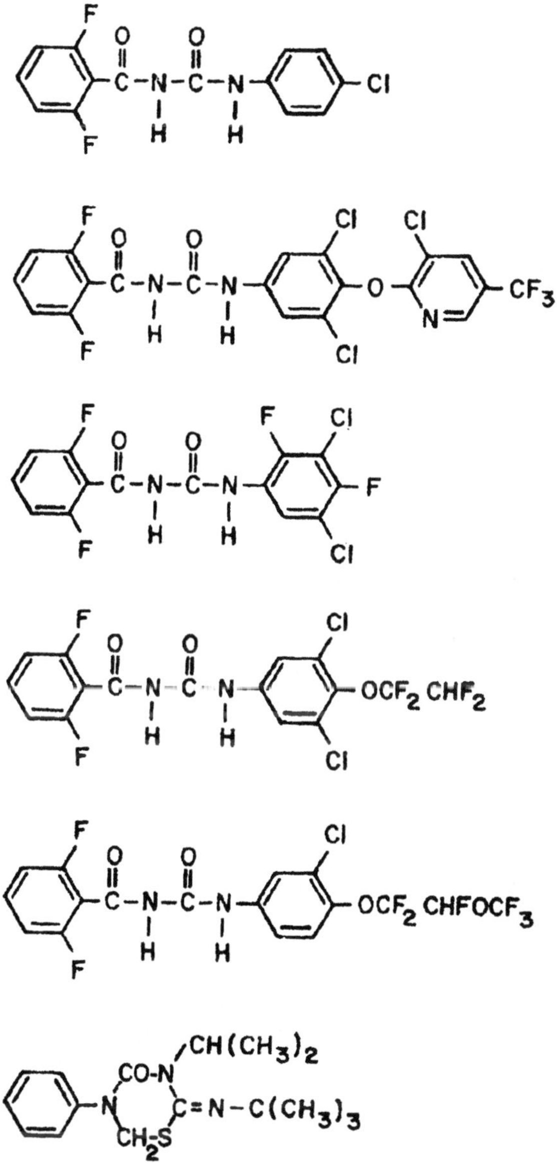

Fig. 20.6. Chitin synthesis inhibitors of agricultural importance. *From top* Diflubenzuron, chlorfluazuron, teflubenzuron, hexaflumuron, novaluron, and buprofezin

epidermis. This, in turn, inhibits chitin synthesis (see reviews by Marks et al. 1982; Cohen 1987).

20.2.1 Compounds of Agricultural Importance

Two important groups of chitin synthesis inhibitors are available today for agricultural use—the benzoylphenyl ureas and buprofezin (Ishaaya 1990b, 1992). Benzoylphenyl ureas act on insects of various orders by inhibiting chitin formation (Ishaaya and Casida 1974; Post et al. 1974; Sowa and Marks 1975; Hajjar and Casida 1979), thereby causing abnormal deposition and abortive molting. Diflubenzuron, the most thoroughly investigated compound of this group, affects the larval stage (Grosscurt 1978). It acts mainly by ingestion, but in some species it suppresses fecundity (Arambourg et al. 1977; Sarasua and Santiago-Alvarez 1983) and inhibits ovicidal and contact toxicity (Ascher and Nemny 1974, 1976b; Holst 1975; Wright and Harris 1976). The main uses of diflubenzuron are against pests in forestry, horticultural, and field crops, and in the home (Grosscurt and Stoker 1991). The search for more potent acylureas has led to the development of new compounds such as chlorfluazuron (Haga et al. 1982), teflubenzuron (Becher et al. 1983), and hexaflumuron (Sbragia et al. 1983; Fig. 20.6). These compounds are very potent against important lepidopteran pests, such as the Egyptian cotton leafworm *Spodoptera littoralis* in cotton and ornamentals (Ishaaya et al. 1986; Ishaaya and Klein 1990), whereas teflubenzuron is used against the grapevine moths *Lobesia bortrana* and *Cryptoblabes gnidiella* (Baum et al. 1990, 1992).

Novaluron. 1-[3-Chloro-4-(1,1,2-trifluoro-2-trifluoro-methoxyethoxy)phenyl]-3-(2,6-di-fluorobenzoyl) urea, a novel benzoylphenyl urea, acts by both ingestion and contact. As such it is a powerful suppresser of lepidopteran larvae such as *S. littoralis* and *Helicoverpa armigera* (by ingestion) and of cotton whitefly larvae *Bemisia tabaci* (by contact). The compound is now under advanced development by Makhteshim Chemicals. Preliminary studies (Ishaaya et al. 1996) indicate that the LC_{50} value of Novaluron on third-instar *S. littoralis* fed on caster bean leaves treated with the compound is ~ 0.1 mg a.i. l^{-1}. This value resembles that of chlorfluazuron and is about ten fold lower than that of teflubenzuron. An application of 250 g a.i.ha^{-1} in a cotton field resulted in 100% mortality of both *S. littoralis* and *H. armigera* upon exposure to treated leaves up to day 8 after application, and about 60 and 30% mortality, respectively, at day 15. Novaluron affects to a much higher extent eggs and larvae of *Bemisia tabaci* than chlorfluazuron and teflubenzuron. Suppression of 100% egg hatch was obtained at a dipping concentration of 0.5 mg a.i. l^{-1} and of first and second instars at a concentration of 1 mg a.i. l^{-1}. Novaluron is far more active in suppressing developing stages of the leafminer *Liriomyza huidobrensis* than teflubenzuron and chlorfluazuron. Suppression of over 80% adult formation was obtained at a concentration of 0.8 mg a.i. l^{-1} and a similar suppression of pupation and mine formation at a concentration of 20 mg a.i. l^{-1}. Addition of 0.1% surfactant to Novaluron solution increased considerably its potency, resulting in 80% and 100% suppression of pupation and adult formation, respectively, at a concentration of 0.8 mg a.i. l^{-1}. The results obtained thus far indicate that Novaluron is a potential compound for controlling lepidopteran pests, whiteflies, and agromyzid leafminers in field crops, vegetables and ornamentals.

Buprofezin. Applaud, 2-tert-butylimino-3-iso-propyl-5-phenyl-3,4,5,6-tetrahydrothiadiazine-4-one, has been developed parallel to the development of benzoylphenyl ureas (Kanno et al. 1981). It acts specifically on some homopteran pests such as the greenhouse whitefly ***Trialeurodes vaporariorum*** (Yasui et al. 1985, 1987), the sweetpotato whitefly ***Bemisia tabaci*** (Ishaaya et al. 1988, De Cock et al. 1990), the brown planthopper ***Nilaparvata lugens*** (Izawa et al. 1985; Nagata 1986) and the citrus scale ***Aonidiella aurantii*** and ***Saissetia oleae*** (Yarom et al. 1988; Ishaaya et al. 1989). Whiteflies are important pests in cotton and vegetables. In some cases they transmit viruses and are considered limiting factors for various agricultural crops. The brown planthopper is an important pest of rice in the Far East. Most of these pests have developed resistance to conventional insecticides and buprofezin is an important addition enabling continuing production of various agricultural commodities. Buprofezin suppresses embryogenesis and progeny formation of ***B. tabaci*** (Ishaaya et al. 1988). The estimated concentration for 50% inhibition of egg hatch was 0.0015% a.i. in the spray solution and 0.006% a.i. for 50% commutative larval mortality. Hence, the compound exerts its effect on the egg hatch and on the larval stage. It has no ovicidal activity, but suppresses embryogenesis through adults. The length of exposure of the whitefly females to buprofezin corresponds well with the suppression of egg hatch. Adult females exposed for up to 5 h to cotton seedlings treated with 62.5 ppm laid eggs with fertility similar to that of the control, but those exposed for a period of more than 24 h laid infertile eggs (Ishaaya et al. 1988). A good correlation between the length of adult exposure to buprofezin and egg fertility was observed also with ***T. vaporariorum*** (Yasui et al. 1987). The compound affects in some cases oviposition (Yasui et al. 1987). The effect on oviposition seems to result from the inhibition of biochemical processes leading to prostaglandin formation; an injection of the latter compound into females of ***Nilaparvata lugens*** antagonizes this effect (Uchida et al. 1987).

Part of buprofezin's efficiency resulted from its vapour phase (De Cock et al. 1990). Cumulative larval mortality of 90, 83 and 65% was obtained when infested seedlings were introduced at 1, 9, and 16 days, respectively, in a cotton field treated with 500 g a.i. ha^{-1} (Table 20.3). These results indicate that vapour phase toxicity of buprofezin persists at least 10 days after application, enabling control of whitely larvae which are present on the lower surface of the leaves and are difficult to control under standard spray conditions of other insecticides. Buprofezin is harmless to aphelinid parasites such as ***Encarsia formosa*** and ***Cales noaki*** (Garrido et al. 1984; Wilson and Anema 1988; Mendel et al. 1994) and to predacious mites (Anonymous 1987) and as such it is a potential component to be used in integrated pest management programs. The compound has been introduced, in 1989, for controlling whiteflies in cotton fields in Israel (Ishaaya 1990a; Horowitz and Ishaaya 1992) accompanied with insecticide resistance management programs aiming at lessening resistance development to this important insecticide (Horowitz and Ishaaya 1994; Horowitz et al. 1994).

Table 20.3. Effect of buprofezin vapor phase on *Bemisia tabaci* larvae under field conditions. (After Ishaaya 1992)

Infested cotton seedlings were introduced into a treated cotton field at various periods after application (days)	Cumulative mortality until pupation (%) (± SE)
1	90 ± 3
9	83 ± 2
16	65 ± 63

Cotton seedlings infested with first instars were introduced for 7 days into a treated or untreated cotton field, at various periods after application. Cumulative mortality until pupation (correlated according to Abbott's formula) was then determined. Each treatment was done with 10 replicates of 50–200 larvae each and presented as means with their SE values.

CHAPTER 21

Insect Sterilants

21.1 Insect Sterilization

The control and eradication of insect pests by means of sterile insects has been a revolutionary idea in entomology. The method consists of breeding insects in the laboratory in large numbers, sterilizing them by chemicals or by radiation (X-rays or gamma-rays), and releasing them in the open so that they can compete for mates with the natural population. If successful, this will reduce the reproductive potential of the natural population in proportion to the ratio of sterile to fertile insects present in the population after the sterile insects are released. If the ratio is 1:1, and the sterilized insects are fully competitive, the reproductive potential of the natural population will be reduced by 50%. If the ratio is 9:1, the reproductive capacity will be reduced by 90%.

This method known as the "sterile male technique," pioneered and described by Knipling (1959, 1960, 1968), was implemented by him and his associates in the early 1950s in the successful eradication of the screwworm fly *Cochliomyia hominivorax* from the island of Curacao (Bushland and Hopkins 1953; Baumhover et al. 1955; Lindquist 1955). Subsequent successes of the sterile male technique against the screwworm fly included eradication of the insect from Florida and Texas (Knipling 1960; Lindquist 1961) and current attempts are to eradicate it from Mexico.

Screwworms are parasites of warm-blooded animals. If such animals happen to have open wounds then the female screwworm fly lays 200–400 eggs in such wounds. The eggs hatch within 12–24 h giving rise to larvae (the screwworm) which feed on the living muscle tissues of the animal. Screwworms are voracious feeders, hence the effect on the animal can be devastating. Young animals can die fairly rapidly from the attack, and even full grown steers can be killed by a heavy infestation in a few days. Larvae complete their feeding in about 5 days, then they drop to the ground and burrow into the soil to pupate and emerge as adult flies in about 2 weeks. The life cycle of the fly thus requires about 20 days, but under adverse conditions it may take as long as 2.5 months. The flies mate when they are 2 days old and the female lays her eggs after 6–8 days. Males mate repeatedly if they can find virgin females but females mate only once.

The annual loss due to infestations by the screwworm fly in the USA was estimated in 1957 to be US$ 25–40 million. Results of research indicated that it was possible to sterilize the flies with either X-rays or gamma-rays. The most efficient method was to irradiate the pupae with 7500 roentgen 2 days before emergence of the adult. It was determined experimentally that the irradiated-induced sterility was permanent and that the sterilized males were competitive with normal males. If a normal female was mated with a sexually sterile male her reproductive potential was completely destroyed. Further research showed that the sterile males produced sperm that fertilized the eggs, but the larvae failed to hatch because the embryos died during the process of development. One major difficulty with this technique is the flight distance of the fly. Using marked flies it has been shown that they can migrate up to 480 km (300 miles) from the point of release so that there is a constant threat of infiltration of flies from other areas.

Knipling (1955) lists five key factors which are essential before the procedure can be regarded as feasible for eradicating or controlling any given pest. These are:

1. A method of mass rearing of the insect must be available.
2. Adequate dispersion of the released sterile males must be obtained.
3. The sterilization procedure must not adversely affect the mating behavior of the males.
4. The female of the insect to be controlled must normally mate only once, or, if more frequent matings occur, the sperms from gamma-irradiated males must be able to compete with those from fertile males.
5. The population density of the insect must inherently be low or it must be reduced by other means to a level which will make it feasible to release a dominant population of sterile males over an extended period of time.

Taking into account the above considerations, attempts have been made to control or eradicate the Mexican fruit fly, melon fly, Mediterranean fruit fly, oriental fruit fly, several species of mosquitoes, and others with some degree of success. In the island of Okinawa, Japan, after the release of about 50 000 million sterile melon flies from 1972 to 1992, the fly was completely eradicated (Yamagishi et al. 1993). Less successful was the attempt to eradicate the cotton boll weevil because of the lethal effect of irradiation on the insect (Davich and Lindquist 1962).

21.2 Chemosterilants

The possible use of chemical compounds as sterilants for insect control parallels the sterility principle with gamma-radiation described in Section 21.1. Knipling (1959, 1968) has considered the desirability of utilizing the natural population of an insect or other noxious animal for its own destruction. In theory, the advantage of a chemical sterilant over a killing agent (an insecticide) can be realized if the chemosterilant is as effective as the insecticide in producing the intended effect. The advantage can be twofold: (1) the insects sterilized cannot reproduce, which has the same effect as that of killing the insect; and (2) the sterile insects are still present in the population to compete with the remaining fertile insects for mating, thus producing the same potential effect that would be achieved by releasing sterile insects. The latter effect represents a distinct advantage of sterility over the killing method.

Chemosterilants are chemicals capable of causing sexual sterility, thus preventing reproduction. Insect chemosterilants may act in three ways: (1) they may cause the insects to fail to produce ova and sperm; (2) they may cause the death of ova and sperm after they have been produced; (3) they have the effect shown by radiomimetic compounds, that is, producing lethal mutations or severely damaging the genetic material in the sperm and ova. Thus, although the sperm and ova remain alive, the zygotes, if formed, do not complete development.

A chemosterilant might be used in two ways: (1) to sterilize insects reared in large numbers for release in infested areas; (2) to induce sterility in the natural population, thus avoiding the necessity of rearing and releasing large numbers of insects, both of which are costly operations. Also, some species, are not adaptable to mass rearing in the laboratory. The first method is not suitable for insects which are mechanical vectors of pathogens because the addition of such vectors to the existing natural population may cause annoyance by biting or invading premises and even temporarily increase the disease potential. For control or eradication of such species, chemical induction of sterility in the natural population would be a distinct advantage. This can be accomplished by using chemosterilant baits with or without the addition of chemical attractants or sex lures (pheromones). Chemosterilants can also be applied by other means but with much caution concerning their effects on other organisms and the problem of residues in the environment.

Several thousand compounds have been evaluated as chemosterilants on several insect species. LaBreque (1968) lists over 200 compounds comprising alkylating agents, antimetabolites, and others as affecting the reproductive system in more than 30 insect species. By far the largest amount of experimental work was done with the housefly. The alkylating agents apholate, aphomide, and aphoxide (tepa) induced sterility in both males and females when the compounds were introduced in their diet at a concentration of 0.5–1%. Using cornmeal baits containing 0.5% aphoxide, excellent results were obtained in reducing housefly populations in an isolated refuse dump in Florida (LaBreque et al. 1962). Other alkylating chemosterilants of interest include thiotepa, metepa, methiotepa, tretamine, morzid, busulfan, and the analogs of alkylating agents hempa and hemel. Alkylating agents replace hydrogen with an alkyl group in basic genetic material with an effect similar to that of radiation. It is of interest to note that the search for effective chemosterilants parallels the search for cancer chemotherapeutic agents since the object in both cases involves the prevention of cell division. Hence, many anticancer agents have been screened as chemosterilants. A comprehensive review of the chemistry and biochemistry of chemosterilants has been written by Turner (1968).

So far no compound has yet been found that will sterilize an insect but is safe enough to apply in the same way that most insecticides are applied. All of them must be considered hazardous, and in experimental use the possibility of their contact with man or beneficial animals should be avoided, although some of them have received very limited field tests, mainly by combining them with baits to minimize the total amount released (Smith et al. 1964; Hayes and Laws 1991). A greater degree of safety would permit perhaps the use of such compounds in more general but highly selective methods of application.

CHAPTER 22

Attractants, Repellents, and Antifeedants

22.1 Attractants

The earliest record of an attractant used for economic purposes was an attempt in 1885 to control grasshoppers in California by means of attractive poisoned baits. Later, wine growers in Europe used traps baited with stale beer, sugar water, and old cider to control the grapevine moth. Other baits included lemons and oranges in the formulation, later changed on a chemical basis by substituting amyl alcohol for citrus fruit. A fly bait consisting of a mixture of old casein, brown sugar, and water was considered an excellent attractant. Molasses–yeast baits for peach moth control were added to the list. In the late 1920s, hundreds of aromatic compounds were screened for their attractant properties. Additional impetus was given by the discovery of geraniol as an attractant for the Japanese beetle. A list of many of the old type attractants is given by Dethier (1947).

Plants defoliated by the cabbage butterfly were found to contain volatile compounds known as mustard oils which were the basis for the attraction of the larvae. This indicated that specific natural attractants present in plants influenced phytophagous insects in their food selection. Other than chemical stimuli, such as odor, taste, and humidity, physical stimuli, such as heat, light, and pressure, also contribute to the attractance of the insect to its host (Dethier 1947). Attractants, in a strict definition, are chemicals which cause insects to make oriented movement towards its source (Dethier et al. 1960) or chemicals acting in a vapor phase, causing an insect to behave in ways which result in its moving towards the source of the material or towards a zone of preferred concentration (Browne 1977).

22.1.1 Poison Baits

Attractants may be used in a bait either alone or in conjunction with a poison. When used alone the attractant is the bait itself. Usually, attractants are added to a bait containing an insecticide and a palatable ingredient such as molasses.

22.1.2 Traps

A trap is a device for immediate killing of insects or retaining them for later identification or destruction. Different types are available for different insect species. In public health work, the New Jersey light trap for mosquitoes is commonly used. In addition to light, carbon dioxide (CO_2) in the form of dry ice is used as an attractant for *Aedes aegypti* and *Culex tarsalis* mosquitoes.

In nature, mosquitoes are attracted and stimulated by chemicals evolved by the human body, such as blood, sweat, and urine. The attractiveness is enhanced by moisture, temperature, and carbon dioxide. Humans also secrete lactic acid in their sweat and this probably accounts for the appeal of sweat to mosquitoes (Fletcher 1974). Dark hands were found to be more attractive than pale hands, and warm hands more than cold hands. Men were more attractive than women to *Aedes aegypti*. Estrogens and certain amino acids enhanced the attractiveness.

Housefly traps constitute an effective adjunct to standard methods of fly control. Many models have been designed, all operating on the same principle. Flies are attracted to molasses, ripe fruit, milk, meat, or fermenting baits placed in shallow pans beneath the traps. Upon completion of feeding, they fly upward, passing through a small opening and enter the screened enclosure from which they are unable to escape. Blowflies are also constant visitors. The conical type traps are the most efficient.

One of the most successful bait traps containing a synthetic attractant (siglure) was used in the eradication of the Mediterranean fruit fly, *Ceratitis capitata*, from Florida in 1956. Other, more potent synthetic attractants include medlure and trimedlure. Other baits consisting of protein hydrolysate fortified with the insecticide malathion are successful to this date in controlling the medfly. A similar success story was the annihilation of males of the Oriental fruit fly *Dacus dorsalis* from an island in the Pacific and the whole Okinawa island, Japan, using the powerful attractant methyl eugenol together with the insecticide naled (Steiner et al. 1961, 1965).

The use of chemical attractants in public health does not appear to be very promising since many insects such as bedbugs, fleas, certain fly species, and others do not seem to possess highly developed olfactory organs which are the basic receptors of the chemical stimuli.

22.1.3 Ovipositional-Type Attractants

Oviposition lures are substances that attract gravid females of a species and induce them to lay eggs. In nature, the external stimuli which govern the time and location of oviposition include temperature, humidity, light, currents, type of surface, odors, and contact with chemical substances. Of these, odors are probably the most important factor with many insect species (Richardson 1925). Houseflies are attracted to manure and ammonia, and green bottle flies are attracted to ammonium carbonate for oviposition purposes. Of particular interest is the attraction of female mosquitoes (*Aedes aegypti*) to oviposit in special type containers (with preference to texture, shape, and color of the container), as well as the odors of methyl butyrate, methyl propionate, and ethyl acetate (Fay and Perry 1965; Perry and Fay 1967). Methyl butyrate was distinctly more attractive than most other fatty acid esters tested.

22.2 Pheromones

Pheromones are chemicals released by one individual organism and affect the physiology or behavior of another individual of the same species. In some cases they cause an immediate behavioral response such as sex attraction, while in others they act primarily on the physiology of the receiving organisms with time lag before causing certain effects. Pheromones are classified according to their effect on the target organism or their functions. Accordingly, there are a variety of pheromones: alarm, trail, aggregation, masking, oviposition marking, mating, caste recognition, caste keeping, deterrent, masking pheromones, etc. Among them, sex attractant pheromones have particular interest from the standpoint of insect control.

22.2.1 Sex Attractants

It has been known since the 1800s that insects release a sex attractant scent into the atmosphere in order to call a mating partner. Insects detecting the specific scent respond by flying or moving upwind in the direction of the emission and mate when contact is made with the partner. The majority of sex attractants are emitted by females although, in some cases, males also release sexual odors which are primarily for the purpose of sexually exciting the females and as a stimulus to copulation. The latter are more in the nature of "aphrodisiacs" (Jacobson 1965).

One of the earliest sex attractants was a crude extract of the tips of the abdomens of female gypsy moths *Lymantria* (=Porthetria) *dispar* which was used in traps to monitor the distribution of the insect over large areas. In 1970, the gypsy moth pheromone, called disparlure, was isolated, identified, and synthesized (Bierl et al. 1970; Table 22.1). The sex attractant can lure males of this species to come to the female from a distance of more than 3 km away. This gives an indication of the potency of the chemical and the high sensitivity of the male's antennae to receive the signal. The successes encountered with disparlure and other sex attractants prompted an extensive research in the isolation, identification, and synthesis of a large number of pheromones of many other species (Table 22.1). Included in these advances were the development of better means of using the pheromones, better means of timing insecticide sprays with pheromone-baited monitoring traps, improved controlled release of pheromone formulations, and more effective methods of pheromone distribution for disorienting insects in order to reduce or prevent their reproduction.

In the use of insect sex attractants major emphasis has been placed on the following (Kydonieus and Beroza 1982): (1) monitoring with traps; (2) mass trapping; (3) trap cropping; (4) mating disruption.

Methyl eugenol of plant origin, though it is not a pheromone, attracts only males of the fruit fly, *Dacus dorsalis*, and it is utilized as a sex attractant.

Table 22.1. Insect pheromones important in agricultural practices

Species	Pheromone	Structure
Silkworm *Bombyx mori*	(E, Z)-10, 12-hexadecadien-1-ol (bomby-ol)	$CH_3(CH_2)_2CH{=}CH{-}CH{=}CH(CH_2)_8CH_2OH$
Gypsy moth *Porthetria dispar*	(Z)-7, 8-epoxy-2-methyloctadecane (disparlure)	$CH_3(CH_2)_9CH{-}CH(CH_2)_4CH(CH_3)_2$ (epoxide bridge O)
Smaller tea tortrix *Adoxophyes orana*	(Z)-9-tetradecenyl acetate (Z)-11-tetradecenyl acetate (9:1)	$CH_3(CH_2)_3CH{=}CH(CH_2)_7CH_2OCOCH_3$ $CH_3CH_2CH{=}CH(CH_2)_9CH_2OC(O)CH_3$
Pink bollworm *Pectinophora gossypiella*	(Z, Z)- and (Z, E)-7, 11-hexadecadienyl acetates (1:1) (gossyplure)	$CH_3(CH_2)_3CH{=}CH(CH_2)_2CH{=}CH(CH_2)_5CH_2OC(O)CH_3$
Tobacco budworm *Heliothis virescens*	(Z)-9-tetradecenal (Z)-11-hexadecenal (1:16)	$CH_3(CH_2)_3CH{=}CH(CH_2)_6CH_2CHO$ $CH_3(CH_2)_3CH{=}CH(CH_2)_8CH_2CHO$
Indian meal moth *Plodia interpunctella*	(Z, E)-9, 12-tetradecadienyl acetate	$CH_3CH{=}CHCH_2CH{=}CH(CH_2)_7CH_2OC(O)CH_3$
Japanese beetle *Popillia japonica*	(Z)-5-(1-decenyl)-dihydro-2(3H)-furanone	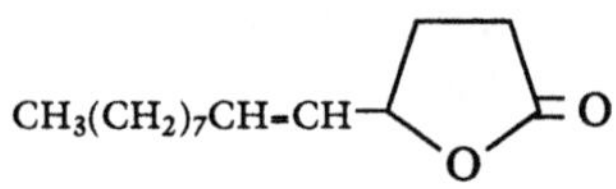
Cotton boll weevil *Anthonomus grandis*	(+)-(Z)-2-isopropenyl-1-methylcyclobutane ethaanol	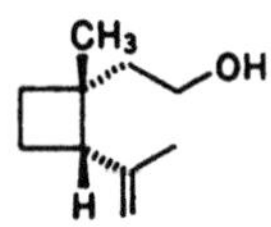
	(Z)-3, 3-dimethyl-$\Delta^{1,\beta}$-cyclohexane ethanol	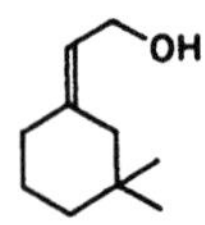
	(Z) and (E)-3, 3-dimethyl-$\Delta^{1,\alpha}$-cyclohexane acetaldehydes	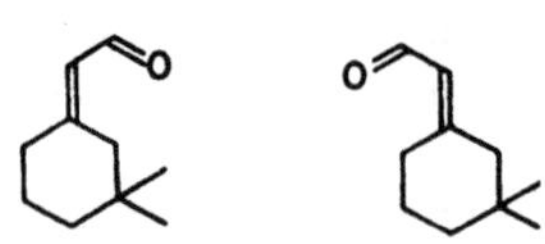
Ips paraconfusus	(R)-(+)-ipsdienol (S)-(−)-ipsenol (Z)-verbenol	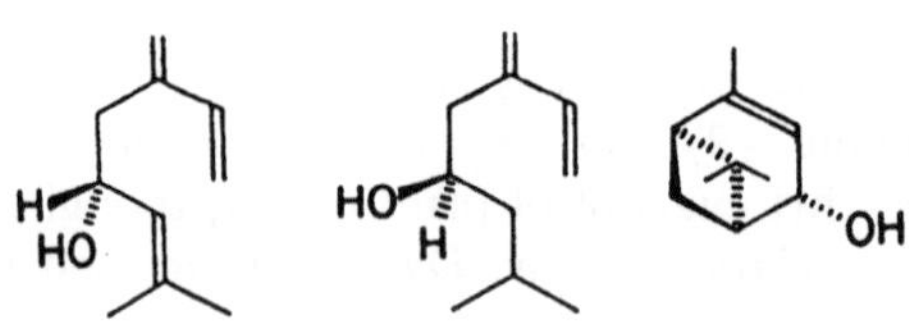

22.2.2 Monitoring

Traps may be used with sex lures for detection as well as estimation of certain species in specific areas and for indicating when and where control measures should be applied. This procedure is useful in limiting insect penetration into new areas, preventing damage from excessively high populations, and in timing of insecticide sprays.

22.2.3 Mass Trapping

It was suggested that pheromone traps can be used in the direct control of insects, thereby minimizing the dependence on insecticides alone. It was realized that large scale trapping would be effective only against low-level insect populations. In this approach, insects can be suppressed to low levels by using an insecticide (especially against the larval stage) and further suppressed by mass trapping the adults that manage to survive. The major difficulty in mass trapping is the large number of traps needed and the high cost of deploying and maintaining them. An advantage of the mass trapping is that it can also be used effectively for monitoring purposes. Trap design, type of pheromone dispenser, height of trap above ground, and distance between traps are important parameters which need to be considered.

22.2.4 Trap Cropping

In trap cropping, several rows of a crop are planted to emerge before the main planting, in which case insects attacking the plants are killed with an insecticide applied to the plant. The method can be improved by treating the trap crop with pheromone, thereby attracting the insects to the lethal trap crop in larger numbers and from a greater distance. In this case less insecticide is needed to achieve good control and the effect on beneficial insects is kept to a minimum.

22.2.5 Mating Disruption

Mating disruption is one of the most promising methods of controlling insects with sex attractants. It consists of permeating the atmosphere with sex-lure-emitting particles. With attractant everywhere, the insect becomes confused because it cannot distinguish between pheromone from the lure-emitting particles and that from a potential sex partner. Thus, it finds it difficult in locating a mating partner.

As with mass trapping, mating disruption works best with low-level populations, that is, against insects that are sufficiently separated to require the lure for mate-finding. If insects are in close proximity to each other, as occurs with dense populations, this method will not be very effective. Here too, suppression of the population with an insecticide or by other means is a prerequisite for maintaining a low population level throughout the season using the mating disruption technique.

22.2.6 Controlled Release

Release of the pheromone in a controlled manner is essential for its success. Releasing too high a concentration can inhibit the insect's entry into the trap and is also wasteful. Releasing too low a concentration can lower the trap catch by reducing the distance from which insects respond. Therefore, various controlled technologies have been devised and such products are now commercially available (see Hercon products listed by Kydonieus and Beroza 1982).

22.2.7 Other Types of Pheromones

In the field of public health the sex pheromone of virgin females of the American cockroach *Periplaneta americana* has been isolated and characterized. It consists of two active components, periplanone A and periplanone B. These components stimulate males to sexual excitement which is characterized by wing fluttering and copulatory attempts (Persoons et al. 1974, 1979), and, more accurately, by antennal response using an electroantennogram technique (Nishino et al. 1983).

A sex pheromone of the Oriental cockroach *Blatta orientalis* was isolated from the feces of virgin females. It elicits sexual response in males of the Oriental cockroach and also in males of the American cockroach (Warthen et al. 1983). Likewise, males of the Oriental roach respond to the sex pheromone of the female American roach. These pheromones may be similar in composition and they show a high degree of interspecificity. The sex pheromone of the housefly *Musca domestica* called muscalure consists of (*Z*)-9-tricosene and other hydrocarbon components which potentiate its action (Carlson et al. 1971). The female housefly begins producing these components 2 days after emergence, but they are absent or present in very small quantities in the male (Dillwith et al. 1981). Sex pheromones that induce mating behavior in males have also been isolated from females of the facefly *Musca autumnalis*, the lesser housefly *Fannia canicularis*, from *Fannia puso*, *Fannia femoralis*, and from the stable fly. A sex pheromone of female tsetse flies *Glossina pallidipes* elicits sexual response in adult males of this species. The pheromone, a C_{35} compound, is present in the hydrocarbon fraction of the adult female cuticle. A similar pheromone, although slightly different in composition, is produced by female *Glossina morsitans* (Langley et al. 1982; Carlson et al. 1984).

Callosobruchus chinensis and *C. maculatus* use several pheromones for different purposes. In oviposition, under low population density conditions, the females oviposit evenly among beans, being guided by marking pheromone. Under high density conditions, the number of eggs on each bean becomes high, but only a few eggs can hatch and grow normally inside the bean, leaving the remaining eggs to die for reducing competition. It is due to the ovicidal action of higher doses of the marking pheromone (lipids in nature). Treatment of beans with the higher doses of the marking pheromone (100 μg per bean) or certain edible oils can protect beans. For mating, a female sex pheromone attracts the nearby male. Then the second sex pheromone (erectin) induces the male to copulation behavior. These pheromones were partially or completely identified (Tanaka et al. 1981; Yamamoto 1990; Yamamoto et al. 1996).

22.2.8 Toxicity

To date, no pheromone has been found to be truly toxic to warm-blooded animals. Most pheromones are fatty acid derivatives or similar to chemicals that make up food or food constituents. Furthermore, the amounts intended for use in insect control are so small that it takes highly specialized equipment to detect them in the atmosphere. They are also of very low toxicity to beneficial insects. Hence, pheromones are distinctly suitable for use in integrated pest management programs.

As of recent years, pheromones and attractants of more than 700 insect species have been characterized and chemically identified. It is beyond the scope of this chapter to discuss pheromones in detail. Numerous articles and several books have been published on this subject (refer to books and chapters by Jacobson 1965; Wood et al. 1970; Birch 1974; Roelofs 1975, 1978; Roelofs et al. 1976; Beroza 1976; Shorey and McKelvey 1977; Blum 1978; Kydonieus and Beroza 1982; Prestwick and Blomquist 1987; Mandava and Morgan 1988; Jutsun 1989; and many others).

22.3 Repellents

Historically, repellents have included a variety of materials such as fresh plants or plant brews rubbed on the skin, pitch rubbed on animal bodies to repel ticks, volatile oils, tar, and various earths applied to the body. Smoke was used to repel the malaria mosquito *Anopheles gambiae*. The Bedouins are known to have used camel urine as hair tonic and medicine, and when applied to the face it left an oily residue which, supposedly, acted as an insect repellent. Camphor crystals sprinkled on woolens have been used for decades to repel clothes moths. The earliest truly chemical repellent is attributed to Bordeaux mixture [$CuSO_4$ + $Ca(OH)_2$]. However, repellents in a strict definition are chemicals which cause insects to make oriented movement away from the source of the chemical (Dethier et al. 1960) or chemicals acting in a vapor phase, preventing an insect from reaching a target to which it would otherwise be attached (Browne 1977).

In more modern times, the bulk of research on repellents was undertaken by the military requirements of World War II. Prior to World War II there were only four principal repellents for use against biting arthropods:

1. Oil of citronella, used as hair-dressing fragrance by Eastern cultures, probably for control of lice and other head ectoparasites. This was the most widely used repellent.
2. Dimethyl phthalate, originally reported as a fly repellent.
3. Indalone (2,2-dimethyl-6-carbobutoxy-2,3-dihydro-4-pyrone).
4. Rutgers 612 (2-ethyl-1,3-hexanediol).

The formulation 6-2-2, a mixture of 6 parts dimethyl phthalate, 2 parts indalone, and 2 parts Rutgers 612, was used as a standard repellent (Table 22.2). Extensive reviews on repellents and the principles underlying their mode of

Table 22.2. Important insect repellents. (Modified from Metcalf et al. 1962; Metcalf and Metcalf 1993)

Repellent	Toxicity LD_{50} to rat (mg/kg)	Species
Dimethyl phthalate	8200	Mosquitoes, *Anopheles, Aedes, Culex* spp., ticks
2-Ethyl-1,3-hexanediol (Rutgers 612)	2400	Mosquitoes, flies, fleas
N, N-diethyl *m*-toluamide (Deet)	2000	Mosquitoes, flies, fleas
2-Ethyl-2-butyl-1, 3-propanediol	5040	Mosquitoes, flies, fleas
N-butylacetanilide	2830	Ticks, fleas
Benzyl benzoate	1900	Mites (chiggers), *Trombicula* spp.
Benzil		Mites (chiggers)
Dibutyl phthalate	21 000	Mites (chiggers)
Dibutyl succinate	8000	Flies on cattle, cockroach, ants
Dipropylpyridine-2,5-dicarboxylate		Flies on cattle
Butoxypolypropylene glycol	11 200	Flies on cattle, houseflies
2-Hydroxyethyl octyl sulfide		Flies on cattle

action have been published by Dethier (1947, 1956). A review of testing techniques for the evaluation of insect repellents has been published by Schreck (1977). Failure of compound 6-2-2 to provide adequate protection to military personnel, especially in the tropics, prompted the screening of thousands of other compounds.

The criteria for an ideal repellent would be that the candidate material should be nontoxic and nonirritating to humans, and long lasting (12 h or more) against mosquitoes, biting flies, ticks, fleas, and chiggers. Unfortunately, such a compound has not yet been found. Some of the newer repellents have unpleasant odors, require large dosages to be effective, are oily, and effective only for short durations, irritate the skin, or soften paints and plastics (Ware 1983). Insect repellents are formulated in a variety of ways: undiluted, diluted in a cosmetic solvent with added fragrance, aerosols, creams, grease sticks, powders, suntan oils, and laundry emulsions for impregnating clothes. The reason that repellents are not long lasting is that they evaporate, are absorbed by the skin, are lost by abrasion with clothing or other material, and are diluted by perspiration (Ware 1983). The most commonly used repellents today are shown in Table 22.2. Of these, diethyl toluamide (Deet) is by far the most effective and the most commonly used repellent against biting flies, certain species of mosquitoes, and chiggers.

Natural pyrethrum is well known for its insect repellent property. Although it is highly effective when applied to livestock to repel tsetse flies and ticks, its complete protection lasts only 2 days. Simkin and Galun (1983) showed that microencapsulation of natural pyrethrum increased significantly the persistence of its repellency (up to 11 days) and its toxicity against tsetse flies *Glossina morsitans*, and prolonged the repellency and toxicity toward the tick *Ornithodoros tholozani*. Microencapsulation of the synthetic pyrethroids bioresmethrin, resmethrin, and permethrin was ineffective in prolonging the repellency.

Another class of repellents may be termed as physical repellents, for example, dust, granules, water, oils and waxes which may occur naturally on the surface of plants or animals being attacked. In this class are also included auditory repellents. Various types of amplified sound producing devices have been reported as being effective in repelling mosquitoes, moths, sand fleas, and other insects.

Many insects and other arthropods produce their own chemical repellents which they use for their defense against predators. Cockroaches, earwigs, stink bugs, caterpillars, scorpions and millipedes discharge their repellents as a spray, often to a distance of several meters.

22.4 Antifeedants

Antifeedants are substances which are not necessarily repellent in action, but which make the crop plant distasteful to the insect attempting to feed on it. The insect might take a bite, then, finding the food unpalatable, it may wander off to another part of the plant. In the presence of the antifeeding compound the insect may starve to death if it remains on the host plant because feeding is inhibited. Alternatively, the insect may move to a nearby untreated plant. If the untreated plant happens to be a weed which is attractive to the insect, the crop may be protected with the added advantage that the weeds might be destroyed. This would be an ideal situation but is seldom the case.

The potential of antifeedants, both natural and synthetic, has been recognized in insect control. The Bordeaux mixture mentioned in Section 22.3 is in fact a feeding deterrent to flea beetles, leaf hoppers, and the potato psyllid. Synthetic compounds like 4'-(dimethyltriazeno)acetanilide and triphenyltin acetate were attempted to be used. Organotin compounds are used as antifeedants on some crops (Ascher and Ishaaya 1973; Ishaaya and Casida 1975). Some commercial insecticides exhibit repellent/antifeedant activity like pyrethroids, chlordimeform, cartap, and pymetrozine.

Several drimane sesquiterpenoids (including warburganal and polygodial) have shown potent activity against the African armyworm. Some clerodane diterpenoids (like clerodin and ajugarin I) were antifeedant to a number of insect species. These natural products were not commercialized by limited spectrum, toxico-

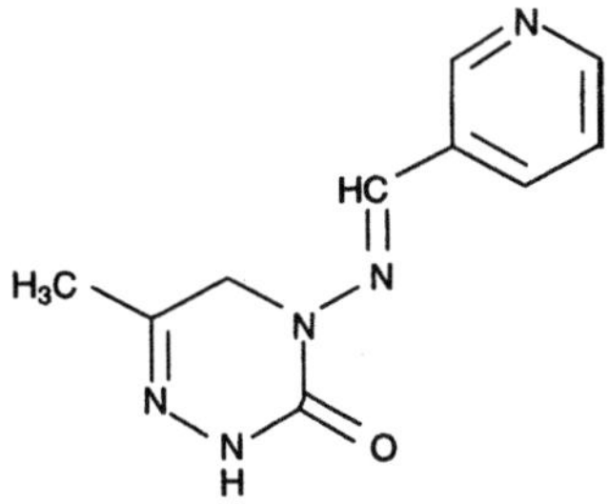

Pymetrozine (CGA 215'944)

logical and/or cost problems. However, of particular interest are components of the neem tree ***Azadirachta indica*** (also called margosa tree or Indian lilac) (see Sect. 13.4 for details).

Pymetrozine, a pyridine azomethrine, is an insecticide of novel structure and mode of action. It is effective on aphids, whitefly, and hoppers, it is systemic, has long residual activity, and is not injurious to beneficial arthropods. It is unique in affecting nervous control of the salivary gland on which it shows abrupt action and irreversible blocking of feeding, resulting in slow mortality due to starvation. It is considered safe to natural enemies and the environment (Kristinsson 1994).

CHAPTER 23

Microbial Insecticides

Microbial insecticides are those microorganisms or their products that are capable of attacking and killing pest insects. This may be considered to be an aspect of biological control, but when one is dealing with the products of such microbes the control may be termed chemical. In the latter case, the products are natural insecticides or microbe-origin insecticides.

23.1 Protozoa

A large number of protozoan species have been isolated from mosquitoes, but most of them were found to be of low pathogenicity. Most studies have concentrated on the microsporidia because of their wide distribution, common occurrence in a wide variety of mosquito hosts, and their greater pathogenicity to many species (Andreadis 1990). The most abundant and widespread genera affecting mosquitoes are *Amblyospora* and *Parathelohania*. Other genera include *Nosema*, *Pleistophora*, and *Stempellia*. *Nosema* species have been extensively studied in association with the European corn borer, *Heliothis (Helicoverpa)*, grasshoppers, and other species (Henry 1990). Some authors have suggested that mosquitoes infected with microsporidia were less likely to transmit malaria. In addition, it was reported that the longevity of *Anopheles albimanus* females infected with *Nosema* was half of that of healthy females (Chapman 1974). Levels of microsporidian infections in field-collected larvae amounted to 1–50%. Release of microsporidia in a field population of mosquitoes showed that although the pathogen was still active in parasitizing mosquitoes after 2 years, the level of infection was very low.

23.2 Bacteria

Among the bacteria, most promising results have been obtained with *Bacillus thuringiensis* (B.t.) of which some 30 different isolates having insecticidal activity have been described (Dulmage 1981; Navon et al. 1990). B.t. was first isolated in 1911 from diseased flour moth larvae *Ephestia kuhniella* and described as a new species in 1915. Later, several varieties were obtained from different species of infected caterpillars, but identification of the different varieties and their use against insect pests began only in the early 1960s.

This group of bacteria is composed of aerobic, gram-positive, spore-forming rods. When spores are introduced into a suitable growth medium, they germinate to yield rod-shaped vegetative cells. After a period of vigorous vegetative growth, the cells sporulate forming nongrowing, heat-resistant spores. Simultaneously a proteinaceous delta-endotoxin crystal is synthesized. Most isolates of this group of bacteria produce toxic protein crystals at the time of sporulation. The most prevalent toxin is the delta-endotoxin (protoxin) which is activated in the insect's midgut by proteolytic enzymes. The active principle of the toxin disrupts the epithelial membrane of the midgut causing disturbances in the osmotic balance of the cells, leading eventually to paralysis and death of the insect (Forsberg et al. 1976).

Delta-endotoxin is toxic primarily to larvae of *Lepidoptera*, although some species of *Diptera* and *Coleoptera* are also affected. The grasshopper *Schistocerca gregaria* showed resistance to the toxic action of this toxin (Yunowitz and Yawetz 1989). Further studies on the mode of action of the toxin (Thomas and Ellar 1983; Yunowitz and Yawetz 1988) showed that it interacts with liposomes (principally phosphatidylcholine) of the epithelial cell membrane in addition to disrupting the epithelial membrane of the midgut.

Some B.t. varieties produce other toxins, such as alpha-exotoxin and beta-exotoxin. Alpha-exotoxin is a thermolabile toxin produced and secreted by vegetative bacterial cells and is not found in sporulated cultures. Hence, it is not present in commercial preparations of B.t. because it is inactivated during the fermentation process. It is claimed to be toxic to several insect species including honeybees and to mice, hence the designation "mouse factor." Beta-exotoxin is a heat-stable toxin and is characterized by its ability to produce malformations in susceptible insects. These effects are thought to be due to mutations caused by beta-exotoxin which can be passed on from parent to offspring. In addition, beta-exotoxin causes derangement or inhibition of pupation in the housefly, hence it is called the "fly factor." Beta-exotoxin is toxic to a wide range of insects such as bees, flies, mosquitoes, grasshoppers, termites, and many Lepidoptera. In addition, it is toxic to spider mites and to some vertebrates, such as white mouse and chicken (Angus 1971). Its action has been determined to be inhibition of RNA formation and inhibition of protein and nucleic acid synthesis.

Commercial preparations of *Bacillus thuringiensis* include Thuricide, Dipel, Agritol, Biotrol, Larvatrol, and several others. These are composed of viable spores and delta-endotoxin (the variety *B.t. kurstaki* does not produce beta-exotoxin). They are effective against many important pests of agriculture and forestry. In Israel, B.t. preparations of variety *kurstaki* are used to control avocado pests. Also, control of *Spodoptera littoralis* in alfalfa and *Heliothis armigera* in tomatoes is effective in bioorganic agriculture. Other B.t. varieties proved to be more effective against many lepidopterous insects (Navon et al. 1990). Laboratory and field experiments are being conducted to test B.t.'s effectiveness against cotton pests, other field crops, and stored product insects.

B.t. preparations have a long shelf life but are not persistent under field conditions (24–48 h), and their activity is affected by temperature, light, plant exudates, etc. (Navon 1993). On the other hand, they may persist on foliage of coniferous trees for 30–42 days (Forsberg et al. 1976). Improvements in the efficacy and host range of B.t. insecticidal activities have been made in recent years by various genetic engineering manipulations and transfer of the δ-endotoxin genes (*cry* genes) to B.t. strains, *E. coli, Pseudomonas clavibacter*, algae, and protozoa, and some new and more potent commercial preparations have already been registered for use as bioinsecticides.

B.t. is not toxic to mammals (it has little or no effect on mice, rats, dogs, hogs, cattle, birds. fish, humans, and some beneficial invertebrates after ingestion of B.t. spore-crystal complex), and no deleterious effects on most predators and parasites of pest insects have been noted. Some food avoidance has been observed in birds fed B.t., and some varieties of B.t. may be implicated in a disease of earthworms. No environmental hazard can be anticipated from its use, hence, it is ecologically desirable as a pest control agent and is therefore compatible with integrated pest management programs, though precaution must be taken for use near sericulture facilities. Current reviews on B.t. can be found in Aronson and Wu (1990); Feitelston et al. (1990); Leemans et al. (1990); McGaughey (1990); Rothnie and Geiser (1990); Yamamoto and Powell (1993); Schnepf (1995).

Bacillus thuringiensis var. *israelensis* (B.t.i.) was isolated from a pond in the southern part of Israel (Goldberg and Margalit 1977) and was shown to be toxic to larvae of *Anopheles sergentii, Uranotaenia unguiculata, Culex univitattus*, and *Culex pipiens*. de Barjac and Coz (1979) added to the list *Aedes caspius, Aedes albopictus, Aedes polynesiensis*, and *Anopheles gambiae*, the principal vector of *Plasmodium falciparum* which is the most virulent form of malaria. Other species have since been added. B.t.i. is used worldwide as a larvicide for mosquito control in ponds, ditches, tidal water, sewage areas, and other water-standing areas, as well as in control of black flies *Simulium damnosum* in the onchocerciasis control program of the World Health Organization in Africa (Lacey and Undeen 1986; Davidson 1990). B.t.i. has virtually no toxicity to lepidopterous larvae.

B.t.i. is commercially available (as Bactimos, Skeetal, Teknar, Vectobac, and others) as wettable powder, flowable concentrate, granules, briquets, etc. It may be applied by ground or air to the surface of the water. It is especially effective against first and second instar larvae at lower application rates, but higher rates are required against third and fourth instar larvae. It is not effective against pupae or adults. Brackish waters do not require higher rates of application. Since the organisms released into the environment do not self-perpetuate (although some studies have demonstrated germination, vegetative growth, sporulation, and crystal production in living and dead mosquito larvae, under laboratory conditions), frequent reapplication is necessary. The main reason for required frequent application is due to the fact that B.t.i. is adsorbed to suspended silt particles and, organic matter (Rashed and Mulla 1989) and consequently sinks to the bottom of the water body (Margalit and Bobrolgo 1984; Ohana et al. 1987), thus disappearing from the feeding zone of the larval mosquitoes. Good control of *Aedes albopictus* larvae in automobile tires for periods of up to eight weeks has been obtained (Lee and Cheong 1987).

The activity of B.t.i. is influenced by the type of formulation, and by mosquito habitat factors such as pollution, water depth, turbidity, temperature, microflora, presence of larval food and canopy (Lacey and Undeen 1986). Organic material and solids in the water lower the persistence of B.t.i. (Margalit and Bobroglo 1984). Culicine larvae are more susceptible than anopheline larvae. The efficacy of B.t.i. against culicine larvae is highest in habitats with clear, shallow, warm water protected from sunlight. Rapid settling of the formulation but not denaturation of the toxin account for its low residual activity. Flowable concentrates are more effective than wettable powders because they settle less. Encapsulated formulations are even better (Lacey et al. 1984; Margalit et al. 1984).

The mode of action of B.t.i is similar to other B.t. varieties. After ingestion by the larvae, proteolytic enzymes cleave the protoxins to yield the active δ-endotoxin, disrupting the membrane permeability and causing the midgut cells to vacuolize and lyse (Porter et al. 1993). Death usually occurs within the first 20 min after ingestion. Detailed description of the pathogenesis of mosquito midgut cells is given by Charles and de Barjac (1981).

B.t.i. is harmless to humans, animals, fish, plants, and most nontarget organisms. Only filter-feeding chironomids and some other related nematocerous midge larvae were found to be sensitive to B.t.i. at high rates of application (Yiallourous et al. 1995)

Efficacy, specificity, biodegradability, absence of resistance, and long shelf-life makes B.t.i. a desirable agent for mosquito and black fly control in pest management programs.

Bacillus sphaericus (B.s.) is a common bacterial species found in many soils and aquatic habitats. A number of isolates with insecticidal activity toward mosquito larvae have been found. There appear to be two or more toxins associated with the various strains. Larvicidal activity may be due to one or more toxins acting on the larval midgut, however, the exact mode of action is not known.

Many factors influence the larvicidal activity of *B. sphaericus.* Among these, species susceptibility is one of the most striking. For example, *Culex quinquefasciatus* larvae are susceptible to very low concentrations of bacterial spores, while *Aedes aegypti* larvae are virtually unaffected by high concentrations. Another factor involves larval specificity to different strains of B.t. Physical factors such as temperature, sunlight, and pH also affect toxicity. Additionally, organically rich water and natural aquatic microflora reduce larvicidal activity (Yousten 1984). Rapid settling of the toxic components also reduces residual activity. The lack of toxic effects on nontarget organisms including vertebrates, coupled with a high level of insecticidal activity against several culicine and anopheline larvae make *B. sphaericus* an added alternative to chemical control under certain conditions.

Bacillus popilliae, the agent of the milky disease of the Japanese beetle *Popillio japonica,* has been a recognized factor in the beetle's control for many years. The infective spores can persist in the soil in large quantities for long periods of time and can exhibit marked resistance to heat, desiccation, and insolation. Young larvae are the most susceptible stage to the disease.

23.3 Avermectins

Avermectins form a new class of compounds having nematicidal, miticidal, and insecticidal activity (Fig. 23.1). A detailed account of the role of avermectins in arthropod pest control has been published by Lasota and Dybas (1991). They are a mixture of natural products produced by a soil microbe *Streptomyces avermitilis* which was isolated from a soil sample in Japan in 1976. Tests with mice infected with a gastrointestinal nematode demonstrated good antihelminthic activity of avermectins against nematodes at concentrations that were nontoxic to the rodent. Avermectins are effective against all stages of various mites by acting as GABA agonist (see Chap. 10).

The avermectins consist primarily of four major (A_1a, A_2a, B_1a, B_2a) components

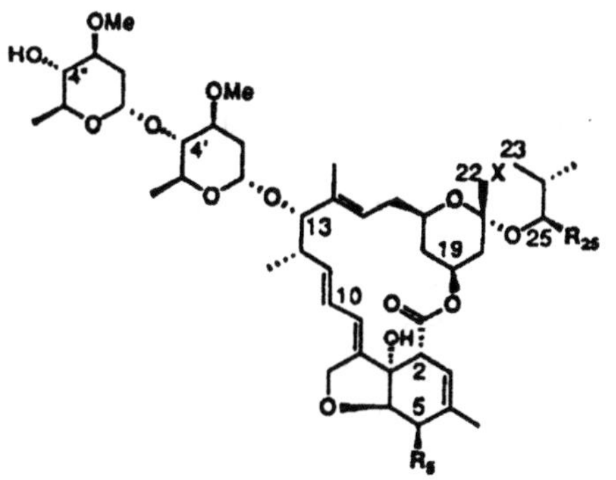

Avermectins

A series: R_5 = OMe
B series: R_5 = OH
1 series: X = CH = CH
2 series: X = CH_2–CH(OH)
a series: R_{25} = s–Bu
b series: R_{25} = i–Pr
Ivermectin
R_5 = OH
X = CH_2–CH_2
R_{25} = s–Bu and i–Pr

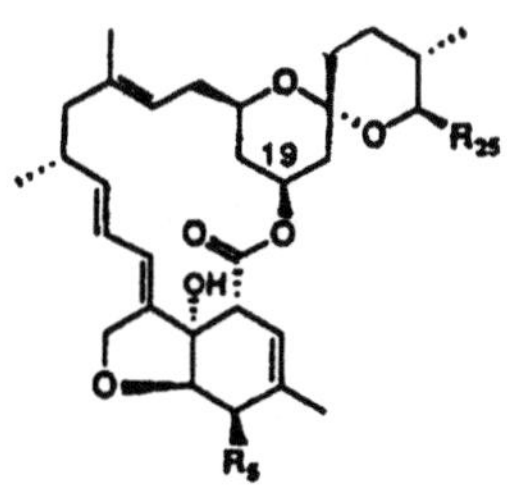

Milbemycins

R_5 = OH
A_3: R_{25} = Me
A_4: R_{25} = Et
D: R_{25} = i-Pr

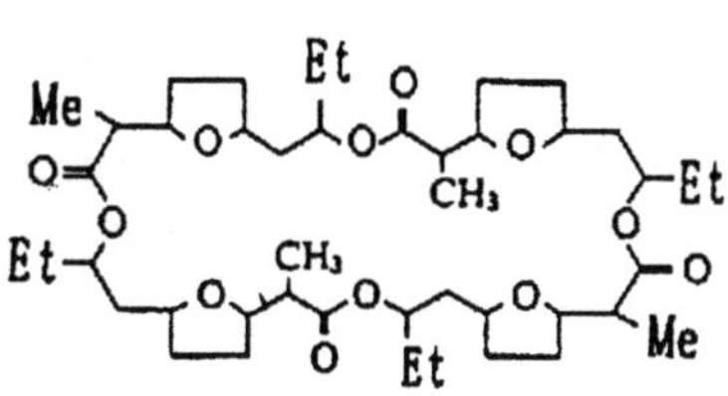

Tetranactin

Fig. 23.1. Structures of avermectins, milbemycins, and tetranactin

and four homologous minor (A_1b, A_2b, B_1b, B_2b) components. The A series has a methoxy group at the C_5 position of cyclohexene moiety, while the B series has an hydroxyl group. In the 1 series a double bond exists between C_{22} and C_{23}, while in the 2 series the double bond is hydrated, having an hydroxyl group at C_{23}. The a series has a secondary butyl group at C_{25}, while the b series has an isopropyl group. The disaccharide substituent at C_{13} contains two indentical α-L-oleandrose units. The B series has more biological activity than the A series.

Abamectin (Avermectin B_1). A mixture of B_{1a} ($\geqq 80\%$) and B_{1b} ($\leqq 20\%$), utilized for control of agricultural and household arthropod pests. It rapidly breaks down in the presence of ultraviolet light. In water, upon exposure to sunlight, abamectin has a half-life of 12 h or less. It is also rapidly photooxidized in organic solvents. In soil exposed to sunlight, abamectin has a half-life of 21 h. It does not bioaccumulate in plants, fish, or in animal tissues, and the degradation products are much less toxic than the parent compound. It is not selectively toxic to birds and other nontarget organisms, but honeybees are very susceptible to abamectin following direct contact, or contact via their food, or their exposure to treated foliage.

Abamectin was found effective in the control of the fire ant *Solenopsis invicta* by causing death or sterility of the queen as well as causing high mortality of workers and brood. Abamectin baits and sprays gave good control of cockroaches *Blattella germanica* in indoor urban situations. It affects nymphs and adults and has an adverse effect on reproduction. Reduced oothecal egg hatch and feeding inhibition are characteristic of its action. In agriculture, the most widespread use of abamectin has been against mites on various food crops and ornamentals. It was found to be effective also against leafminers, psyllids, and lepidopteran pest insects. No injury or slight toxicity to predatory mites has been observed.

Ivermectin (22, 23-Dihydroavermectin). A semisynthetic avermectin B_1 for veterinary and medical use against a variety of helmintic species. It is produced by reduction of the double bond between C_{22} and C_{23}. It is important to control *Onchocerca volvulus*, a microfilaria transmitted by *Simulium* black flies, which causes onchocerciasis (African river blindness).

4"-Epi-methylamino-4"-deoxyavermectin B_1 (EMA). It is produced by introducing an amino substituent at the 4" position in the terminal disaccharide unit of abamectin. Abamectin has a widely different efficacy to lepidopterans, while EMA has a high intrinsic activity against a wide range of species, in some instances over 1500-fold, and is considerably more toxic than commercially used carbamate and pyrethroid insecticides against several pest species.

23.4 Milbemycins

Milbemycins are a mixture of macrocyclic lactones, discovered in Japan in 1967 from a soil actinomycete *Streptomyces hygroscopicus* subsp. *aureolacrimosus* as having remarkable miticidal activity. Milbemycins consist 24 or more components similar to avermectins but without disaccharide moiety (Fig. 23.1). They are effective against all stages of various mites by acting as GABA agonists (see Chap. 10).

Milbemectin. A mixture of milbemycin A_3 and A_4 (ratio: 3/7), is more miticidal among the homologues to all developmental stages of various phytophagous mites such as *Tetranychidae, Eriophysidae,* and *Tarsonemidae* all year-round because the activity is not affected by the ambient temperature. Although nonpersistent, it shows a long-lasting effect on mite population by decreasing oviposition with sublethal doses.

Milbemycin D. It controls fatal canine heartworm disease, caused by *Dirofilaria immitis.*

Mode of action of avermectins. Both are effective to all stages of various mites by acting as GABA agonists (see Chap. 10).

23.5 Polynactins

Polynactins are miticidal antibiotics, produced by *Streptomyces aureus,* which consist of tetranactin, trinactin, and dinactin. The major component is tetranactin. Polynactin acts as an uncoupler of oxidative phosphorylation in mitochondria, apparently forming a complex with potassium ions, and destroys the concentration gradient between intra- and extracellular fluids. The resultant leak is considered to be the cause of the uncoupling (Dekeyser and Downer 1994).

23.6 Fungi

The early history of the discovery of parasitic fungi on insects and the principles underlying their use as insect control agents have been documented by Steinhaus (1956). Approximately 700 species of fungi are pathogenic to insects. They cause mycosis in many different arthropods and in almost every order of insect. They infect all stages of the life cycle of the host and they are commonly found in aquatic, terrestrial, and subterranean habitats. Their mode of infection is different from that of other microbes. Whereas most other pathogens such as bacteria, viruses, protozoa, and the like infect their hosts through the gut following ingestion, fungi penetrate the insect's cuticle; therefore, fungi can infect insects with sucking mouthparts (orders Hemiptera and Homoptera).

Level of infectivity depends largely on host density and on moisture. The life cycle of the fungus begins with spore germination and penetration of the host's cuticle. This is followed by proliferation of fungal cells within the hemocoel resulting in death of the insect. Production of infective spores can be initiated again following death of the insect and the cycle is repeated, or resting spores may be produced which require a period of dormancy; thus they persist in the environment for long periods of time (Carruthers and Hural 1990). The disease of silkworms was shown as early as 1835 to be caused by the fungus *Beauveria bassiana* which also attacks a number of insect pests including mosquito larvae. Recently it was shown to be effective for the biological control of fire ants and several agricultural pests and is produced commercially under the trade names of Biotrol, Boverin, and ABG-6178 (McCoy 1990).

Three genera of fungi, *Coelomomyces, Lagenidium*, and *Culicinomyces,* have demonstrated potential as control agents against mosquitoes. They have the ability to self-perpetuate and are efficacious under field conditions. The last two have an advantage in that they can be cultured on artificial media. A detailed account of the life cycles and infectivity of these three genera of fungi can be found in a review by Lacey and Undeen (1986). The current status of several other species of fungi being tested as potential microbial agents against agricultural pests has been reviewed by McCoy (1990).

Coelomomyces. Coelomomyces are obligate parasites found mostly in mosquito larvae. About 40 species have been described the world over. They are effective as microbial control agents against a broad range of mosquito species. However, most evaluations have been made under laboratory conditions. The most successful field trials have been made with *C. iliensis* in the former USSR (Dubitskii 1985).

A significant breakthrough in the mass production of *Coelomomyces* (at least in one species) was the discovery that a copepod intermediate host (*Cyclops vernalis*) is required for completion of the life cycle. After penetration of the fungus into the mosquito, hyphal bodies develop in the hemocoel. Subsequently, extensive growth of mycelia fills the hemocoel and spores are formed at their tips. As the dead larva decomposes, the spores are released into the medium and these invade the intermediate host. After several stages of development, thousands of haploid uniflagellate gametes are released which mate and fuse to produce the mosquito-infective diploid zygote. Infected blood-fed adult females of *Aedes trivittatus* were unable to oviposit. This fungus shows little or no potential for control of black flies.

Lagenidium giganteum. *Lagenidinm giganteum* is a facultative pathogen of mosquitoes with both sexual and asexual cycles. The pathogen is effective in suppressing several species of mosquitoes under field conditions with the benefit of recycling and persisting at certain levels throughout the mosquito breeding season. In some cases it has maintained endemic infection for several years, for example, in Californian rice fields that were inoculated. Successful mosquito control with *L. giganteum* depends on the type and quality of water in the mosquito habitat. High (>35 °C) and low (<15 °C) temperatures, salinity and low levels of organic matter inhibit the production of spores and larvicidal activity. Younger larvae are considerably more susceptible than older instars, and complete elimination of the target population without a subsequent hatch within 1–2 days could prevent recycling. Several insecticides such as methoprene, malathion, fenthion, and temephos are not toxic to *L. giganteum* at recommended rates and would be compatible with the fungus in pest management programs.

Culicinomyces clavisporus. ***Culicinomyces clavisporus*** has been isolated from laboratory mosquito colonies and from mosquito natural habitats. This fungus has a broad range of mosquito hosts and is the easiest to culture and mass produce on artificial media. Several factors influence the level of pathogenicity of *C. clavisporus* to mosquito larvae. Species and age of larvae have a distinct effect on the degree of susceptibility. Anophelines are less susceptible than culicines. Apparently, feeding behavior plays a role in the quantity of conidia that are ingested; anophelines filter feed from the surface of the water and, most likely, have little or no contact with conidia that settle beyond their feeding zone. Culicines, on the other hand, feed deeper into water and come in greater contact with fungal conidia. Early instar larvae are more susceptible than older instars. Temperature influences germination and growth of the fungus. Infection of larvae by the fungus will readily take place at temperatures between 15 and 27.5 °C but not at 30 °C. High salinity and organic pollution can inhibit infection of larvae.

C. clavisporus has been screened against a large number of nontarget organisms. Although *Nematocera* and mosquitoes are the most susceptible hosts, certain other dipterous larvae as well as certain simuliids, chironomids, chaoborids, and ceratopogonids are also affected. However, strains of the fungus are not sufficiently pathogenic to black fly larvae to warrant their consideration as microbial control agents for these insects. There is no evidence that the fungus is pathogenic to vertebrates. A disadvantage of *C. clavisporus* is the high dosage required for effective control, and persistence of infective levels of the fungus in the environment is short-lived.

A detailed account of the life cycles and infectivity of these three genera of fungi can be found in a review by Lacey and Undeen (1986). The current status of several other species of fungi being tested as potential microbial agents against agricultural pests has been reviewed by McCoy (1990).

23.7 Nematodes

Several species of nematodes pathogenic to insects are highly regarded as biological control agents because they possess a broad range of insect hosts, are safe to vertebrates and plants, are highly virulent, killing their hosts within 48 h, can be easily and inexpensively mass-produced, are easily applied with conventional equipment, and are compatible with other control strategies. Nematodes are best suited for control of soil-inhabiting insects. They occur naturally in many soils and have the advantage of great motility and attraction to hosts (Kaya 1990). Improved efficacy by genetic manipulation, improved formulations, and standardization of bioassay procedures for virulence are currently under study.

23.8 Viruses

Virus diseases of insects have been known for a long time, the first record being a virus disease of silkworms observed in 1527. Over 700 species of insects and mites are reported to have viral diseases, and more than 320 viruses have been isolated from many insect and mite species of agricultural importance, representing important pests of food, fiber, and forest crops. Arthropod viruses are represented by six genera: ***Baculovirus***, comprising the nuclear polyhedroses (NPVs) and granuloses (GVs); ***Poxvirus; Iridiovirus; Parovirus; Enterovirus;*** and ***Rhabdovirus***. The species of ***Baculovirus*** are the best known arthropod viruses and some NPVs have been used for pest control on a commercial basis (such as VIRON/H or ELCAR), especially against lepidopteran pests of agriculture and forestry. A list of viruses successfully used for biological control of insects was published by Fuxa (1990). Most of the above viruses have also been isolated from a large number of mosquito species (David 1975). However, their potential in the biological control of mosquitoes has not been ascertained.

Until recently, all the viruses used in field tests or sold commercially have been produced from living insects. In the past few years progress has been made in obtaining good yields of polyhedral viruses produced in tissue culture. Research results have indicated that arthropod viruses: (1) are selective, safe, environmentally nondisruptive control agents that are suitable for use in pest management programs; (2) may aid in the control of certain pests in cases where the insecticide is no longer effective, or in cases in which the use of an effective insecticide is restricted; and (3) could help reduce the current heavy reliance on chemical insecticides (Falcon 1976).

Baculoviruses. The NPV and GV baculovirus subgroups have received the most attention as viral insecticides for several reasons: viruses of the family Baculoviridae cause lethal infections only in invertebrates and are regarded as inherently safer than other insect viruses that are more apparently related to vertebrate or plant viruses; members of the Baculoviridae usually have a relatively narrow host range; some baculoviruses are highly pathogenic in insects and can produce sufficient progeny virus per insect; the virus particles of NPVs and GVs are occluded in proteinaceous crystals, thus making them more stable in the environment and compatible with pesticide formulations (Miller et al. 1983).

Numerous observations have been made on the persistence of baculoviruses in the ecosystem. In soil, they are resistant to leaching and can persist from one season to the next. When exposed to foliage, they are rapidly inactivated by ultraviolet light. For field use, these viruses are formulated with various additives that screen ultraviolet light. One of the most effective additives is activated carbon.

Recombinant Baculoviruses. The major limitation to the use of baculoviruses as biological insecticides is their slow kill. Such slow kill is often of benefit to the virus, which leads to the production of a large amount of virus for subsequent infection. However, extensive damage in annual crops often occurs before the infected insect dies. The use of recombinant DNA technology has made it possible to insert a gene into virus or delete a gene from virus. Recombinant viruses speed the death of pest insects and reduce crop damage by the expression of toxic peptides or proteins or by expressing one of the pest insect's own proteins at an inappropriate time (Hammock 1993). An overview of arthropod toxins for potential use in recombinant baculoviruses was made by Zlotkin (1991).

CHAPTER 24

Miscellaneous Compounds

24.1 Arsenicals

Arsenic occurs in two valence forms: trivalent (As_2O_3) reacting with water to give arsenious acid and forming arsenite salts, and pentavalent arsenic (As_2O_5) yielding arsenic acid and the corresponding arsenate salts. The arsenite compounds are more toxic to insects, mammals, and plants. Paris green (a complex between copper acetate and copper arsenite) was used in the USA since 1867. It is quite toxic to plants and animals; therefore, its use is limited to insect poison baits. Lead and calcium salts of arsenic have been the commonest arsenicals in use. Acid lead arsenate ($PbHAsO_4$) and basic lead arsenate [$Pb_2(AsO_4)_2$] are no longer used in the USA, but they still find some use in other countries for control of various beetles, weevils, leaf rollers, codling moths, and other insects damaging fruit trees. Several organoarsenicals are used as herbicides. In view of the high mammalian toxicity of lead, preparations of calcium arsenate are often preferred. Environmental pollution by lead presents a serious problem since lead is a cumulative poison.

The arsenicals are stomach poisons that release water-soluble, toxic, inorganic arsenic ions after ingestion by insects. A large portion of the ingested material goes through the gut unchanged and is excreted or regurgitated. Toxic symptoms in mammals include abdominal pain, vomiting, and rapid fall in blood pressure leading to a state of shock. In lethal cases, death may be delayed by up to 14 days. Sublethal effects include polyneuritis with pain and tenderness of the limbs (O'Brien 1967; Hayes 1982; Hayes and Laws 1991). In insects poisoned by arsenic, eating stops, regurgitation occurs, and activity diminishes. Death occurs due to inhibition of respiratory enzymes. The mode of action of arsenical compounds is similar to that of rotenone in that they inhibit respiration. In addition, they interact with various enzymes containing SH groups, and they cause gross protein precipitation. Due to the highly poisonous nature of arsenic and the potential dangers of environmental contamination, sodium arsenite has been banned and the use of inorganic arsenicals has markedly diminished. A comprehensive review of arsenical pesticides has been published by Woolson (1975).

24.2 Fluorine Compounds

24.2.1 Inorganic Fluorine Compounds

Inorganic fluorines are generally stomach poisons. The degree of their toxicity is related to their fluorine content. The important series of fluorine salts which have been used as insecticides comprise the fluorides (as sodium fluoride, NaF), the fluosilicates (as sodium fluosilicate, Na_2SiF_6), and the fluoaluminates (as sodium fluoaluminate or cryolite, Na_3AlF_6).

Sodium fluoride is soluble in water to the extent of 4.3% at ordinary temperatures. The pure compound is a white powder, but, in actual use, federal regulations require it to be colored blue or green to reduce the possibility of its being confused with edible or medicinal substances, as has happened in some instances. Before World War II it was used to a large extent for the control of cockroaches, ants, and other household pests, and against biting lice (Malophaga) which infest poultry. It is not used on plants because it is phytotoxic. Sodium fluoride is toxic to all forms of life. It has been used as an insecticide, rodenticide, herbicide, and wood preservative. The oral LD_{50} to rats is 180–200 mg/kg. For adult man death occurs by ingestion of about 4000 mg of sodium fluoride (about 57 mg/kg in man) (Peters 1948; Sollman 1957; Hayes and Laws 1991).

Sodium fluosilicate has been used considerably in poison baits as insecticide against cockroaches, grasshoppers, and crickets, and as a mothproofing agent. It is highly phytotoxic and cannot be applied to foliage. The oral LD_{50} to rats is 125 mg/kg. Sodium fluoaluminate or cryolite occurs either as a natural mineral or in the synthetic form. It has been used against a wide variety of agricultural insects but is not effective in poison baits. The oral LD_{50} to rats is 200 mg/kg.

24.2.2 Organic Fluorine Compounds

Fluoroacetate (FCH_2COOH). In the early 1930s sodium fluoroacetate was reported to be a systemically active compound in plants against the bean weevil, that is, when applied to the roots or to the leaves it was transported to other parts of the plant. Its high mammalian toxicity prevented its widespread usage as an insecticide, but under the designation of compound 1080 it was widely used and is still being used as a rodenticide. During the years of World War II the Germans were searching for toxicants to be used as warfare agents but for which no antidote was available. One such compound was fluoroacetic acid which was found to be the active principle of a poisonous South African plant ***Dichapetalum cymnosum*** or gifblaar. As a consequence, the biochemical basis of fluoroacetate poisoning was discovered and was termed "lethal synthesis" (Peters 1963a,b).

Sodium fluoroacetate (FCH_2COONa) is an odorless, white, nonvolatile powder, very water soluble and hygroscopic, but of low solubility in alcohol, acetone, and petroleum oils. It is formulated as an aqueous solution containing 0.5% dye as a warning color. It is a quick acting poison causing symptoms after a short duration of 20–60 min exposure. It is rapidly absorbed by the gastrointestinal tract. It is not well absorbed by the intact skin, but absorption is greater in the presence of dermatitis or other skin injury. The mode of action of fluoroacetate is similar both in mammals and insects (Peters 1963b; Matsumura and O'Brien 1963).

Fluoroacetate was found to have no effect on all enzymes tested in vitro. Its toxicity in vivo is due to its conversion in the body to fluorocitric acid. With the mediation of acetyl coenzyme A and condensation with oxaloacetate, fluoroacetate is converted to fluorocitrate which is the toxic principle, hence the term "lethal synthesis." Fluorocitrate inhibits the enzyme aconitase in the tricarboxylic acid cycle (Krebs cycle). The enzyme aconitase is involved in the metabolism of citrate to isocitrate; hence, its inhibition causes the accumulation of citrate in the tissues resulting in the blockage of the Krebs cycle and a marked reduction in oxygen consumption. The toxic effect may not be directly due to the accumulation of citrate per se, but rather to the blockage of energy metabolism. As might be expected, the heart and central nervous system are the most critical tissues involved in poisoning by fluoroacetate due to general inhibition of oxidative energy metabolism. Hence, the symptoms following fluoroacetate poisoning, in addition to nausea and vomiting, include cardiac irregularities, cyanosis, convulsions, and death from ventricular fibrillation and respiratory failure. Estimates of the mean lethal dose of fluoroacetate in man range from 2 to 10 mg/kg, and there have been a number of human fatalities, mostly children (Hayes and Laws 1991). The oral LD_{50} to rats is 0.22 mg/kg and there are qualitative and quantitative species differences among animals (see Table 24.1).

Monoacetin injected intramuscularly, as well as treatment with acetamide

Table 24.1. Single dose LD_{50} for sodium fluoroacetate

Species	Route	LD_{50} (mg/kg)	Reference
Rat	Oral	0.22	Dieke and Richter (1946)
Rat	Oral	2.5	Ward and Spencer (1947)
Rat	Oral	1–2	Phillips and Worden (1957)
Rat	Intraperitoneal	3–5	Ward and Spencer (1947)
Mouse	Subcutaneous	19.3	Hutchens et al. (1949)
Mouse	Subcutaneous	17.0	Tourtellotte and Coon (1951)
Mouse	Intraperitoneal	10.0	Ward and Spencer (1947)
Mouse	Intraperitoneal	16.5	Tourtellotte and Coon (1951)
Mouse	Intraperitoneal	14.7	Raasch (1958)
Guinea pig	Oral	0.4	Ward and Spencer (1947)
Guinea pig	Intraperitoneal	0.37	Hutchens et al. (1949)
Rabbit	Subcutaneous	0.28	Hutchens et al. (1949)
Dog	Oral	0.06	Tourtellotte and Coon (1951)
Cow	Oral	0.39	Robinson (1970)
Calf	Oral	0.22	Robinson (1970)
Opossum	Oral	0.79	Bell (1972)
Mallard duck	Oral	4.8	Hudson et al. (1972)
South African clawed toad (*Xenopus laevis*)	Oral	500	Chenoweth (1949)

injected intravenously as a 10% solution in glucose, can afford protection against poisoning in man. A 10% solution of ethanol in normal saline administered subcutaneously at a rate of 800 mg/kg gave protection against poisoning in rodents but was not effective as antidote in dogs and monkeys (Hutchens et al. 1949; Chenoweth et al. 1951). Commercially available monoacetin containing about 60% glycerol monoacetate was found to be superior to any other substance tested in mice, rats, rabbits, dogs, and rhesus monkeys as therapy against fluoroacetate poisoning (Chenoweth et al. 1951). Barbiturates may be of help in reducing convulsions.

2-Fluoroacetamide (FCH_2CONH_2). Fluoroacetamide, also known as compound 1081, is a crystalline solid, freely soluble in water, moderately soluble in acetone, and little soluble in chloroform. At one time fluoroacetamide was used as a systemic insecticide for scale insects, aphids, and mites on fruit trees. It is now used mostly as a rodenticide, but still finds some use as an insecticide in Britain and Japan. Fluoroacetamide is a compound of moderate to high acute toxicity depending on the species (see Table 24.2). It is absorbed by the skin. Animals poisoned by this compound show listlessness, irritability, clonic convulsions, and irregular respiration (Araki 1972).

There is general agreement that fluoroacetamide is less toxic than fluoroacetate. This is due to the fact that fluoroacetamide is converted in the body to fluoroacetate and this conversion is rather slow (Matsumura and O'Brien 1963). However, Phillips and Worden (1957) reported that rats treated with fluoroacetamide excreted 62% of the dose in the urine unchanged, and concluded that this compound may have inherent toxicity of its own and does not depend entirely on its conversion to fluoroacetate. Fluoroacetamide has a specific effect on spermatogenesis causing degeneration of the testis in rats and other species (Egyed 1973). Evidence that the mode of action of fluoroacetamide is the same as that of fluoroacetate is the finding that mammals poisoned with fluoroacetamide contain large amounts of citrate in the tissues (Matsumura and O'Brien 1963; Egyed and Brisk 1965; Allcroft et al. 1969; Egyed and Miller 1971; Egyed and Shlosberg 1973). The same is true for insects (Matsumura and O'Brien 1963).

Table 24.2. Single dose LD_{50} for fluoroacetamide

Species	Route	LD_{50} (mg/kg)	Reference
Rat	Oral	15	Phillips and Worden (1957)
Rat	Oral	13	Bentley and Greaves (1960)
Rat	Dermal	20[a]	Phillips and Worden (1957)
Mouse	Oral	31	Araki (1972)
Mouse	Subcutaneous	34	Araki (1972)
Mouse	Intraperitoneal	85	Matsumura and O'Brien (1963)
Rabbit	Oral	1.5–2.0	Phillips and Worden (1957)
Rabbit	Intravenous	0.25	Buckle et al. (1949)
Chicken	Oral	4.25	Egyed and Schlosberg (1977)

[a]Lowest lethal dosage.

Acetamide and L-cysteine hydrochloride were found to be useful antidotes if given within 20–60 min after poisoning, depending on the animal species (Phillips and Worden 1957; Egyed 1971; Egyed and Shlosberg 1977). Treatment of poisoning in man is the same as that with fluoroacetate. Removal of the poison as fast as possible and treatment with acetamide is recommended.

24.3 Organothiocyanates

Lethane 384 was the first widely used synthetic organic insecticide which was marketed in 1932 as a rapid knockdown agent for the control of household insects. A wide variety of organic compounds incorporating the thiocyano group (-SCN) or the isothiocyano group (NCS) were shown to have useful insecticidal properties. In addition to Lethane 384, lauryl thiocyanate and isobornyl thiocyanoacetate (Thanite) have been marketed as insecticides (Metcalf 1971).

Mode of Action These compounds have a rapid paralytic action on insects which is attributable to the thiocyano group. The activity depends on the length of the alkyl chain, maximum activity being attained with lauryl thiocyanate (C_{12} chain length) beyond which the activity drops rapidly. Certain organothiocyanates release hydrogen cyanide by metabolism with glutathion S-transferases-glutathion system (GST-GSH), which contributes to the insecticidal activity (Ohkawa et al. 1972).

$$RCH_2SCN \xrightarrow[GSH]{GST} HCN + [RCH_2SSG] \xrightarrow[GSH]{} RCH_2SH + GSSG\,.$$

Toxicology. The thiocyanates are of relatively low toxicity to mammals with oral LD_{50} values to the rat of 250 mg/kg for Lethane and 1600 mg/kg for Thanite. Being somewhat volatile they should always be used with good ventilation.

Uses. The thiocyanates are used principally as household aerosols and sprays for control of mosquitoes, flies, and cockroaches, and as dairy and livestock sprays for fly control. Their rapid knockdown makes them particularly useful for control of flying insects, but they tend to have unpleasant odors (Metcalf 1971). In the last decade the thiocyanates have declined in appeal and have largely been replaced by the synthetic pyrethroids.

CHAPTER 25

Molluscicides

Molluscicides are compounds used to control snails and slugs in vegetables, flower gardens, and citrus groves. Medically, aquatic snails of several genera such as *Bullinus, Oncomelonia*, and *Ostralorbis* are extremely important because they act as intermediate hosts for *Schistosoma* parasites that cause the disease schistosomiasis (bilharzia) in humans. Other snails serve as alternate hosts of parasitic lung and liver flukes of humans, dogs, cats, and domestic animals.

There are specific pesticides that control these snails and slugs: metaldehyde is a specific attractant and toxicant for garden snails and slugs and is used at 1.5% with moist bran as bait. It is not soluble in water but is soluble in organic solvents. Calcium arsenate, sodium arsenate, copper sulfate, carbon disulfide, and several insecticides are often added to improve the toxicity of the bait. Metaldehyde has been used commercially and around the home since its discovery in 1936. Clonitralid (Baylucide) is one of the most promising moluscicides discovered. It is especially toxic to freshwater snails that serve as intermediate hosts for the organisms that transmit bilharzia and persists for several months. It is also useful as a fish poison, but it has undesirable toxic side effects on other aquatic organisms such as frogs, crayfish, and clams. It has no effect on plankton and aquatic vegetation.

Triphenyltin acetate and other trialkyltins are very effective against aquatic snails and their eggs at concentrations of 0.1–0.4 ppm and they persist in the environment for several months. However, they are toxic to fish, amphibia, and other aquatic life. Pentachlorophenol (PCP) is also used as a molluscicide in Egypt to control aquatic snails. Other molluscicides used to control snails and slugs around the home include the carbamates methiocarb (Mesurol), carbaryl (Sevin), and mexacarbate (Zectran, production discontinued). The organophosphorus compound azinphosmethyl (Gusathion) is highly effective in controlling the European brown scale *Helix aspersa* on citrus plants, but is not effective against the common garden slug.

CHAPTER 26

Avicides

The vast majority of the more than 8000 species of birds are very beneficial and a great pleasure to observe. Some species, however, can seriously compete with human interests. The areas of conflict include:

1. Destruction of agricultural foodstuffs and depredation of crops;
2. Contamination of foodstuffs or defacing buildings with their feces;
3. Transmission of diseases, directly and indirectly, to man, poultry, and dairy animals;
4. Creating hazards at airports and freeways;
5. As a general nuisance affecting human comfort.

Three birds in particular are of interest to public health. These are the pigeons, starlings, and English sparrows. These three birds create health hazards because of their potential for transmission of many microbial, parasitic, and viral disease to humans and domestic animals, the pollution they produce, and the great economic losses they can cause. Some of the bacterial diseases include fowl cholera, fowl typhoid, paratyphoid, salmonellosis, spirochetosis, tuberculosis; mycotic diseases include aspergillosis, blastomycosis, histoplasmosis, and others; protozoal diseases include American trypanosomiasis, coccidiosis, toxoplasmosis; viral diseases include encephalitis, Eastern equine encephalitis, St. Louis encephalitis, West Nile encephalitis, meningitis, and gastroenteritis. In addition, these birds transmit rickettsial, parasitic trematode, and parasitic cestode diseases. Associated with these birds also are about 50 species of insects, mites, and ticks capable of transmitting disease. A complete list and detailed descriptions of the diseases transmitted as well as the biology and control of these birds are given by Weber (1979).

The basic requirements for effective chemical control of pest birds are:

1. The chemical must pose no hazard to humans and nontarget animals or birds.
2. It must be selective for the species to be controlled.
3. It must not produce secondary hazard to animals or birds which consume the affected birds.
4. It should act rapidly and with a minimum amount of pain to the target pest bird (Weber 1979).

The methods of control include repellents, perch treatments (sticky chemicals on ledges and roosts), toxic baits, soporifics (stupefacients), and chemosterilants (Ware 1983).

26.1 Repellents

Probably, the most prominent of the avicides is Avitrol (4-aminopyridine). The repellent effect results from the distress calls made by the birds. This material has a low LD_{50} of 20 mg/kg, hence some mortality of birds is inevitable. The treated corn kernels must be eaten to be effective, and only a small number of birds need be affected to alarm the rest of the flock. After one alarming exposure, the birds will not return to the treated area. Methiocarb (Mesurol), a carbamate insecticide, is also used as a bird repellent on cherries, blueberries, grapes, and some cereal crops. The chlorinated cyclodiene insecticide endrin has been used for many years as an elevated, out of reach, perch treatment for control of pigeons, starlings, and English sparrows. It should be used only by licensed pest control operators trained in bird control. Another compound is the organophosphorus insecticide fenthion. This compound acts quickly after its absorption through the feet. It should be applied only by licensed operators. Glucochloralose is registered as a bird repellent and hypnotic. The affected birds are stunned and unable to fly for several hours.

Chemically treated baits work well in areas where pest birds are actively feeding. Starlicide (3-chloro-p-toluidine hydrochloride) is a slow acting avicide against starlings and blackbirds. It is not effective against English sparrows. Since the chemical acts slowly, large numbers of dead birds do not appear in the treated area but rather die in flight or in their roosts. The old general-purpose poison strychnine is used to control pigeons and English sparrows. It is highly toxic: the LD_{50} is 4 mg/kg for English sparrows and 8 mg/kg for pigeons. It is fast acting, the normal time for mortality being from 12 to 30 min. Prebaiting for several days prior to application of the poison is necessary.

Toxic baits are hazardous to the applicator, to other individuals and to nontarget species and should be used sparingly and with great caution.

26.2 Chemosterilants

Ornitrol is a pigeon chemosterilant designed to control population growth rather than eradication. It was found effective in inhibiting ovulation and inhibiting reproduction for about 6 months, thus limiting the population size without injuring the birds. The pigeons die of old age, but during their lifetime they are able to transmit disease and cause pollution. The chemical is selective for pigeons and does not affect mammals. It is impregnated on whole corn kernels, too large for other birds to feed on, and is formulated as a 0.1% bait.

CHAPTER 27

Rodenticides

Rodenticides are substances used to control rodent pests such as rats, mice, gophers, ground squirrels, and others. They consist of a wide variety of chemicals differing in their chemical structure, biological activity, and mode of action. Rats and mice compete with man for food. In some developing countries, where it is necessary to store grain in the open, as much as 20% of the grain is consumed by these mammals. Rats are also important pests in various field crops. In addition to widespread destruction of food and fiber, rodents are also intermediate hosts in the transmission of vector-borne diseases such as bubonic plague, leishmaniasis, leptospirosis, tick-borne encephalitis, tularemia, and others. Rodents also do other damage mainly by gnawing. Although rodents resemble man physiologically and biochemically, certain of their physiological characteristics differ from those of primates and other animals. The degree of selectivity in favor of man and domestic animals is reflected in part in the inability of rats to vomit. Thus, many stomach poisons that have emetic properties are likely to be more toxic to rats than to other animals.

Since rodenticides can be used in baits and placed in inaccessible locations they do not contaminate the environment and they pose little or no hazard to man if used with caution and according to instructions. Toxicological problems arise mainly from accidental or suicidal ingestion of the toxicants. Historically, plant toxins such as strychnine and red squill have been used as rodenticides for many years. Added to the list were inorganic compounds such as white arsenic, yellow phosphorus, barium carbonate, zinc phosphide, and thallium sulfate. More recently, synthetic organic rodenticides have proven more effective and specific and have largely replaced the older compounds (Fig. 27.1).

27.1 Plant-Derived Rodenticides

Red Squill. Mediterranean red squill is obtained from the bulbs of the plant *Urginea (= Scilla) maritima*. It has been used for many years as a relatively safe rodenticide. The plant contains several cardiac glycosides of which scillirocide is highly toxic to rats with an LD_{50} of 0.43 and 0.70 mg/kg to females and males, respectively. Symptoms which are associated with ingestion of large doses of red squill include vomiting and abdominal pain, blurred vision, cardiac irregularity, convulsions, and death from respiratory failure. The emetic action of scillirocide

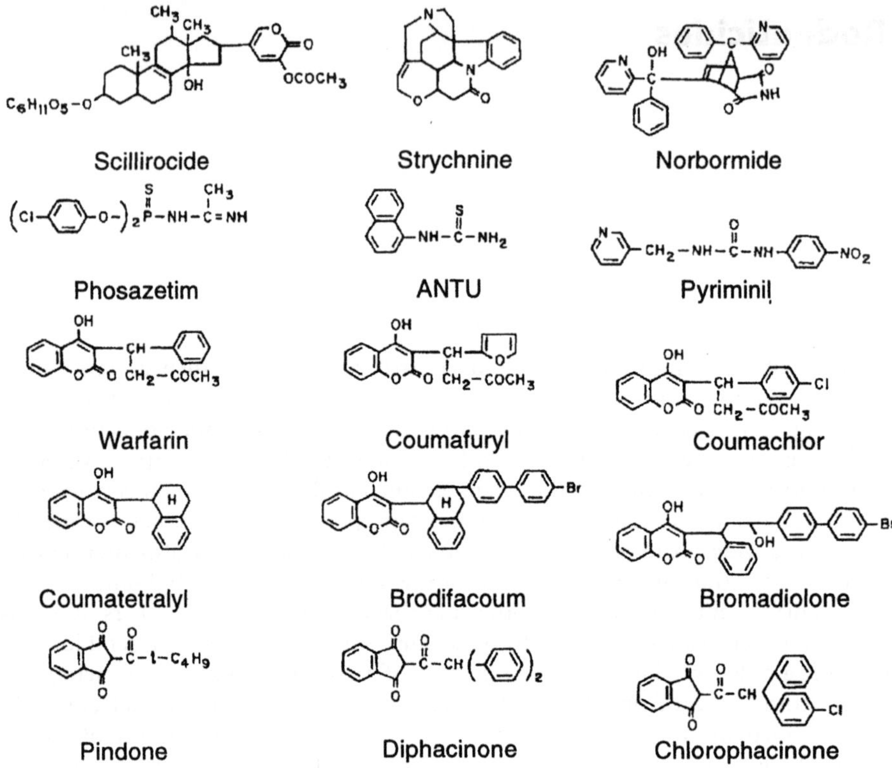

Fig. 27.1. Organic rodenticides

which reaches a concentration of 0.037% in red squill is the main factor that contributes to the safety of the rodenticide to humans. The specific toxicity of this poison to rats, therefore, is due to the rat's inability to vomit.

Strychnine. Strychnine is an alkaloid obtained from the seeds of the Asiatic plant *Strychnos nux-vomica*. It is converted to strychnine sulfate for use as a rodenticide. Strychnine is highly effective against mice and rats (oral LD_{50} for rats is 16 mg/kg), but the compound is generally toxic to all warm-blooded animals. It acts by paralyzing specific muscles, resulting in cessation of breathing and cardiac arrest. It has been largely replaced by other rodenticides (see also Chap. 26).

27.2 Inorganic Rodenticides

Phosphorus. Phosphorus comes in two forms: the relatively harmless red and the highly toxic white or yellow phosphorus. White or yellow phosphorus has caused many poisonings because of the practice of spreading a paste containing this element on bread as a rodenticide bait (Lisella et al. 1971). A dose of 15 mg phosphorus can cause severe poisoning in humans and 50 mg can be fatal. Shortly after ingestion, phosphorus causes severe gastrointestinal irritation with hemorrhage and cardiovascular failure which can be fatal within 24 h. If a person survives the initial poisoning effect, liver necrosis and kidney damage may ensue. Phosphorus causes vomiting in warm-blooded animals but not in rats because rats are unable to vomit. Phosphorus has been largely replaced by safer rodenticides.

Barium Carbonate ($BaCO_3$). Barium carbonate has been used as a rodenticide in the past but is not currently used for that purpose. Barium carbonate is a white powder insoluble in water but soluble in hydrochloric acid in the stomach. It is readily absorbed from the gasrointestinal tract. Barium stimulates striated, cardiac, and smooth muscle and may cause vasoconstriction, tremors, cramps, and arrhythmia. Treatment of poisoning includes induced vomiting or gastric lavage followed by administation of sodium or magnesium sulfate (30 g in 250 ml water) and artificial respiration, preferably with oxygen.

Arsenic Trioxide (As_2O_3). Arsenic trioxide is a powerful compound for controlling rodents. For more details see Chapter 24.

Zinc Phosphide (Zn_3P_2). Zinc phosphide is used against mice, rats, gophers, squirrels, and prairie dogs. It is highly toxic to mammals and birds. It has an unpleasant garlic odor which, apparently, is not offensive to rodents. The rat oral LD_{50} is 45 mg/kg. Zinc phosphide reacts with water and HCl in the gastrointestinal tract to produce phosphine gas (PH_3) which causes severe gastrointestinal irritation. In the presence of moisture it evolves phosphine gas which, if inhaled in sufficient quantity, can cause fatal pulmonary edema. Insensitivity of dogs and cats to this compound may be due to the emetic qualities of zinc. Treatment of poisoning is symptomatic (Hayes 1982).

Thallium Sulfate (Tl_2SO_4). Thallium sulfate has been used for a long time as a rodenticide and also to control some bird pests, squirrels, and prairie dogs. It is a general cellular toxicant being lethal to most animals in doses of 10–20 mg/kg. It acts by inactivating some vital enzymes, especially those containing sulfhydryl (SH) groups. The toxicant is readily absorbed through the skin as well as by the respiratory and gastrointestinal tracts. It has cumulative, secondary, and acute toxicity. Acute poisoning is accompanied by gastrointestinal irritation, motor paralysis, and death from respiratory failure. The most serious effect is on the nervous system. Sublethal doses cause reddening of the skin, loss of hair, and pathologic changes in brain, liver, and kidneys. The oral rat LD_{50} is 15.8 ± 9.0 mg/kg (Dieke and Richter 1946). Antidotes for thallium sulfate poisoning in man are discussed by Hayes and Laws (1991).

27.3 Anticoagulant Rodenticides

Spoiled sweet clover (*Meliotus* spp.) has long been known to be highly poisonous to cattle. In 1941, Link of the University of Wisconsin, USA identified the active ingredient in "sweet clover poisoning" as dicoumarol (Link 1944). The compound was synthesized in 1948 but was found to be impractical for use as a rodenticide. Of the many analogs of dicoumarol which were synthesized and screened for toxicity, warfarin was found to be the best candidate as a rodenticide and an anticoagulant drug. The anticoagulant rodenticides are classified as coumarin and indandione types. The coumarin type compounds are one of the most successful groups and are represented by the classical compound warfarin and several others.

27.3.1 Coumarin-Type Anticoagulant Rodenticides

Warfarin. It has been one of the most widely used rodenticides. Its safe usage is due to the fact that it requires repeated doses for toxicity to develop. Hence, when placed in baits accessible only to rodents, repeated ingestion results in fatalities to rodents with little likelihood that pets and children would be repeatedly exposed. Rats are very susceptible to this toxicant, the LD_{50} being 10 mg/kg, and they do not develop "bait shyness." However, there have been reports of rats developing resistance to warfarin (Hayes and Laws 1991). Warfarin acts as an anticoagulant. It interferes with the action of vitamin K in the body, and inhibits the synthesis of prothrombin, the blood clotting factor. Multiple doses are required until prothrombin levels are depleted to result in hemorrhage, which is the cause of death. In addition to its anticoagulant action, warfarin causes direct blood capillary damage and enhances the hemorrhage throughout the body. Symptoms of poisoning in man which begin after a few days or weeks of repeated ingestion include pallor, bleeding gums, rash leading to hematomas, blood in the feces and urine, and sometimes paralysis due to cerebral hemorrhage. Markedly reduced prothrombin activity is the principal diagnostic test for repeated ingestion of warfarin. Therapy is directed by administration of vitamin K_1 (Hayes and Laws 1991).

Coumachlor was introduced in 1953 but was not successful in competing with warfarin. Coumatetralyl was developed in Germany and introduced in 1957 achieving some success in cases where rats had already shown bait shyness to warfarin. Coumafuryl is very similar to warfarin and was the last of the early coumarin compounds introduced as an anticoagulant rodenticide. Two newer coumarin-type rodenticides were introduced later, namely brodifacoum (Talon) and bromadiolone (Maki). Although these are similar in action to other anticoagulants, their advantage is that only one dose is sufficient to kill the rats within 4–7 days after ingestion. These compounds are very useful against rats that have developed resistance to other anticoagulants. Rat oral LD_{50} for brodifacoum is 0.27 mg/kg; for bromadiolone the rat oral LD_{50} is 1.125 mg/kg. Antidote for both compounds is vitamin K_1.

27.3.2 Indandione-Type Anticoagulant Rodenticides

Three compounds, pindone, diphacinone, and chlorophacinone, belong to the indandione-type anticoagulants which differ chemically from the coumarin-type compounds. Pindone which appeared in 1942 was the first truly anticoagulant rodenticide. It is odorless and tasteless, but rodents develop bait shyness after a number of feedings. Several feedings are required to kill rodents. The rat oral LD_{50} is 50 mg/kg. Diphacinone was introduced in the early 1950s and was the first anticoagulant to require only one dose to kill rats. In most cases, however, two or three feedings are necessary with death occurring in 5–7 days. The acute rat oral LD_{50} is 1.86–2.88 mg/kg, whereas that for mice is 340 mg/kg, and for rabbits 35 mg/kg. Like with other anticoagulants, rats develop bait shyness to diphacinone. Chlorophacinone was introduced in 1961. It does indeed require only a single dose of a bait containing 0.005% chlorophacinone to control rats 5 days after feeding. It does not induce bait shyness as do the others. In addition to its anticoagulant action, chlorophacinone inhibits energy production which contributes to its single dose effectiveness. The oral rat LD_{50} is 2.1 mg/kg.

27.4 Other Rodenticides

Norbormide. Norbormide is a highly selective rodenticide, lethal to rats but much less toxic to other rodent species, and is essentially nontoxic to cats, dogs, chicken, ducks, swine, sheep, primates, etc. The oral LD_{50} for Norway rats ranged from 5.3 to 15 mg/kg while that of mice was 2200 mg/kg and that of rabbits about 1000 mg/kg. This compound acts directly on the smooth muscle of peripheral vessels causing irreversible constriction, resulting in the slowing of blood flow and death.

Organochlorine Compounds. DDT 50% dust was used for many years by pest control operators as a tracking powder. The dust was sprinkled in the runways of mice, and, after tracking through the dust, the mice stopped to preen themselves and clean their feet, thus obtaining a lethal dose of DDT. Convulsions, paralysis, and death ensued. DDT was especially useful against bats. DDT is no longer used after it was banned due to its high persistence in the environment and its cumulative nature in ecosystems.

Endrin was used as a spray on fruit tree trunks and orchard soils in heavy concentrations. Field mice attempting to eat the bark of trees and tracking across sprayed soil ingested lethal doses of the insecticide. This practice is no longer used because of endrin's high toxicity, its persistence in the environment, and the hazard to other species.

Organophosphorus Compounds. Phosazetim (Gophacide) was developed in Germany and proved to be an effective rodenticide. It controlled gophers, mice, and rats after a 24-h exposure to the bait. As with other organophosphorus compounds it is a cholinesterase inhibitor. The manufacture of this compound was discontinued.

Sodium Fluoroacetate and Fluoroacetamide. See Chapter 24.

Alpha-Naphthyl Thiourea (ANTU). ANTU was the first synthetic rodenticide introduced in 1946. It is a selective poison lethal to the Norway rat. It was synthesized following the observation that phenylthiourea was toxic to rats but had no effect on humans. However, rats detected and rejected phenylthiourea effectively, so it proved to be of no use as a rodenticide. A search for other thiourea derivatives with high toxicity to rats and little or no taste produced the compound ANTU which proved to be a good rodenticide. A disadvantage of ANTU is that young Norway rats and roof rats of all ages are resistant to the compound. Furthermore, tolerance and bait refusal in adult Norway rats develop quickly and last a long time. It has been largely displaced by other rodenticides.

Some species of rats are not susceptible to this compound, while others develop resistance to it. There is a wide range of susceptibility to ANTU among mammals. The oral LD_{50} to rats is about 8 mg/kg, for dogs 10 mg/kg, for pigs, horses, and cows 30–50 mg/kg, for guinea pigs 400 mg/kg, and for monkeys 4000 mg/kg (hence, the safety margin for humans). ANTU induces vomiting in dogs and some of the impurities in it might be carcinogenic. It produces its toxic action by causing massive pulmonary edema due to action on pulmonary blood capillaries. It is not absorbed through the skin. Symptoms of poisoning include heavy breathing, cyanosis, lowered body temperature, and death from asphyxiation. Treatment of poisoning is symptomatic (Hayes 1982).

Pyriminil (Vacor). Pyriminil was introduced in 1975 as an acute toxicant which killed rats in one feeding. It proved to be effective against rats which had developed resistance to anticoagulant rodenticides. Norway rats are especially susceptible to this compound, with an oral LD_{50} of 4.75 mg/kg. Other animal species are much more resistant. Pyriminil kills in 4–8 h after ingestion of a single dose of 0.5 g of a 2% bait. Its mode of action in rodents is to inhibit niacinamide metabolism causing paralysis and pulmonary arrest. In humans, pyriminil produces diabetogenic effects (destruction of pancreatic beta-cells and severe hyperglycemia) and autonomic nervous system dysfunction. Repeated doses of nicotinamide soon after ingestion of the compound is beneficial in the treatment of poisoning. Because of its high toxicity to humans, pyriminil was removed from the market by the manufacturer.

CHAPTER 28

Hazards Associated with Pesticide Use

There are several types of potential hazards associated with the use of pesticides. People exposed to some highly toxic compounds may suffer short-term or long-term health problems. Excessive residues in the environment may contaminate water supplies and lead to lower water quality. They may contaminate our food through excessive residues on sprayed crops. Pesticides may cause injury to nontarget organisms such as bees, birds, other wildlife, and natural enemies of pest insects. Improperly applied pesticides may cause damage to treated surfaces, or through drift to surfaces adjacent to treated areas. Some pesticides may be phytotoxic, that is, injurious to crops and ornamental plants. The mode of action of pesticides, the symptoms of poisoning they manifest, and therapy are discussed in the text under each functional group of pesticides.

28.1 Human Injury

Human injury may be caused by breathing or swallowing pesticides, and spilling or splashing pesticides into the eyes or onto the skin. Some pesticides are very toxic and can cause poisoning at low doses, while others, of low toxicity, require substantial amounts to cause injury. At this stage, it is necessary to distinguish between toxicity and hazard. Toxicity is the innate capacity of a substance to cause injury to living organisms and is characterized in terms of acute, chronic, local, or systemic effects. These are arrived at by animal experiments. Hazard is a function of toxicity. It involves the likelihood of injury or danger that may arise from the degree of exposure to the chemical under defined circumstances.

Hazard = toxicity × exposure.

Therefore, hazard to humans must be regarded strictly in relation to the chemical's usage. Some highly toxic compounds may be considered safe if their usage involves little or no exposure of the applicator or the people in the vicinity. For example, highly toxic rodenticides are considered safe if used properly, because of little likelihood that people may come in contact with them.

Exposure also depends on the rate of contamination and time. Thus, actual exposure is the absolute amount of pesticide landing on the skin (not on clothing), and therefore available for absorption, or present in the air and available for inhalation (possibly also ingestion of large particles or droplets) in a given time.

Time is very important, because pesticide which is absorbed from the skin is usually distributed to other organs in the body during which it can be metabolized and excreted. However, if the pesticide is not broken down rapidly, it may accumulate in the body and be stored in certain tissues, such as body fat, as is the case with DDT.

Because potential hazards exist, anyone working with pesticides should avoid exposure to the skin, lungs, digestive tract, and eyes. All pesticides should be treated with respect, regardless of their level of toxicity. It is impossible to predict accurately what effects can result from long-term repeated exposure even to the least hazardous pesticides. Much attention in recent years has focused on the social and environmental risks associated with pesticides. Improper use of pesticides by untrained workers has led to many poisonings during application. Also, workers who entered treated areas too soon after treatment with pesticides have been contaminated by brushing against sprayed foliage. An outbreak of poisoning from the mercurial compound ethyl mercury *p*-toluene sulphonanilide occurred in Iraq in 1972 (Damluji and Tikrity 1972). The poisoning occurred among farmers whose wheat grain had been treated with the fungicide. More than 5500 people were admitted to the hospital and some 280 people died. A similar outbreak, on a smaller scale, occurred in that country in 1961 (Jalili and Abbasi 1961). In Minamata, Japan, an epidemic (ca. 1956) was associated with the ingestion of seafood contaminated by methylmercury originating from industrial waste, known as Minamata disease. Many adults suffered from brain damage and infants whose mothers had been exposed to this chemical were born defective.

Worldwide, the World Health Organization in 1972 estimated that annually about 9200 deaths occurred and the number of cases of accidental human poisoning was 490000 but that the limits of error were 250000–1453000. More recent estimates put the figure at about 20000 deaths, but this may be on the high side and some suicides may be included (Copplestone 1977). Pimentel et al. (1980) discussed environmental and social costs of pesticides and estimated that there were annually about 45000 cases of human poisonings by pesticides in the USA with 3000 hospitalizations and 200 deaths (two-thirds intentional death). The importance of pesticide poisoning is that it is 100% preventable if the factors that make up the hazard can be controlled.

28.2 Routes of Entry into the Body

There are several ways in which people come in contact with pesticides. For pesticide applicators the most common is during mixing and loading, during application, and when entering treated areas soon after application. Usually, the highest levels of pesticide exposure result from accidents, many due to carelessness. Pesticide spills and leaks during application and transportation of pesticides are also a cause of risk and injury. Protective clothing and prompt emergency procedures can greatly reduce the extent of injury resulting from an accident. People may also be exposed to small amounts of pesticides if they live adjacent to treated areas, or touch recently treated livestock, foliage, store products, clothing, or furnishings.

The general public, however, is exposed in a different way, namely, to very small amounts of pesticides which might occur as residues in food crops or in the meat of treated animals which we eat. While exposure of workers is generally to high concentrations of the chemical for short periods of time, the public tends to be exposed to very low concentrations for long periods. For example, a chemical which is used successfully to treat cereals against damage by insects or fungi may be ingested in the diet in small amounts every day, by large segments of the population, for several years or even a lifetime (Sharratt 1974). The subtle effects of very small doses over a very long period are not known and are difficult to measure. The most problematic substances are heavy metals and some persistent insecticides such as the organochlorine compounds (the use of which has been largely discontinued), or compounds that are not easily biodegradable.

In summary, human exposure to pesticides may be accidental, work-related, due to food contamination, improper use in the home, and through contact with nonfood items.

There are several ways by which pesticides can enter the body. It can be through the skin (dermal); through the mouth (oral); through the lungs (respiratory); and through the eyes (ocular).

Dermal Exposure. Dermal exposure, or skin contact, is the most frequent type of exposure. Pesticides on the skin may cause irritation, dermatitis, or more severe injury upon prolonged exposure. When absorbed through the skin they enter the blood stream and are transported to various organs in the body where they can cause internal poisoning. The ability of a pesticide to penetrate through the skin depends on the chemical characteristics of the compound and its formulation. Pesticides that are more soluble in oil or formulated in petroleum solvents are more readily absorbed than those that are water soluble.

Oral Exposure. Oral ingestion may occur by accidentally drinking a pesticide, by splashing spray material into the mouth, or by eating contaminated food or beverage. Smoking while using pesticides may also cause ingestion. Chemicals are absorbed into the blood stream through the linings of the mouth, stomach, and intestines. Food and beverages must be kept away from areas where pesticides are being used or stored. Pesticides should be kept in their original packages and should never be placed into anything that can be mistaken for a food or beverage container. Small children are the major group of pesticide poisoning victims. Keep away pesticides from children's reach.

Respiratory Exposure. Pesticide dusts or vapors may be inhaled during mixing and loading, during aerosol spraying, and during spray applications in general. Upon entering the lungs, they are quickly absorbed and transported to all parts of the body. For this reason, respirators should always be worn while mixing and loading pesticides, and during their application.

Ocular Exposure. Serious damage can result from pesticides entering the eyes. Both the pesticide and the organic solvents used to dissolve it can cause serious damage. Besides eye injury itself, the eyes provide another route of pesticide entry into other parts of the body via the blood stream. The eyes should be protected by wearing goggles or a faceshield.

28.3 Groundwater Contamination

Potential contamination of groundwater with pesticides and other chemicals is a serious matter and of utmost concern. The extent of this potential hazard is just beginning to be realized. Recognition of the vulnerability of aquifers to chemical pollution has intensified over the last decade as more and more identification of contaminants in groundwater in many parts of the world, together with more sensitive analytical instrumentation for their detection, comes to our attention. Until recently, it was generally believed that groundwater was relatively free of toxic chemicals as compared with surface waters (such as lakes, streams, rivers, and ponds), and in cases where contamination was detected, it was assumed that it posed no hazard to human health.

To be sure, groundwater contamination is not a direct process. The pollutant has to percolate through a top layer of soil, then penetrate through a protective layer (the unsaturated zone), eventually reaching groundwater either in its original form or as a breakdown product. This process is slow and time consuming during which the pollutant may undergo various physical, chemical, and biological modifications (Perry et al. 1989). Very likely, the pollutants which are detected in groundwater today originated from carelessness and indifference to environmental sanitation many years ago.

It is estimated that only 1% or less of the total water on the planet is in a form available to people for drinking, irrigation, household use, and manufacturing purposes. Two-thirds of this fresh water (about 4.1 million km^3) is groundwater, trapped beneath the soil; the rest is surface water (Marer et al. 1988). In Israel, groundwater is the major source of water supply (about 65% of the total) with the coastal plain aquifer supplying about 30% of the overall water consumption. This aquifer, underlying a large, heavily populated area, intensively cultivated, is subject to a high degree of pollution. This potential hazard is enhanced by agricultural practices involving irrigation with sewage effluents as a water conservation measure (Muszkat et al. 1988).

The organochlorine insecticides are, by and large, lipophilic compounds of limited water solubility. It is, therefore, predictable that they will remain in the upper layers of the soil with little downward movement. Lichtenstein et al. (1971) demonstrated that DDT, lindane and aldrin, although partly metabolized in the soil, are still present in the 15–25-cm layer 15 years after their application.

The insecticides DDT, DDE, and DDD, which are highly insoluble in water and were originally thought not to penetrate deep into the soil, have been detected in the soil in Israel to depths of 3 and 4 m. Given sufficient time, they could well reach groundwater. Several herbicides, such as atrazine, alachlor, prometryn, prometon, propazine, bromacil, and dinoseb have been detected in the Ashdod area, Israel, in soil and in groundwater (Tables 28.1 and 28.2). It has been shown that the spread of pollutants in this area occurs by horizontal

Table 28.1. Pesticides detected in groundwater, Ashdod area, Israel. (Muszkat et al. 1993b)

Pesticides	ppb
Atrazine	30
Alachlor	30
Bromacil	60
Propachlor	3
Terbutryne	40
Triethazine	3
Anthraquinone	1

Table 28.2. Contaminants in soil, Ashdod area, Israel. (Muszkat et al. 1993a)

Compound	Depth (cm) 0.5	6	135	300
	Concentration (ppb)			
Trifluoromethyl aniline	20	36	–	–
Dichloroaniline	770	1700	–	–
Propazine	–	320	40	110
Prometryne	–	176	–	32
p, p'-DDE	400	1980	26	–
p, p'-DDD series	500	11980	86	–
DDT	–	2100	180	–
Diisooctylphthalate	1100	900	480	236

migration of the organic contaminants deep in the aquifer. This migration is the result of lithological conditions, namely the high content of clay in this area (Muszkat 1994). In addition, a large number of industrial chemicals have been found in soils and in groundwater of an effluent-irrigated site (Muszkat et al. 1989, 1993a,b). Irrigation with effluents, which is a common practice in Israel, enhances groundwater pollution by accelerating the downward movement of micropollutants present in the effluents and in the soil. These results have recently been summarized in a review paper by Muszkat (1994).

Pesticides have been detected in groundwater in many parts of the world. As an illustration may be cited the organochlorine residues found in potable water in three locations in the USA where 16 compounds were detected (Table 28.3). Although organochlorines are lipophilic compounds and highly immobile, they nevertheless found their way to groundwater. In California, more than 2500 wells were closed due to contamination of the drinking water with the potential carcinogenic and testicular toxic compound DBCP (dibromochloropropane—a soil nematicide), and a large number of pesticides, including organochlorines, organophosphates, carbamates, as well as soil fumigants and herbicides have been detected in groundwater (Cohen 1986). Many other examples exist, especially with herbicides which are more water soluble (see Perry et al. 1989 for references), and several books have been published on this subject (for example, Garner et al. 1986; Gerstl et al. 1989; Jones 1991).

These chemicals, even though they are present in small quantities (in the parts per billion range), may pose a serious threat to the health of the population in the long run. The long-term health effects of such pollutants is presently

Table 28.3. Organochlorine pesticide residues in potable water supplies of Chesterfield and Hampton counties. South Carolina (Sandhu et al. 1978) and from New Jersey groundwater (Page 1981)

Compound	Chesterfield		Hampton		New Jersey groundwater
	Mean (ng/l)	Maximum (ng/l)	Mean (ng/l)	Maximum (ng/l)	Maximum (ng/l)
BHC-alpha	–	–	–	–	800
BHC-beta	–	–	–	–	8700
Lindane	23	193	143	319	900
Heptachlor	15	159	9	44	1000
Heptachlor epoxide	8	90	18	87	600
Aldrin	17	191	9	56	1200
Dieldrin	65	153	204	771	900
Endrin	–	–	–	–	200
Chlordane	–	–	–	–	400
p,p'-DDT	261	3307	264	812	900
p,p'-DDE	17	200	8	30	–
p,p'-DDD (TDE)	284	4333	129	779	400
o,p-DDT	–	–	–	–	500
o,p-DDE	–	–	–	–	1000
Methoxychlor	33	312	23	100	–
Mirex	2	230	83	437	–
Total residue per sample	725	9168	890	3435	–

unknown, but in laboratory tests some of these compounds have been found to be potential carcinogens.

28.4 Hazard to Nontarget Organisms

Nontarget organisms include all plants and animals in or near a treated area that are not intended to be controlled by pesticide application. Some types of pesticides have a drastic effect on the environment including air, soil, water, and a large variety of living organisms. Not all the pesticide that is sprayed in a given location actually remains in that area. It is estimated that as much as 55% of the applied pesticide may leave the treated area due to spray drift, volatilization, leaching, run-off, and soil erosion. Some of this drift may move to adjacent areas and contaminate residences, water bodies, other crops, forest trees, and wildlife.

28.4.1 Hazard to Wildlife

Prior to the use of the newer synthetic insecticides, calcium arsenate and lead arsenate were used to control insect pests in forests by application from aircraft. No observable wildlife fatalities were recorded except when the concentration of calcium arsenate dust (as 40% arsenic trioxide) at 50 kg/ha was used, in which case it killed many hares, rabbits, deer, and other animals. Likewise, sodium arsenite dust applied from the air for locust control in Africa was fatal to big game animals. Lead arsenate had similar effects with an accumulation of large quantities of lead in the liver and bones of wildlife (Brown 1978). With the advent of DDT, forest spraying with this insecticide replaced the older type insecticides. At the recommended dosages DDT showed no adverse effects on wildlife. Bats were shown to be quite susceptible to DDT, so residual applications of DDT wettable powders were used to eliminate bat colonies carrying rabies infections. Small mammals pick up residues remaining in the forest floors after application of DDT, but not in quantities that produce harmful effects. However, mink that were fed fish contaminated with DDT produced fewer offspring due to death of embryos in utero.

Cyclodiene insecticides have caused considerable mortality among nontarget wildlife. Heptachlor which was used for fire ant control, killed many raccoons, rabbits, and opossums. The use of heptachlor in seed dressing resulted in the death of some 1300 foxes in England. Dieldrin caused high mortality of fox squirrels, woodchucks, voles, and shrews, and practically eliminated ground squirrels, cottontails, and muskrats from treated areas. Many badgers were also affected after feeding on contaminated wood pigeons that had eaten treated seed. Endrin, another cyclodiene insecticide, proved to be even more toxic to large herbivores (Brown 1978). There are no records of wild animals killed from organophosphorus or carbamate insecticides, although some insecticides of both groups are quite toxic to wild mammals.

Amphibians. Frogs, toads, and their tadpoles are considerably less susceptible to insecticide poisoning than fish. However, high dosages of DDT (12 kg/ha) caused 50% mortality of frogs and toads in a sprayed forest area. Secondary poisoning due to frogs eating DDT-contaminated tent caterpillars has also occurred. Tadpoles affected by DDT are more subject to predation by newts. Insecticides such as TDE and methoxychlor are more toxic than DDT to tadpoles. The order of toxicity to toads and frogs is: endrin > dieldrin > aldrin > toxaphene > DDT.

Reptiles. Treatment of ponds with 2 ppm DDT or more killed water snakes, turtles, and salamanders. In cotton fields sprayed with DDT, residues of total DDT, (DDT + DDE + DDD) exceeded 400–800 ppm in four species of snakes. Residues in water snakes were much higher than in terrestrial snakes.

Among the organophosphorus and carbamate insecticides, only carbophenothion and azinphosmethyl caused considerable mortality of tadpoles. Other compounds, including the carbamate insecticide carbaryl, had no effect.

Extensive tables of residues and toxic effects of pesticides in terrestrial and aquatic organisms, as well as other environmental effects, have been well documented in books by Edwards (1973, 1975), Perring and Mellanby (1977), Brown (1978), and Pimentel, (1990), and in numerous articles. In particular, the *Handbook of Toxicity of Pesticides to Wildlife* by Tucker and Crabtree (1970) is a valuable reference since it gives toxicity data of 108 pesticides on 15 bird species as well as ducks, geese, chickens, turkeys, rats, mules, horses, and bullfrogs. See also Stickel (1973) *Pesticide Residues in Birds and Mammals.*

28.4.2 Hazard to Honeybees

Certain types of insecticides may kill honeybees, causing severe economic losses to beekeepers and loss of certain crops due to poor pollination. The activities of honeybees are essential to ensure pollination and, therefore, growth of fruit in orchards. The danger of insecticides to bees results not only from direct contact with the poison, but also from taking poisoned nectar, pollen, and water, and their transport to the hive. Usually, insecticidal dusts are more hazardous than sprays, and oil solutions and concentrates are more hazardous than water emulsions and suspensions.

When arsenical dusts were used many years ago, large numbers of bee hives were poisoned every year. The substitution of DDT for calcium or lead arsenate decreased the danger to honeybees since DDT deposits were less toxic. Also, contact with DDT made the bees excitable and repelled them, so the bees left the area. There was no hive poisoning with this insecticide. On the other hand, benzene hexachloride (HCH) and the cyclodiene insecticides were highly toxic. Most of the organophosphorus (OP) and carbamate insecticides are highly hazardous to bees, including the nonpersistent compounds such as tetraethylpyrophosphate (TEPP) and mevinphos, and the systemic compounds phosphamidon and dimethoate. The carbamate insecticide carbaryl (Sevin) poisoned some 30000 bee hives in California. Malathion appears to be safer to bees than most OP compounds when applied as dilute sprays, but its application in the undiluted form, such as ULV (ultra low volume), is very hazardous to honeybees. The only low risk OP compounds are trichlorfon, a nonpersistent insecticide, and the systemic insecticide schradan.

Based on laboratory acute toxicity tests, the synthetic pyrethroids are considered to be highly or extremely toxic to honeybees. LD_{50} values for the photostable pyrethroids permethrin, cypermethrin, deltamethrin, and fenvalerate fall in the range of 0.02–0.3 μg per bee. The photolabile pyrethroids showed greater variation in toxicity with LD50's of 0.02–10 μg per bee (Hill 1985). However, extensive field studies have demonstrated that synthetic pyrethroids pose little or no hazard to bee populations following agricultural applications due to the low field application rates, partly due to repellency effects, and also because the residues on plant surfaces show a rapid decline in toxic effect with time. The relative toxicities of pyrethroids to honeybees are given in Table 28.4.

The bacterial toxin of *Bacillus thuringiensis* appears to be harmless to honeybees. Wild bees such as *Megachile rotundata* and *Nomia melanderi* were found to be more susceptible than honeybees to 29 pesticides tested, but less susceptible to trichlorfon carbaryl, and dicofol (Johansen 1963; Johansen and Eves 1967). Bumblebees were more resistant than honeybees. Honeybees are more severely damaged if the harmful pesticides are applied when bees are foraging in the field for nectar and pollen, so application of materials toxic to bees should be avoided when crops are in bloom, and when cover crops, weeds, and wild flowers are in bloom in the treatment area or nearby. Treatment over colonies in hot

Table 28.4. Laboratory acute toxicity of pyrethroids to *Apis mellifera*. (Leahey 1985)

Pyrethroid[a]	24-h LD_{50}(μg a.i. per bee)	
	Oral	Topical, Acetone
Permethrin	0.19–0.28	0.05–0.10
Cypermethrin	0.18–0.26	0.013–0.056
Fenvalerate	0.29	0.077–0.34
Deltamethrin	0.079	0.02–0.051
Cyhalothrin	–	0.027
Flucythrinate	–	0.27
Allethrin	4.6–9.1	1.6–3.4
Bioallethrin	–	0.1
Resmethrin	0.069	0.015
Bioresmethrin	0.055	0.06–0.05
Pyrethrum	0.15	0.13–0.29

[a]Test material: technical.

weather when bees usually cluster on the outside of the hive is highly hazardous. Treatments made during the evening or early morning before the bees are foraging are much safer.

The location of bee colonies is important. Colonies located in the treated area may sustain greater losses than those farther away. Usually, losses are not significant if the colonies are located 0.4 km or more from the site of treatment, unless the treated crop is the only attractive field in the area. Pesticide applications from aircraft are more hazardous than ground applications. Covering hives with burlap or dark plastic material for 1 or 2 h during and after treatment in the early morning may give added protection, especially if application is by aircraft. When using pesticides hazardous to bees, beekeepers should be notified so that they may provide proper protection to their colonies. Modern pesticides, in contrast to arsenicals, are less hazardous to bees if the above facts and precautions are taken into consideration (Anderson and Atkins 1968).

28.5 Impact on Bird Populations

Research on the effects of modern pesticides on bird populations began soon after the introduction of DDT in 1943. At a time when DDT was hailed as a compound of great benefit to mankind, both in agriculture and public health, and long before the publication of *Silent Spring* by Carson (1962), many scientists had already recognized the potential danger of the chemical to birds and other wildlife.

28.5.1 Organochlorine Insecticides

Dead birds were frequently seen in fields sprayed with high concentrations of DDT (such as 7 kg/ha), and even at a rate of 2.8 kg/ha applied annually there was a reduction in the density and reproductive capacity of forest birds (Hall

1987). This was corroborated by large scale experiments made at the Patuxent laboratories (US Fish and Wildlife Service) with DDT, DDD, endrin, aldrin, and dieldrin incorporated into the diets of pheasants and quail, which showed lowered reproductive success of these birds (DeWitt and George 1960). More than 500 publications on the effect of pesticides on wildlife have been published since 1964 by scientists at these laboratories. Residue levels of pesticides in dead birds vary greatly even when the birds are killed by the same pesticide and under same conditions. Total pesticide residues in tissues does not always correlate with percent mortality of the birds. Residues of DDT in the brains of birds were found to correlate more accurately with lethality (Hill et al. 1971; Stickel 1973).

The most rapid and convenient indicator of toxicity to birds for most chemicals is the acute toxicity as measured by the LD_{50} (Table 28.5). However, this is not always the most sensitive measure of potential hazard. Compounds that are poorly absorbed or are cumulative in action are better tested by chronic long-term feeding (such as 30-day feeding period) or by repeated oral doses. The 30-day feeding period shows the cumulative nature of the toxicant. A useful index of cumulativeness is the ratio of the single acute LD_{50} dose to the chronic MLD (minimum lethal dose) which is the lowest daily oral intake of a pesticide that kills one or two out of six birds in a 30-day period. A low ratio indicates that the compound is not cumulative, perhaps due to its efficient detoxication and elimination from the body, while a high ratio shows that the compound is cumulative in the tissues of the bird and, likely, more hazardous.

A large number of other reports have appeared in the literature regarding insecticide residues in birds and bird eggs (see Edwards 1975; Moriarty 1975; Brown 1978, for references). Generally, much larger quantities of insecticide residues were found in raptorial and fish-eating birds than in herbivorous and insectivorous species. Residues of organochlorine insecticides and their metabolites in bird eggs have been implicated in the thinning of egg shells with a consequence of egg breakage or reproductive failure. In Israel (Mendelssohn 1972; Mendelssohn

Table 28.5. Acute oral LD_{50} values of insecticides for selected birds. (Tucker and Crabtree 1970)

Insecticides	Mallard (females)	Pheasant (females)	*Coturnix (males)* (quail)
Chlorinated hydrocarbons			
DDT	2240	1296	841
Dieldrin	381	79	70
Endrin	5.64	1.78	–
Organophosphates			
Parathion	1.90	12.4	5.95[a]
Dursban	75.6	17.7	17
Abate	90	31.5	84.1
Carbamates			
Propoxur	11.9	20[a]	28.3[a]
Zectran	3.0	4.5	3.2
Carbaryl	2179	707	2290

[a]The sex is different from the one indicated in the heading.

and Paz 1977; Mendelssohn et al. 1979), as in many other countries, certain insectivorous and omnivorous bird species have declined in numbers as a result of feeding on pesticide-contaminated food in heavily sprayed agricultural areas. Heavy losses were also recorded among several raptorial species due to secondary poisoning from pesticides, as well as from secondary poisoning caused by the frequent application of rodenticides to control the Levant vole *Microtus guentheri* (Mendelssohn 1972).

Residues in eggs of 16 bird species monitored in Israel (Perry et al. 1990; Table 28.6) showed that eggs of all species contained DDE (a metabolite of DDT), with the highest amounts found in kestrels and in fish-eating birds (both carnivores), and the lowest amounts in chukar, a herbivore, and in Palm dove and Turtle dove, both granivores. Varying amounts of DDE were found in all bird tissues analyzed, with highest amounts in heart, liver, and brain (Table 28.7). The effect of organochlorine insecticides on eggshell thickness is well documented. It became clear that DDE was the main factor in eggshell thinning of cormorants, falcons, pelicans, blue herons, herring gulls, sparrow hawks, golden eagles, and many others (Brown 1978). The effect on seed-eating birds was not as pronounced. An example of the relationship between eggshell thickness and DDE content of eggs is shown in Fig. 28.1.

In contrast to DDT, methoxychlor is broken down quite rapidly so that little or no residues are found in birds under field conditions. The cyclodiene compounds aldrin, dieldrin, endrin, heptachlor, and toxaphene have a considerably greater acute and chronic toxicity to birds than DDT. Dieldrin caused the death of many song birds, especially, robins and meadowlarks, but also pheasants, grackles, and starlings. Extensive bird kills were noticed following application of heptachlor for fire ant control. Isobenzan (Telodrin) is perhaps the most toxic to birds of all the organochlorine compounds, causing the death of quail after a few days at 10 ppm in the diet.

Chlordecone and mirex have lower acute toxicity to birds, but they are very persistent in the body, more so than DDE.

28.5.2 Organophosphorus (OP) and Carbamate Insecticides

Organophosphorus (OP) and carbamate insecticides have, generally, high acute toxicity to birds, but species differ greatly in their susceptibility to these compounds. Monocrotophos (Azodrin) is one of the most toxic OP compounds having killed many bird species when applied to control insect pests in cotton and potato fields. In Israel, azodrin was used extensively in alfalfa fields in the Hula Valley to control the Levant vole *Microtus guentheri* that caused heavy damage to the alfalfa crop. As a consequence, many birds feeding in these fields were killed. In addition, azodrin caused a mass mortality of birds of prey by way of secondary poisoning when they fed on poisoned voles and birds. Among the heaviest losses were spotted eagles, common buzzards, kestrels, and barn owls, and to a lesser extent 12 other species of birds of prey (Mendelssohn et al. 1979). Parathion which was the first widely used OP insecticide caused extensive casualties of many bird species (Brown 1978). The acute oral toxicities of OP insecticides to several bird species are shown in Table 28.8.

Table 28.6. Residues of DDE in eggs of 16 bird species. (Perry et al. 1990)

Species	Number of clutches	Number of eggs	DDE (ppm)	Food preference
Chukar *(Alectoris chukar)*	5	45	0.20 ± 0.44	H
Feral pigeon *(Columba livia)*	4	8	0.57 ± 0.21	G
Greenfinch *(Carduelis chloris)*	9	41	0.35 ± 0.21	G
Turtle dove *(Streptopelia turtur)*	19	36	0.23 ± 0.17	G
Palm dove *(S. senegalensis)*	26	48	0.04 ± 0.06	G
Cattle egret *(Bubulucus ibis)*	11	36	0.62 ± 0.38	O
Hooded crow *(Corvus corone)*	6	18	0.61 ± 0.14	O
House sparrow *(Passer domesticus)*	18	79	0.28 ± 0.22	O
Crested lark *(Galerida cristata)*	5	17	0.36 ± 0.15	I/G
Yellow vent bulbul *(Pycnonotus xanthopygos)*	17	38	0.57 ± 0.43	I/F
Blackbird *(Turdus merula)*	13	28	0.66 ± 0.46	I/F
Graceful warbler *(Prinia gracilis)*	9	25	0.96 ± 1.05	I
Kestrel *(Falco tinnunculus)*	3	11	0.46 ± 0.15	C
Common tern *(Sterna hirundo)*	7	16	0.88 ± 0.60	C
Little egret *(Egretta alba)*	10	28	1.61 ± 1.43	P
Night heron *(Nycticorax nycticorax)*	12	32	1.62 ± 1.41	P

H, herbivore; G, granivore; O, omnivore; I, insectivore; F, fructivore; C, carnivore; P, piscivore.

Carbamate insecticides are safer than most OP compounds to birds with some exceptions. Carbaryl was shown to have no effect on birds at high dosages, but carbofuran and aldicarb are quite toxic. Most carbamates are characterized as having low chronic toxicity to birds. The acute oral toxicities of certain carbamates to several bird species are shown in Table 28.8 (see also Kuhr and Dorough 1976).

28.5.3 Pyrethrum and Pyrethroids

Extensive field tests with pyrethrum sprays for control of forest insect pests failed to show any appreciable changes in species composition and relative abundance of birds attributable to spraying. More than 30 bird species were monitored before and after spraying (Pillmore 1973). Birds have been shown to

Table 28.7. DDE residues (ppm) in various tissues of 11 bird species collected in an agricultural area of the coastal plain of Israel. (Perry et al. 1990)

Species	Number	Liver	Heart	Brain	Muscle	Skin and feathers
Cattle egret	5	0.47 ± 0.14	0.48 ± 0.06	0.39 ± 0.18	0.48 ± 0.18	0.56 ± 0.11
Crested lark	12	0.45 ± 0.07	0.68 ± 0.24	0.38 ± 0.17	0.13 ± 0.02	0.45 ± 0.18
Yellow vent bulbul	11	0.36 ± 0.07	–	0.41 ± 0.26	0.11 ± 0.03	1.40 ± 1.10
Greenfinch	5	0.39 ± 0.12	–	0.11 ± 0.04	0.09 ± 0.07	–
Hooded crow	7	0.21 ± 0.08	0.41 ± 0.04	0.25 ± 0.05	0.12 ± 0.04	–
Blackbird	5	0.30 ± 0.09	0.35 ± 0.18	0.18 ± 0.05	0.12 ± 0.04	–
House sparrow	10	0.34 ± 0.10	0.55 ± 0.19	0.29 ± 0.07	0.17 ± 0.06	–
Palm dove	4	0.40 ± 0.20	0.25 ± 0.10	0.18 ± 0.06	0.03 ± 0.02	–
Turtle dove	6	0.30 ± 0.10	0.39 ± 0.18	0.37 ± 0.15	0.03 ± 0.02	0.32 ± 0.15
Feral pigeon	5	0.24 ± 0.05	–	0.06 ± 0.06	0.04 ± 0.04	–
Graceful warbler	5		Whole body 0.15 ± 0.06			

be even less sensitive to pyrethroids than are mammals. In acute and subacute (8-day) oral toxicity tests, the birds in the treated group never consumed and retained a sufficient amount of chemical to provide LD_{50} or LC_{50} figures (Table 28.9; see Leahey 1985). Since pyrethroid insecticides are not used for seed dressing they do not present acute, subacute, or dietary hazard to birds. The newer type insecticides such as juvenile hormone mimics and chitin synthesis inhibitors have no effect on bird populations.

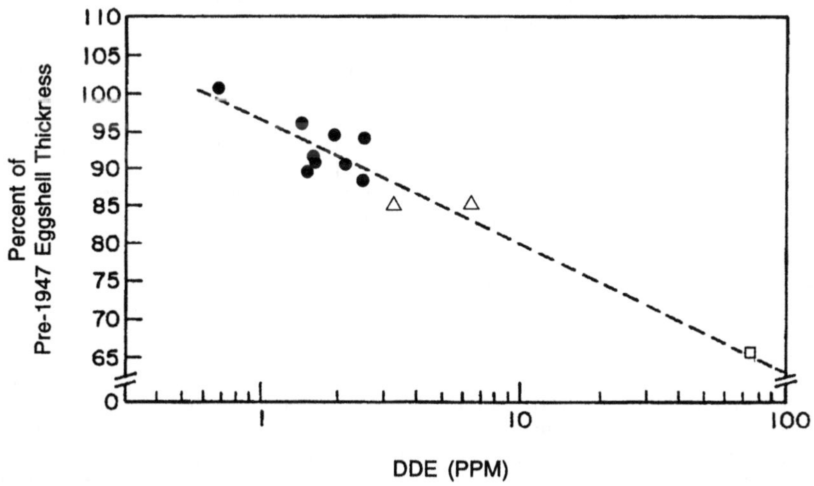

Fig. 28.1. Relationship of DDE and eggshell thickness in brown pelicans, based on field data from Florida (•), South Carolina (Δ), and California (□). (Marco et al. 1987)

Table 28.8. Acute oral toxicities of insecticides to pheasant and other birds: LD_{50} values (mg/kg), 14-day observation period. (Tucker and Crabtree 1970; Tucker and Haegele (1971))

Compound	Ring-necket pheasant	Coturnix (Quail)	Chukar (Partridge)	Mallard (Duck)	Rock dove (Pigeon)
Organophosphorus insecticides					
Monocrotophos	2.8	3.7	6.5	4.8	2.8
Dicrotophos	3.2	4.3	9.6	4.2	2.0
Demeton	8.2	8.5	15	7.2	8.4
Parathion	12	5.9	24	2.0	2.5
Chlorpyrifos	12	17	61	75	27
Fenthion	18	11	26	5.9	4.6
Temephos	32	84	270	90	50
Methyl demeton	42	84	113	54	15
EPN	53	5.2	14	3.1	5.9
Carbamate insecticides					
Mexacarbate	4.5	3.2	5.2	3.0	65
Propoxur	20	28	24	12	60
Landrin	52	71	60	19	168
Mobam	228	668	237	1130	273
Carbaryl	>2000	2290	–	>2180	2000
Organochlorine insecticides					
Dieldrin	79	70	23	381	27
DDT	1300	840	–	>2240	>4000

Table 28.9. Toxicity of pyrethroids to mammals and birds. (Leahey 1985)

Pyrethroid	Mammal		Bird[a]	
	Acute oral toxicity to rat and/or mouse (LD_{50}, mg ai/kg)	Acute dermal toxicity to rat and/or rabbit (24-h LD_{50}, mg ai/kg)	Acute oral toxicity (LD_{50}, mg ai/kg)	Subacute (8-day) oral toxicity (LC_{50}, mg ai/kg/day)
Natural Pyrethrum	200–2500	>1500	>10000 (1,2)	>5000 (1,2,4)
Photolabile pyrethroids				
Allethrin	500–2500			
Resmethrin	700– >5000			
Photostable pyrethroids				
Permethrin	400– >5000	>200– >2000	>13000 (1–4)	>27000 (1–4)
Cypermethrin	100– >4000	>2000– >20000	>10000 (2)	>2000 (2,5)
Fenvalerate	200–3000	>2500–5000	>4000 (1,2,5)	>15000 (5)
Deltamethrin	20–150	>800– >3000	>4000 (2)	

[a]Species tested shown in parentheses: 1, *Coturnix coturnix japonica,* Japanese quail; 2, *Anas platyrhyncos,* mallard duck; 3, *Sturnus vulgaris,* starling; 4, *Phasianus colchicus,* ring-necked pheasant; 5, *Colinus virgianus,* bob-white quail.

28.6 Effects on Aquatic Ecosystems

Inland bodies of water such as groundwater, lakes, streams, reservoirs, and estuaries are, unlike the open seas and oceans, relatively small captive sinks of the by-products of man's activities. Thus, they are especially vulnerable to the accumulation of abnormal levels of nutrients, trace contaminants, pesticides, and other chemicals. Pesticides reach aquatic systems by direct application, spray drift from ground or aerial spraying, atmospheric fallout (rain and duststorms), run-off from agricultural land, discharge of effluent from chemical factories, and from sewage.

As the progressive use of organic insecticides reached massive proportions in the past 40–50 years, it became increasingly evident that they were having detrimental effects on aquatic forms of life. Carried away by natural forces such as wind, rain, duststorms, the flow of rivers, and ocean currents, residues of pesticides began to appear everywhere on the globe from tropical forests to Antarctic snows. The more stable insecticides, such as the organochlorine compounds, slow in decomposing, were taking heavy tolls of many fish species, aquatic invertebrates, and a wide variety of nontarget organisms. A greater long-term threat was the gradual buildup of organochlorine residues in fish, with the consequential hazard to humans who stand at the end of the food chain.

Once in aquatic systems, pesticides must be either broken down to simpler, less toxic compounds, or remain in the medium, or move back into the atmosphere by volatilization and codistillation, in which case they recycle back to water bodies by fallout due to wind, rain, snow, and duststorms. The effects of pesticide pollution on aquatic systems depend on the chemical characteristic of the compound, its stability and persistence in water, its water solubility, its potential for uptake and bioconcentration into aquatic organisms, and the physical characteristics of the ecosystem, such as size, form, and location (Edwards 1977). Such characteristics of pesticides that cause hazards to aquatic systems are shown in Table 28.10. The toxicities of various pesticides to several species of freshwater fish are shown in Tables 28.11–28.14 (see Brown 1978). To marine and estuarine

Table 28.10. Characteristics of pesticides that cause hazards to aquatic systems. (Edwards 1977)

Characteristic	Pesticides			
	Organochlorines	Organophosphates	Carbamates	Herbicides
Toxicity of aquatic fauna	+ + +	+ +	+ +	+
Solubility	–	+	+	+ + +
Potential for uptake and bioconcentration	+ + +	+	+	+ +
Persistence in aquatic systems	+ + +	+	+	+ +
Heavy usage on or close to water	+ + +	+ + +	+ +	+ + +

Table 28.11. Comparative susceptibility of fish families to insecticides: 96-h LC_{50} by static test. (Macek and McAllister 1970)

Fish	Organochlorines (ppb)		Organophosphates (ppm)		Carbamates (ppm)	
	DDT	Lindane	Methyl parathion	Malathion	Azinphos-methyl	Carbaryl
Salmonidae						
Rainbow trout	7	27	2.8	0.17	0.014	4.3
Brown trout	2	2	4.7	0.20	0.004	2.0
Coho salmon	4	41	5.3	0.10	0.017	0.8
Percidae						
Yellow perch	9	68	3.1	0.26	0.013	0.7
Centrarchidae						
Red-ear sunfish	5	83	5.2	0.17	0.052	11.2
Bluegill sunfish	8	68	5.7	0.10	0.022	6.8
Large-mouth bass	2	32	5.2	0.28	0.005	6.4
Ictaluridae						
Channel catfish	16	44	5.7	9.0	3.3	15.8
Black bullhead	5	64	6.6	12.9	3.5	20.0
Cyprinidae						
Goldfish	21	131	9.0	10.7	4.3	13.2
Fathead minnow	19	87	8.9	8.6	0.24	14.6
Carp	10	90	7.1	6.6	0.70	5.3

Table 28.12. Comparative susceptibility of salmonid fishes and three-spine stickleback to insecticides: 96-h LC_{50} by static test in ppb. (Katz 1961; Post and Schroeder 1971)

Compound	Coho salmon	Chinook salmon	Brook trout	Rainbow trout	Cutthroat trout	Three-spine stickleback
Endrin	0.5	1	0.4	0.4	0.1	0.4
Toxaphene	9	2	–	8	–	9
Dieldrin	11	6	–	10	–	15
DDT	18	11	7	2	0.8	18
Aldrin	46	8	–	18	–	40
Lindane	50	40	–	38	–	44
Chlordane	56	57	–	44	–	90
Heptachlor	59	17	–	19	–	112
Metoxychlor	66	28	–	63	–	86
Azinphosmethyl	4	4	–	3	–	12
Malathion	265	23	130	122	150	94
Coumaphos	15000	–	–	1500	–	1862
Carbaryl	1300	–	1070	1470	1500	3990

fish, endrin is the most toxic insecticide followed by DDT. There is a wide variation among the species in their susceptibility to other insecticides (Table 28.15; see Brown 1978).

Due to its intensive agricultural practices, Israel is one of the largest users of pesticides, per unit of cultivated agricultural land, in the world. From the standpoint of pesticide application, especially by aircraft, the line of demarcation between

Table 28.13. Comparative toxicity of pesticides to rainbow trout: 48-h LC_{50} (ppb). Data are from compilation of Pimentel (1971)

Compound	LC_{50} (ppb)	Compound	LC_{50} (ppb)	Compound	LC_{50} (ppb)
Endosulfan	1	Azinphosmethyl	4	Carbaryl	1350
Endrin	1	Chlorpyrifos	20	Mexacarbate	8000
Toxaphene	3	Parathion	47	DNOC	210
Aldrin	3	Diazinon	170	Pyrethrins	54
DDT	3	Malathion	196	Dimethrin	7000
Methoxychlor	7	Naled[a]	240	Silvex	650
Heptachlor	9	Fenthion[b]	930	2,4-D ester	1100
TDE(DDD)	9	Temephos[c]	1500	2,4,5-T acid	1300
Chlordane	10	Methyl parathion	2750	Diuron	4300
Dieldrin	13	Trichlorfon	3200	Simazine	5000
Lindane	18	Monocrotophos	7000	Atrazine	12600
Chlorobenzilate	710	Phosphamidon	8000	Na arsenite	36500
Dicofol	100000	Dimethoate[a]	19000	2,4-D amine[a]	250000

[a]24-h; [b]96-h; [c]brook trout.

Table 28.14. Comparative toxicity of organochlorine and organophosphorus insecticides to freshwater fish: 96-h LC_{50} by static test. (Henderson et al. 1959; Pickering et al. 1962)

Compound	Minnows	Bluegills	Goldfish	Guppies
	Organochlorines (LC_{50} values in ppb)			
Endrin	1.0	0.6	1.9	1.5
Toxaphene	7.5	3.5	5.6	20
Dieldrin	16	7.9	37	22
DDT (technical)	32	16	27	43
Aldrin	33	13	28	33
Chlordane	52	22	82	190
Lindane	62	77	152	138
Methoxychlor	64	62	56	120
Heptachlor	94	19	230	107
HCH	2300	790	2300	2170
	Organophosphates (LC_{50} values in ppb)			
Azinphosmethyl	0.09	0.005	1.3	0.11
EPN	0.25	0.10	0.45	0.03
Parathion	1.3	0.10	2.7	0.05
TEPP	1.9	1.1	21	1.8
Chlorthion	3.0	0.7	2.3	1.2
Demeton	3.2	0.10	11	0.61
Disulfoton	3.7	0.06	6.5	0.25
Methyl parathion	8.0	1.9	9.6	7.8
Dioxathion	10	0.03	32	0.21
Malathion	23	0.09	0.45	0.84
Coumaphos	> 18	0.18	> 18	0.56
Schradan	100	110	610	20
Trichlorfon	140	3.8	99	7.1

Table 28.15. Toxicity of organochlorine and organophosphorus insecticides to estuarine fish[a]: 96-h LC_{50} by static test (ppb). (Eisler 1970)

Compound	Silverside	Bluehead	Mullet	Killifish	Eel	Mummichog	Puffer
Endrin	0.05	0.1	0.1	0.3	0.6	1.0	3.1
p,p'-DDT	0.4	7	1.5	1	4	5	89
Heptachlor	3	0.8	194	32	10	50	188
Dieldrin	5	6	23	4	0.9	10	34
Lindane	9	14	66	28	56	60	35
Aldrin	13	12	100	17	5	8	36
Methoxychlor	33	13	63	30	12	46	150
Dioxathion	6	35	39	15	6	20	75
Malathion	125	27	550	250	82	240	3250
Monocrotophos	320	74	300	75	65	300	800
Dichlorvos	1250	1440	225	2300	1800	2680	2250
Methyl parathion	5700	12300	5200	13800	16900	58000	75800

[a] *Menidia menidia*, Atlantic silverside; *Thalassoma bifasciatums*, bluehead; *Mugil cephalus*, striped mullet; *Fundulus majalis*, striped killifish; *Anguilla rostrata*, American eel; *Fundulus heteroclitus*, mummichog; *Sphaeroides maculatus*, northern puffer.

sprayed agricultural areas and inland bodies of water, including fishponds, is less than desirable. Inevitably, some of the pesticide drift contaminates these water bodies and adds to the burden of the aquatic ecosystem along with other contaminants. Assessment of the presence and bioconcentration of organochlorine insecticides and PCBs (polychlorinated biphenyls, industrial chemicals) in commercial and experimental fishponds on the coastal plain of Israel (Perry et al. 1983) showed that all specimens analyzed contained DDE and PCBs (Tables 28.16, 28.17). In addition, eggs of fish-eating birds contained also DDT and DDD. A study of the kinetics of parathion in an experimental fishpond showed an exponential type decline of the chemical from the water over a period of 3 weeks (Fig. 28.2) and revealed a rapid bioconcentration of parathion in algae, zooplankton, aquatic invertebrates, and fish (Fig. 28.3).

The synthetic pyrethroid insecticides are rapidly and strongly adsorbed in soil and in water-sediment systems. Consequently, they will be present in the water phase for only a relatively short time. Many aquatic organisms, however, may be exposed to bottom sediments or suspended particulate matter to which the

Table 28.16. DDE and PCB residues in various components of a commercial fishpond[a]. (Perry et al. 1983)

Component	Number	DDE (ppb)	Bioconcentration factor	PCBs (ppb)	Bioconcentration factor
Water (unfiltered)	4	0.5 ± 0.1	–	2.3 ± 0.3	–
Sediment	4	35 ± 12	70	78 ± 25	34
Plankton	3	80 ± 22	60	300 ± 85	130
Carp	8	148 ± 38	296	420 ± 140	182
Tilapia	6	160 ± 59	320	530 ± 170	230

[a] Commercial fishponds usually contain a high level of organic matter from added manure or wastewater.

Table 28.17. Organochlorine and PCB residues in eggs of the fish-eating birds night heron *Nycticorax nycticorax* and little egret *Egretta garzetta*. (Perry et al. 1990)

Species	Number	Residues (ppm)			
		DDE	DDD	DDT	PCBs
Night heron	26	2.25 ± 0.73	0.86 ± 0.49	1.0 ± 0.48	0.56 ± 0.30
Little egret	31	1.88 ± 0.98	0.19 ± 0.11	0.43 ± 0.10	0.52 ± 0.30

chemicals had been adsorbed. Desorption from sediments may pose another hazard. On the other hand, it has been conclusively shown that the gills of fish, in particular, are extremely efficient in removing persistent insecticides from water. For example, Atlantic salmon exposed to 1 ppm DDT in solution (greater than its water solubility of 1.2 ppb) concentrated 1.5 ppm DDT in their livers and spleens in 5 min and 31 ppm in 1 h. There are many other such reports. It seems clear that the main uptake of pesticides by fish from water is direct but passive through the gills and the uptake through the food is much less important.

Natural pyrethrum and the synthetic pyrethroids have been extensively studied for their effects on aquatic organisms (Hill 1985). In controlled laboratory experiments with these insecticides in solution, in water free of particulate matter, these insecticides proved to be highly toxic to fish and to many aquatic invertebrates. High toxicity was also shown by direct application of pyrethroids to simulated experimental and natural field ponds. However, spray drifts and run-offs of

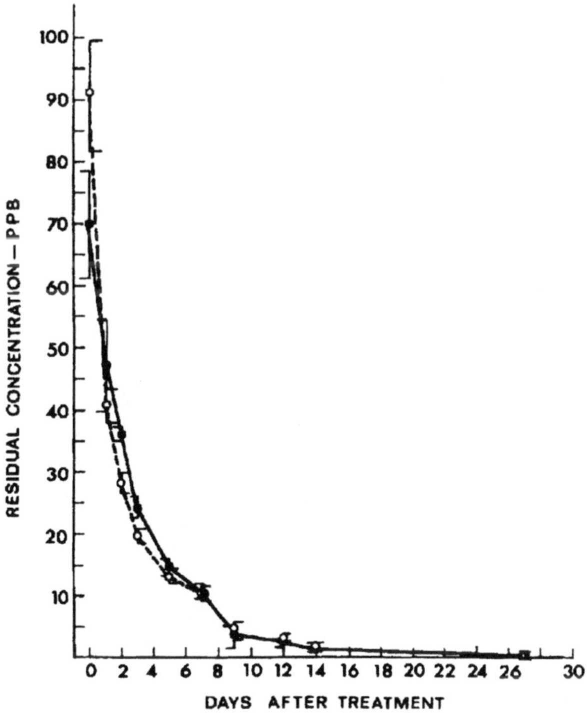

Fig. 28.2. Dynamics of parathion in the water of treated ponds A (*open circles*) and B (*filled circles*). *Bars* indicate standard deviation. (Gasith and Perry 1980)

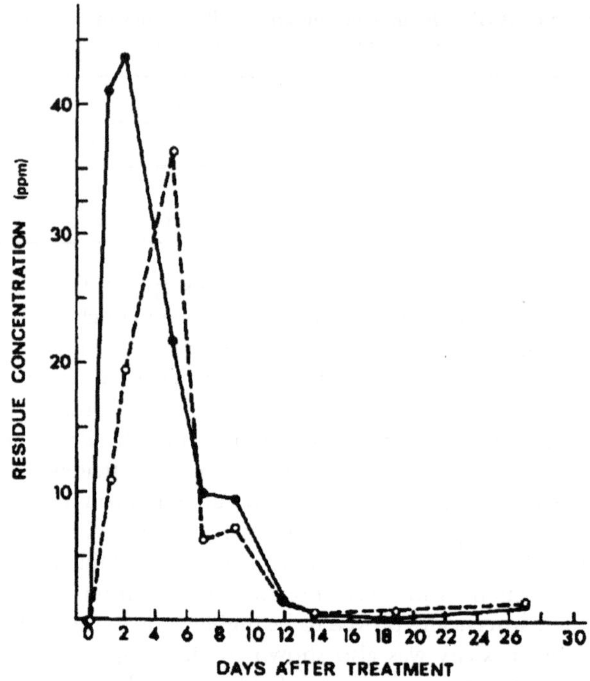

Fig. 28.3. Dynamics of parathion in zooplankton in treated ponds A (*open circles*) and B (*filled circles*). (Gasith and Perry. 1980)

permethrin, cypermethrin, deltamethrin, and fenvalerate resulting from agricultural and forestry applications had no adverse effects on fish populations. The same is true for natural pyrethrum. Toxic effects to other aquatic organisms are greater but are considered to be transient with a potential for fast recovery. Fish are also able to metabolize the pyrethroids efficiently to water soluble metabolites; hence bioaccumulation of these insecticides under field conditions is of minor importance.

In general, the 96-h LC_{50}'s for many warm and cold water fish species were in the range of resmethrin 0.4–30 ppb, cypermethrin 0.4–3.2 ppb; and fenvalerate 0.3–5.4 ppb.

28.7 Hazard to Natural Enemies of Insect Pests

The concept of integrated pest management implicitly emphasizes the maximum use of natural enemies of arthropod pests supplemented with selective insecticides when necessary. There is no doubt that most pesticides, both persistent and nonpersistent types, have had disruptive effects on the balance between pest populations and those of their natural enemies. It is quite evident that both target pest and nontarget natural enemies are killed by insecticides. However, since many more insect pests develop resistance to insecticides than their natural enemies do, the question is asked whether physiological or biochemical differences might exist between the two groups in their responses to insecticides? The scant amount of information available indicates that such differences are more biological than biochemical in that the reproductive potential of the pest is far greater than that of the natural enemy. Hence, greater populations of the pest are subjected to selection pressure, eventually resulting in resistance.

Resurgences. A number of important pests, particularly of perennial crops such as citrus trees, have been controlled for many years by their natural enemies. If a second pest, not so controlled by natural enemies, appears, then its control by chemicals is likely to produce a recrudescence of the first pest. An example of this happened in California where the cottony cushion scale (*Icerya purchasi*) had been successfully controlled since late in the 19th century by the vedalia ladybird beetle (*Rodolia cardinalis*) imported from Australia. When insecticides were used against other pests of citrus, the scale once more caused damage because the natural enemy (its predator) was killed by the chemical sprays.

Another classical example was the use of DDT against spider mites. DDT was especially ineffective against acarina. Mites infesting plants present a peculiar problem in that their life cycle is so short that there is always a sufficient proportion of the population in the egg stage which is not susceptible to DDT. On the other hand, the mite predators lack this ample supply of eggs to replenish their population. Therefore, when the destructive mites hatch from the unaffected eggs, they find no natural enemies to control them and their population explodes and becomes more abundant than before. Thus, DDT has been found to encourage the development of infestations of several mite species in orchards, greenhouses, and crop plants.

The large amount of data for the toxicity of insecticides to predators and parasites of insect pests has been summarized by Croft and Brown (1975). Most of the data pertain to predators such as coccinellids and others. The order of toxicity arranged in five groups of insecticides (Croft and Brown 1975) shows the organophosphorus insecticides to be more acutely toxic than some organochlorine compounds. In general, the predators showed greater tolerance to these insecticides than their prey, and, in most cases, predators were more tolerant to insecticides than were parasites which occurred in the same habitat. Other than direct toxic effects by exposure to sprays or through contact with residues, insecticides may also exert a direct lethal effect on predators and adult parasites by acting as stomach poisons, being taken in the nectar, pollen, honeydews, or in contaminated water. Synthetic pyrethroids are more toxic to spider mite predators than to their hosts. Under field conditions, both predators and parasites of pests of several crops were affected (reduced numbers) by permethrin, fenvalerate, and deltamethrin (Hill 1985). In many instances, predators exhibited pyrethroid sensitivities similar to or less than the pest, and recovery from the effect of the insecticide was rapid.

Numerous studies (for example, Wright and Spates 1972; Vinson 1974; Granett et al. 1975; Smilowitz et al. 1976; Fashing and Sagan 1979; Peleg and Gothilf 1980; Peleg 1983, 1988) indicated that insect growth regulators and chitin synthesis inhibitors have relatively mild effects on parasites and predators of several insect pests; therefore, they can be compatible with insect pest management programs. There are, however, some exceptions. It is worth noting that several predators and parasites have been selected for resistance to insecticides in the laboratory (see Chap. 29).

CHAPTER 29

Insect Resistance

Resistance, as defined by the World Health Organization, is the development of an ability in a given strain of insects to tolerate doses of an insecticide which would prove lethal to the majority of individuals in a normal population of the same species (Brown 1958). The word "development" rules out those cases where a particular insect species is naturally tolerant to an insecticide before any selection pressure has been applied.

The resistance phenomenon is not entirely new, for as long ago as 1911 there was evidence that extensive fumigation of citrus trees with hydrogen cyanide (HCN) to control citrus scales had selected a resistant strain of scale insects to this compound. Melander (1914) is credited with the first publication on insect resistance and the now famous quotation, "Can insects become resistant to sprays?" Prior to World War II there were only a few cases of insect resistance to the then used insecticides.

Perhaps the most significant commentary on the resistance problem was made by the famous American entomologist Stephen A. Forbes (1915) when he said:

> "The struggle between man and his insect enemies began long before the dawn of civilization, has continued without cessation to the present time, and will continue no doubt as long as the human race endures...We commonly think of ourselves as the lords and conquerors of nature, but the insects had thoroughly mastered the world and taken full possession of it long before man began the attempt...We cannot even protect our very persons from their annoying and pestiferous attacks, and since the world began, we have never yet exterminated...we shall probably never exterminate...so much as a single species."
>
> "It is true that through the centuries man has fought indefatigably to restrain, control or eradicate his insect enemies, but there is little doubt that, through intensification of this gallant fight with formidable new chemical weapons, man has been instrumental in the past several decades in accelerating certain microevolutionary processes which otherwise might have gone unnoticed. For many insect species, the environmental changes of recent years have been most sudden and overwhelming, as one insecticide after another, with complete disregard for selectivity, has been poured in their paths...so the insects have responded valiantly with a weapon of their own...a phenomenon well known to us as resistance."

In view of the steady increase in the use of synthetic organic insecticides following the introduction of DDT after World War II it is not surprising that there was also a commensurate increase in the number of species becoming resistant to insecticides. From a single case of resistance to lime sulfur which was detected in a scale insect in 1905, the number of resistance cases grew to 14 in the early 1940s and by the end of 1986 it proliferated to 490 species of insects and mites that had developed resistance to one or more types of insecticides (Georghiou 1987). This number is possibly much larger today, especially since many resistance cases are not reported.

The frequency of occurrence of resistance in various orders of arthropods (Table 29.1) (Georghiou and Saito 1983; Georghiou 1987) reflects the extent to which members of each order have been subjected to chemical stress. The large number of resistant species in the order Diptera (36% of the total) is an indication of the widespread use of insecticides by international organizations to control flies and mosquitoes that are vectors of human diseases. Even larger numbers of resistant species are found among orders of insects and mites that are pests of agricultural crops.

When arranged by chemical groups of insecticides from more recent origin (Georghiou 1990; Table 29.2) 58% of the species show resistance to the cyclodiene class, 52% can resist DDT, 52% are resistant to organophosphorus compounds, 17% to carbamates, 9.5% to pyrethroids, and about 10% to miscellaneous compounds. Obviously, some species are resistant to more than one class of insecticides. These percentages reflect not only the extent of usage of each type of insecticide, but also the relative persistence of the chemical in the environment, for example, the organochlorine-type insecticides is the most persistent.

29.1 Origin and Development of Resistance

It is widely accepted that the development of insecticide resistance by insects is due to the selection of variants in the population carrying preadaptive genes for resistance. The rate at which resistance develops in a population depends on: (1) the frequency of resistant genes present in the population; (2) the nature of these genes (either single or multiple, dominant or recessive); (3) the intensity of selection pressure, that is, the magnitude of the population exposed to the chemical and the proportion killed; and (4) the rate at which the species breeds (number of generations per year). Experience has shown that factors (3) and (4) are the most important. In some instances, geographical isolation, confined populations, or highly inbred strains of insects may not contain the prerequisite resistant genes; therefore, these may fail to develop resistance.

Toxicological observations indicate that there exists a dynamic variation in tolerance to a poison among individuals of a given population. However, truly sublethal doses of insecticides do not induce resistance in susceptible populations, or increase the level of enzymes responsible for the metabolism of the chemical. On the contrary, the cumulative effect of daily sublethal doses of insecticide may render the insect more susceptible to the poison.

Table 29.1. Number of species of arthropods with reported cases of resistance to pesticides through 1980. (Georghiou and Saito 1983)

Order	Pesticide group[a]							Importance[b]		Total (%)
	DDT	Cyclod.	OP	Carb.	Pyr.	Fumig.	Other	Med./Vet.	Agr.	
Acarina	17	15	42	6	1		30	15	38	53 (12.4)
Anoplura	4	4	2	1				6		6 (1.4)
Coleoptera	24	55	26	9	3	14	5		64	64 (14.9)
Dermaptera	1	1							1	1 (0.2)
Diptera	106	107	60	11	6		1	130	23	153 (35.8)
Ephemeroptera	2								2	2 (0.5)
Hemiptera/Heteroptera	8	16	6					4	16	20 (4.7)
Hemiptera/Homoptera	13	13	28	9	3	3	1		42	42 (9.8)
Hymenoptera	1	3							3	3 (0.7)
Lepidoptera	40	40	31	14	8		2		64	64 (14.9)
Mallophaga		2						2		2 (0.5)
Orthoptera	3	3	2	1	1			3		3 (0.7)
Siphonaptera	7	5	2					8		8 (1.9)
Thysanoptera	3	5	1				2		7	7 (1.6)
Total	229	269	200	51	22	17	41	168	260	428
(%)	(53.5)	(62.9)	(46.7)	(11.9)	(5.1)	(4.0)	(9.6)	(39.3)	(60.7)	(100)

[a] Cyclod, cyclodiene; OP, organophosphate; Carb, carbamate; Pyr, pyrethroid; Fumig, fumigant.
[b] Med./Vet, medical/veterinary; Agr, agricultural.

Table 29.2. The occurrence of resistance in insects and mites according to pesticide chemical group. (Georghiou. 1990)

Chemical group	Number of species	Percent of total
Cyclodiene	291	57.7
DDT	263	52.2
Organophosphate	260	51.6
Carbamate	85	16.9
Pyrethroid	48	9.5
Fumigant	12	2.4
Other	40	7.9

Many factors are involved in insect resistance to insecticides. The most important of these are increased detoxication of the insecticide (physiological factor), reduced penetration of the chemical (morphological factor), and altered site of action (insensitivity of target site). Resistance has also been claimed to be associated with changes in behavior following insecticide treatment, especially in houseflies and mosquitoes. However, these changes could not be proven to be inheritable characteristics; hence, they might constitute a deterrent action due to irritability caused by certain insecticides such as DDT.

Morphological factors are associated with the penetration of the insecticide through the insect's cuticle, and many reports indicated that slower penetration of the insecticide might account for resistance. However, this is not of universal occurrence, and unless another mechanism, such as enhanced detoxication, is present, slower penetration can only contribute to a delaying action of the toxic effect. Hence, penetration might be said to supplement another, more efficient, mechanism for survival.

Behavioral resistance has been shown with certain strains of anopheline mosquitoes that will not settle on surfaces sprayed with DDT or enter buildings that have been treated with DDT. Certain strains of houseflies have been shown to avoid resting on surfaces treated with DDT or malathion. Similarly, codling moth larvae, *Carpocapsa pomonella*, have evolved the habit of discarding the first bite of the apple when they bore into it, to avoid ingesting the poison which the apples have been sprayed with.

By far the most important factor underlying resistance is physiological or biochemical, that is, the ability of the resistant insect to breakdown the chemical more efficiently and at a faster rate than its susceptible counterpart. Ordinarily, the breakdown products are less toxic or nontoxic and many can be conjugated and excreted. Evidence for this is quite universal among resistant insects and numerous publications have appeared on this subject as well as on the genetics of insecticide resistance (see reviews by Oppenoorth 1965, 1985; Brown and Pal 1971; Agosin and Perry 1974; Perry and Agosin 1974; Oppenoorth and Welling 1976; Plapp 1976; Georghiou and Saito 1983; and many others).

At the biochemical level, the principal enzymes associated with metabolism of insecticides include: (1) the cytochrome P-450 mediated mixed-function oxidases (MFO) which introduce oxygen, often as an hydroxyl (OH) group, into the molecule; (2) esterases which hydrolyze esters and amides rendering the molecule less toxic and more polar for faster excretion; (3) glutathione S-transferases which catalyze the conjugation of the molecule with the thiol (SH) group of glutathione and the demethylation of *O,O*-dimethyl phosphorus esters; (4) the enzyme DDT-dehydrochlorinase which converts DDT to the noninsecticidal metabolite DDE. These primary metabolic changes are often complemented by secondary processes involving conjugation of the metabolized molecule with sugars, amino acids, or sulfate groups to make the compound even more water soluble and facilitate its excretion. Examples of these reactions have been given under the heading "metabolism" in the discussions of the respective classes of insecticides.

Microsomal mixed-function oxidases play a vital role in both the activation (making the molecule more toxic) and detoxification (making the molecule less toxic) of many organophosphorus, carbamate, and pyrethroid insecticides. Resistance to DDT and related compounds in certain insect species is also due to detoxification via this enzyme system. Esterases play a major role in resistance to certain organophosphorus insecticides (for example, carboxyesterase in malathion metabolism; carboxyamidase in breakdown of dimethoate), certain carbamates (for example, carbaryl metabolism in the German cockroach) and, especially, in many cases of resistance to the newer pyrethroid insecticides, such as permethrin, cypermethrin, and deltamethrin. Recently, an esterase B_1 was shown to be responsible for the extremely high resistance to temephos (as high as 800-fold resistance) in the mosquito *Culex quinquefasciatus* (Georghiou and Pasteur 1978; Georghiou et al. 1980). Studies of the molecular genetics of this type of resistance known as "gene amplification" resulted in the cloning of esterase B_1 and the demonstration that the gene encoding the esterase enzyme is amplified 250 times (Mouchès et al. 1986, 1987) and that the esterase protein represents 6–12% of the total soluble protein present in the resistant mosquito (Fournier et al. 1987). Increased levels of esterase production due to gene amplification occurs in other resistant species of *Culex* and in the resistant peach aphid *Myzus persicae* (Devonshire and Field 1991). It was earlier shown that gene amplification occurs in cultured mammalian cells that develop resistance to cytotoxic substances (review by Devonshire and Field 1991).

29.2 Target Site and Metabolic Resistance

29.2.1 Altered Cholinesterase

The first indication that resistance might be associated with changes in acetylcholinesterase (AChE) was the observation that in organophosphorus (OP)-resistant spider mites, *Tetranychus urticae*, the affinity of AChE toward acetylcholine (ACh) was only one-third that of the susceptible counterpart (Smissaert 1964). Subsequent investigations into the rate of inhibition of AChE by various OP compounds confirmed this and pointed out that these differences were due to alterations in the enzyme itself. A good correlation was found between resistance to malathion in seven strains of mites and the "insensitivity index" of AChE, which is the ratio of the dose required for 50% inhibition in the resistant strain divided by that of the susceptible strain (Zahavi and Tahori 1970a,b). Insensitivity of AChE is also a major factor in OP resistance in the cattle tick *Boophilus microplus*, in the resistance of the green rice leafhopper *Nephotettix cincticeps* to carbamate insecticides, in strains of houseflies resistant to OP insecticides, in *Anopheles albimanus* resistant to parathion and propoxur, and in other species as well.

The existence of numerous resistant species with target insensitivity toward cycodiene insecticides (such as dieldrin) and their cross-resistance to heptachlor, aldrin, chlordane, endrin, toxaphene, and HCH is well documented (Brown 1971). Among these, dieldrin resistance in mosquitoes and houseflies is well known. Target insensitivity in this case is of a different nature in that it does not involve the enzyme AChE, but results from the lower sensitivity of the central nervous system, such as reduced binding of the chemical to nerve components.

29.2.2 Resistance Due to kdr Gene

Resistance due to kdr gene was thoroughly studied genetically and was found to be due to a recessive gene called kdr for "knockdown resistance." The kdr factor confers resistance to DDT and pyrethroids through a target insensitivity mechanism (Milani and Travaglino 1957). This type of resistance conferred by a recessive gene residing on chromosome 3 in the housefly is not associated with enhanced detoxification of DDT by the enzyme DDT-dehydrochlorinase (residing on chromosome 2) or with oxidative enzymes involved in the metabolism of DDT (carried on chromosome 5). The kdr factor confers cross-resistance to natural pyrethrins and to synthetic pyrethroids. Reduced nerve sensitivity due to the kdr factor has been demonstrated in DDT- and pyrethroid-resistant houseflies, in pyrethroid-resistant *Anopheles albimanus*, and in DDT- and permethrin-resistant German cockroaches. Nerve preparations from houseflies carrying the kdr gene were found to bind lower amounts of radiolabeled pyrethroids than susceptible houseflies. The inheritance of a specific type of resistance is generally simple and often depends on a single gene, although the influence of the principal gene may sometimes be modified by secondary genes. This might be one reason why resistance develops so quickly in the field.

29.2.3 DDT-dehydrochlorinase (DDT-ase)

DDT-ase is responsible for the in vitro conversion of DDT to the noninsecticidal metabolite DDE. The same reaction occurs in vivo. A correlation between resistance and DDT-ase activity was shown in many strains of houseflies, the Mexican bean beetle, in many American (but not Asiatic) strains of the mosquito *Aedes aegypti*, in certain *Culex* species, but not in several anopheline species (reviews by Perry 1974; Oppenoorth and Welling 1976). The correlation of resistance with enzymatic detoxification of insecticides can, in many cases, be ascertained by the use of synergists. The compound 1,1-bis(p-chlorophenyl) ethanol (DMC), a noninsecticidal analog of DDT, was found to inhibit the conversion of DDT to DDE in DDT-resistant houseflies and in *Ae. aegypti*, thus confirming the role of detoxification in resistance. A large number of compounds have been screened for this purpose and the only one which attained commercial use albeit with limited success was N,N-dibutyl (p-chlorobenzene) sulfonamide, designated as WARF-antiresistant. The lack of success in the practical use of these synergists is due to the fact that selection of the insects with DDT–DMC or with DDT–WARF antiresistant confers resistance to the combinations and cross-resistance to other compounds.

29.2.4 Mixed-Function Oxidases

The important role of mixed-function oxidases in insecticide metabolism is well established. Their role in resistance to insecticides was shown from the action of certain synergists known to inhibit these enzymes. In particular, the group of compounds containing the methylenedioxyphenyl moiety, such as piperonyl butoxide, piperonyl cyclonene, sulfoxide, and sesamex, were found to inhibit the enzymatic oxidations of DDT, carbamates, pyrethrins, and pyrethroids, and the oxidation of organophosphorus insecticides and oxidative metabolism of other compounds, extending to juvenile hormones, hormone mimics, precocenes, etc. Numerous examples exist in the literature (reviews by Casida 1970; Wilkinson, 1971, 1976a,b; Yamamoto 1973; Agosin and Perry 1974; Oppenoorth and Welling 1976; and others).

29.2.5 Esterases

It is well established that esterase enzymes play important roles in detoxification of organophosphorus (OP) insecticides. The most clear-cut case of the importance of esterase activity in resistance to OP compounds is malathion resistance. This was first shown in *Culex tarsalis* resistance to malathion to be due to the higher activity of carboxyesterase in the resistant strain than in the susceptible counterpart (Matsumura and Brown 1961). Malathion resistance is unique in that it does not confer cross-resistance to other OP insecticides. A similar resistance was subsequently shown in the green rice leafhopper, in several housefly strains, spider mites, flour beetle, blowfly, bedbug, and other insects. In general, malathion resistance was due, mainly, to this carboxy-esterase-mediated detoxification mechanism. Other esterases, such as phosphatases (which are enzymes hydrolyzing OP insecticides at the P-O-C or P-S-C bonds), have been suggested as being important in resistance to phosphates (such as "oxons"), but resistance levels to "thions," such as parathion and diazinon, are much higher than to the corresponding "oxons." At the present time, most scientists agree that phosphatases play only a minor role, if any, in OP resistance. Esterase inhibitors such as n-propyl paraoxon, EPN-oxon, triphenyl phosphate (TPP), tri-ortho-cresyl phosphate (TOCP), and S,S,S-tributyl-phosphorotrithioate (DEF) act as synergists for malathion and other compounds and are therefore useful in determining the type of resistance.

Pyrethroid esterases of *Trichoplusia ni* (Ishaaya and Casida 1980) and *Spodoptera littoralis* (Ishaaya et al. 1983) consistently hydrolyze the *trans*-isomer of permethrin and cypermethrin more extensively than the *cis*-isomer, whereas the lacewing larva, *Chrysopa carnea*, has a unique specificity for hydrolyzing the *cis*-isomer (Ishaaya and Casida 1981). Profenofos fed to *T. ni* larvae at a level inhibiting the gut pyrethroid esterases by 65% with *trans*-permethrin, and by >90% with *cis*-cypermethrin, increased the toxicity of topically applied *trans*-permethrin fourfold and *cis*-cypermethrin 20-fold, resulting in a good correlation between the in vivo larval gut pyrethroid esterase(s) inhibition and synergized toxicity of *cis*-cypermethrin (Ishaaya and Casida 1980). Addition of monocrotophos to a cypermethrin formulation at a ratio of 1:1 and of profenofos at a ratio 1:4 strongly synergized the toxicity of cypermethrin and prolonged its activity on ***Bemisia tabaci*** (Ishaaya et al. 1987). According to log dose-probit mortality plots, monocrotophos synergized the toxicity of cypermethrin about 50- and 38-fold at 3 and 10 days after treatment, respectively (Ishaaya et al. 1987). The toxicity of permethrin and cypermethrin was synergized considerably (up to 30-fold) by addition of DEF against susceptible and resistant strains of *B. tabaci* (Horowitz et al. 1988). Similarly various OP compounds synergized the toxicity of cypermethrin, fenvalerate, and deltamethrin against *S. littoralis* larvae (Ascher et al. 1986).

OP-sensitive esterase activity seems to be an important means of pyrethroid detoxification in lepidopterous species such as *S. littoralis* (Ishaaya et al. 1983), ***Spodoptera eridania*** (Abdel-Aal and Soderlund 1980) and *T. ni* (Jao and Casida 1974; Ishaaya and Casida 1980), and in whiteflies (Ishaaya et al. 1987). However, the esterase inhibitor profenofos was not synergetic with any of the cyano-pyrethroids for ***Tribolium castaneum*** larvae. In contrast, the oxidase inhibitors piperonyl butoxide, *0,0*-diethyl-*0*-phenylphosphorothioate (known as SV-1), and *0*-(2-methylpropyl) *0*-2-propynyl phenylphosphonate (known as MPP and Niagara 16824) synergized considerably the toxicity of pyrethroid against ***Tribolium*** and housefly larvae (Ishaaya et al. 1983, 1984). Thus, the predominant pathway for pyrethroid detoxification, whether hydrolytic or oxidative, depends largely on the insect species involved.

29.2.6 Glutathione S-Transferase and Glutathione System

This enzyme (GST) catalyzes glutathione conjugation and is important in the detoxication of many xenobiotics. The involvement of GST in the metabolism of organophosphorus insecticides was first indicated by Shishido and Fukami (1963) and Fukami and Shishido (1966), implying the importance of GST in the insecticide resistance mechanism. GST plays an important role in the resistance of a predacious mite, *Neoseiulus fallacis* (Motoyama et al. 1971) and housefly (Motoyama and Dauterman 1972) to azinphosmethyl. On the other hand, in the resistance of *Sitophilus granarius* to methyl bromide, GST activity is not altered, but the endogenous glutathione is considerably increased (Starratt and Bond 1981)

29.2.7 Cross-Resistance

When a specific mechanism confers resistance to two or more compounds, the phenomenon is termed "cross-resistance" since this involves the same gene for the different chemicals. For example, when an organism becomes resistant to DDT, it generally shows resistance to related compounds such as DDD, although not to the cyclodiene (dieldrin) and HCH (lindane) insecticides which fall under a different cross-resistance group. Organophosphorus insecticides may be divided into two major cross-resistance groups exemplified by parathion and malathion, and insects resistant to these groups are often cross-resistant to carbamate insecticides. Pyrethroids offer another cross-resistance group. In general, cross-resistance involves compounds having the same mode of action or similar metabolic pathway. When different resistant mechanisms exist in an insect it is said to show multiple resistance. This can happen when an insect population has been selected with insecticides having different modes of action.

29.2.8 Use of Synergists

It is sometimes difficult to obtain a clear-cut indication of the role played by detoxification of the chemical in vitro as a factor in resistance. An example is shown by DDT-resistance in the body louse in which both the resistant and the susceptible strains metabolized DDT in vitro with equal facility, but only the resistant strain accomplished this in vivo (Perry and Buckner 1958). The reason for this discrepancy has never been established, but it appears that another factor is involved in this resistance. The use of synergists (see Chap. 15) often can establish the type of resistance, whether by oxidase attack, by the enzyme DDT-dehydrochlorinase, or by esterase hydrolysis.

29.3 Management

The resistance problem, of all other factors, probably has had the greatest impact on pest control. The development of resistance to pesticides means control failure, economic loss, increased hazards due to higher quantities of pesticides needed for control of pests, and the uncertainty that any new chemical introduced for that purpose, and at a very high cost to the manufacturer, would not suffer the same fate after a short period of use. The financial costs of resistance are difficult to assess, but it is estimated to be at least US $ 1 billion on a worldwide basis (Georghiou 1986). Hence, management of resistance so that its development can be avoided or minimized is of paramount importance.

Resistance is an inevitable consequence of the intensive selection pressure with an insecticide against an insect population. Literally thousands of cases of resistance to the commonly used insecticides have been reported (Brown and Pal 1971; Georghiou and Saito 1983). From more recent laboratory experiment applying selection pressure with some of the newer type insecticides (see Georgniou 1987), it is clear that resistance appeared in each case, but the degree of resistance to each chemical and the number of generations required for its selection differed markedly. For example, resistance developed most rapidly and to the highest level with permethrin and most slowly and at the lowest level with the toxin of *Bacillus thuringiensis israelensis* (B.t.i.).

As a result of research in a few laboratories over the past several years, a set of principles for resistance management has emerged. The aim of these principles is to maintain the level of susceptibility of the pest population within limits that permit the continued use of the insecticide. These strategies have been grouped under three categories, that is (1) management by moderation; (2) management by saturation; and (3) management by multiple attack (Georghiou 1983).

The idea behind "management by moderation" is that susceptibility genes are a valuable resource and that the frequency of resistance genes in a population can be prevented from rising by selecting less rigorously by using lower doses of insecticides. This would maintain sufficient numbers of susceptible phenotypes. The use of insecticides with short environmental persistence is preferred in this case, since residual compounds of long duration would continue to select resistance genes for longer periods. Other measures in this category include less frequent applications of insecticides; leaving part of the population of a particular life stage of the pest untouched with insecticide, thus allowing many susceptible refugees to survive; treatment of only one stage of the life cycle, with emphasis on adulticides rather than larvicides where possible. These measures are rather conservative and must be supplemented by other components of nonchemical control, consistent with the principles of integrated pest management.

While management by moderation is desirable from an environmental standpoint and is less destructive to natural enemies of pests, it may not be acceptable where high value crops must be protected or newly introduced pests eradicated. "Management by saturation" is an alternative. This does not imply saturation of the environment with pesticides. It rather indicates saturation of the insect's defense mechanisms by insecticide dosages high enough to overcome resistance. The aim is to kill off most heterozygous individuals, by means of suppressing their detoxification mechanisms, for example, during the early stages of selection of a population where resistance genes are rare.

This approach may require great caution, for it is not absolutely certain that all heterozygotes would be killed even at very high doses. Chances are, too, that a high dose treatment will also eliminate the natural enemies of the pest. Management by saturation can succeed if the heterozygotes in the population are only slightly more resistant than the susceptible individuals so that a small increase in dose can result in a large gain. This can be followed by a "saturation" treatment of a different type, such as pheromone traps, which would be more compatible with integrated pest management.

"Management by multiple attack" is a more practical approach when the moderation approach is not feasible. This is based on the premise that control can be achieved through the action of several independently acting stresses, including insecticides, each exerting selection pressure that is below the level that might lead to resistance. In this approach, a number of different strategies are available:

1. "Rotation" or alteration of treatment with different insecticides.
2. The use of "mixtures" of insecticides.
3. The use of "synergists" to suppress the insect's detoxification mechanisms.
4. The use of negatively correlated resistance.
5. The use of pesticide-resistant predators and parasites.
6. The use of multiple components of the integrated pest management system in combination with insecticide treatments.

The experimental basis of the rotation approach was provided by Lagunes

(1980) who demonstrated that a strain of the mosquito *Culex quiquefasciatus* that contained resistance genes for temephos, propoxur, and permethrin at low frequencies did not develop resistance when a rotation scheme was used to alternate these three chemicals with different actions. Since selection of the same strain with one chemical always developed a high degree of resistance within five generations, the above experiment showed the usefulness of the rotation approach. Among these three insecticides, propoxur showed a propensity for causing resistance development, while temephos and permethrin suppressed resistance development for each other. The strategy of using rotation is based on the knowledge that resistant individuals have lower biotic fitness than susceptible counterparts, so that their frequency tends to decrease when the alternative insecticide is being used.

29.3.1 Insecticide Mixtures

Mixtures of certain insecticides with different modes of action can nullify each other's resistance when selection is made with a combination of the chemicals. As an example, when temephos, propoxur, and permethrin were used in various combinations for selection of resistance in a *Culex* mosquito, resistance to propoxur was always evident, but no resistance occurred with either temephos or permethrin after eight generations of selection. When the chemicals were used singly, a high degree of resistance to each compound appeared after five generations of selection. While combinations of insecticides might be useful in suppressing resistance, one must be cognizant of the possibility that resistance to insecticide mixtures could also develop.

Restriction of insecticide use, especially pyrethroids and insect growth regulators (IGRs), to "windows" when they are most needed, alternation of insecticides with different modes of action, and preserving the natural enemies when they are most active served as components in an insecticide resistance management strategy aimed at preventing or delaying resistance development in Israeli cotton pests to IGRs and other novel compounds (Horowitz and Ishaaya 1992; Horowitz et al. 1993, 1994).

29.3.2 Use of Synergists

Once the mechanism of resistance is known, the use of synergists can have the same effect as mixtures of insecticides have in suppressing resistance. The mode of action of synergists has already been discussed in Chapter 15. Their function is to inhibit the breakdown of the chemical. Since physiological resistance, in many instances (but not in all), is associated with enhanced breakdown of the chemical, inhibition of this process by synergists might prevent the buildup of resistance. As discussed in Section 29.2, several enzyme systems are involved in the metabolism of insecticides. Prominent among these are mixed-function oxidases, esterases, glutathione-S-tranferases, and DDT-dehydrochlorinases. There are specific synergists for each one of these enzyme reactions that lead to inhibition of detoxification and, hence, to suppression of the resistance mechanism. It should be emphasized that synergist–insecticide mixtures may also lead to the development of resistance to the combination, as has happened with mixtures of DDT–DMC and DDT–WARF-antiresistant compound. Another disadvantage is that the combination may eliminate the selectivity of the insecticide by increasing the danger to natural enemies and to other nontarget organisms.

29.3.3 Negatively Correlated Insecticides

It has been shown that in certain circumstances resistant insects develop higher sensitivities to certain insecticides than their susceptible counterparts. The green rice leafhopper, *Nephotettix cincticeps*, has developed cross-resistance to whole aryl N-methylcarbamates due to alteration of the acetylcholinesterase (AChE). The altered AChE became insensitive to the N-methylcarbamates, while it became more sensetive to N-propylcarbamates. For the field resistant strain, the combination of N-methyl- and N-propylcarbamates was effective, while N-propylcarbamate was powerful in controlling the strain highly resistant to N-methylcarbamates (Yamamoto et al. 1983). Oxadiazolone compounds are also negatively correlated with the resistant development (Yamamoto et al. 1993). Other instances of such cases have been reported by Brown (1971).

29.3.4 Resistant Predators and Parasites

From the standpoint of resistance management it is advantageous to restrict the use of pesticides to a minimum. Because natural enemies are less likely to develop resistance to pesticides in the field (Croft 1977), it is of interest to develop pesticide-resistant strains of predators and parasites by selection for resistance in the laboratory and then releasing them in the field. A number of such attempts have been made, some of which have proven to be highly successful (Havron 1983; Hoy and Conley 1987; Hoy and Cave 1988; Rosenheim et al. 1989). This subject has been reviewed by Croft and Strickler (1983).

29.3.5 Impact of Resistance on Control of Disease Vectors

The control of disease vectors has, for many years, depended largely on the use of organochlorine insecticides, especially DDT, HCH, and dieldrin, and later on organophosphorus, carbamate, and pyrethroid insecticides. Resistance to these insecticides has impeded the control of many insect vectors and the diseases they transmit. Notable among these are many anopheline vectors of malaria, the mosquito vector of yellow fever, *Aedes aegypti*, the common house mosquito *Culex fatigans* which spreads filariasis, the louse vectors of typhus, the tropical rat flea which carries plague, the triatomid bug which transmits Chagas' disease, and other species such as black flies (*Simulium*), tsetse flies (*Glossina*), sandflies (*Phlebotomus*), ticks and mites, snails, etc., which transmit other diseases but in which resistance is not as marked. Resistance of houseflies (Brown and Pal 1971; Keiding 1977), which are mechanical transmitters of the viral disease trachoma causing human blindness, and cockroaches, *Blattella germanica*, which have been strongly implicated in the mechanical transmission of hepatitis virus, nematodes, typhoid, and dysentery, is of worldwide occurrence.

Resistance in arthropods that are characteristically not vectors of disease in man (blow flies, flesh flies, horn flies, stable flies, *Drosophila* flies, bed bugs, midges, etc.) (Brown and Pal 1971) is also widespread. Rodent resistance to certain rodenticides (for example, warfarin) has also occurred in many countries. Among the veterinary pests, the beef industry in Australia is hampered by the high level of resistance of cattle ticks, and sheep farming is plagued by insecticide-resistant blowflies (Drummond 1977).

To counteract resistance in disease vectors, insecticides should always be used in combination with other feasible control measures. Some insect vectors of disease can be eliminated from human dwellings by improved housing, improved water-borne sewage systems, better drainage and sewage disposal, and improved general sanitation. These, of course, may be too expensive to be implemented in developing countries at the present time, but health measures should be given high priorities in any country and under any political system.

References

Abbott WS (1925) A method of computing the effectiveness of an insecticide. J Econ Entomol 18: 265–267

Abdel-Aal YAI, Soderlund DM (1980) Pyrethroid-hydrolyzing esterases in southern armyworm larvae: tissue distribution, kinetic properties and selective inhibition. Pestic Biochem Physiol 14: 282–289

Agosin M, Perry AS (1974) Microsomal mixed-function oxidases. In: Rockstein M (ed) The physiology of insecta, vol 5. Academic Press, New York, pp 538–597

Allcroft R, Salt FJ, Peters RA, Shorthouse M (1969) F-acetamide poisoning II. Toxicity in dairy cattle—confirmation of diagnosis. Vet Rec 84: 403–409

Aller HE, Ramsay JR (1988) RH-5849—a novel insect growth regulator with a new mode of action. In: Brighton Crop Protection Conf, Pests and Diseases. British Crop Protection Council, Farnhamy pp 511–518

American Conference of Governmental Industrial Hygienists (ACGIH) (1989–1990). Threshold limit values and biological exposure indices. ACGIH, Cincinnati, Ohio

Ames BN (1974) A combined bacterial and liver test system for detection and classification of carcinogens as mutagens. Genetics 78: 91–95

Ames BN, McCann J, Yamasaki E (1975) Methods for detecting carcinogens and mutagens with the Salmonella/mammalian-microsome mutagenicity test. Mutat Res 3: 347–364

Anderson LD, Atkins EL Jr (1968) Pesticide usage in relation to beekeeping. Annu Rev Entomol 3: 213–238

Andreadis TG (1990) Polymorphic microsporidea of mosquitoes. Potential for biological control. In: Baker RR, Dunn PE (eds) New direction in biological control: alternatives for suppressing agricultural pests and diseases. Liss, New York, pp 77–78

Angus TA (1971) *Bacillus thuringiensis* as a microbial insecticide. In: Jacobson M, Crosby DG (eds) Naturally occurring insecticides. Dekker, New York, pp 463–493

Anonymous (1987) Applaud, a new pesticide (insect growth regulator), technical information. Nihon Nohyaku, Tokyo

Araki S (1972) Studies on organofluorine pesticides from the medico–legal viewpoints. Jpn J Leg Med 26: 203–219

Arambourgh Y, Pralavario R, Dolbeau C (1977) Premières observations sur l'action du diflubenzuron (PH 6040) sur la fécondité, la longévité et la viabilité des oeufs de *Ceratitis capitata* Wied (Diptera: Trypetidae) Rev Zool Agric Pathol Veg 76: 118–126

Aronson A, Wu D (1990) Specificity in vivo and vitro of *Bacillus thuringiensis* δ-endotoxins active on Lepidoptera In: Baker RR, Dunn PE (eds) New directions in biological control: alternatives for suppressing agricultural pests and diseases. Liss, New York, pp 547–560

Ascher KRS (1993) Nonconventional insecticidal effects of pesticides available from the neem tree *Azadirachta indica*. Arch Insect Biochem Physiol 22: 433–449

Ascher KRS, Ishaaya I (1973) Antifeeding and protease and amylase inhibition activity of fentin acetate in *Spodoptera littoralis* larvae. Pestic Biochem Physiol 3: 326–336

Ascher KRS, Nemny NE (1974) The ovicidal effect of PH 60-40 [1-(4-chlorophenyl)-3-(2,6-difluorobenzoyl)-urea] in *Spodoptera littoralis* Boisd. Phytoparasitica 2: 131–133

Ascher KRS, Nemny NE (1976a) Toxicity of the chitin synthesis inhibitors, diflubenzuron and its dichloro-analogue, to *Spodoptera littoralis* larvae. Pestic Sci 7: 1–9

Ascher KRS, Nemny NE (1976b) Contact activity of diflubenzuron against *Spodoptera littoralis* larvae. Pestic Sci 7: 447–452

Ascher KRS, Eliahu M, Ishaaya I, Zur M, Ben-Moshe, E (1986) Synergism of pyrethroid–organophosphorus insecticide mixtures in insects and their toxicity against *Spodoptera littoralis* larvae. Phytoparasitica 14: 101–110

Awake (1983) Is man winning the war against insects? Watchtower Bible and Tract Society, New York, vol 64, pp 20–23

Barthel WF (1973) Toxicity of pyrethrum and its constituents to mammals. In: Casida JE (ed) Pyrethrum: the natural insecticide. Academic Press, New York, pp 123–142

Baum D, Yablonski S, Ishaaya I (1990) Biological mode of action of benzoylphenyl ureas on the grapevine moth *Lobesia botrana* Den & Schiff (Lepidoptera: Tortricidae). Abstr 7th Int Congr Pestic Chem, IUPAC, Hamburg, vol 1, p 375

Baum D, Yablonski S, Ishaaya I (1992) The ovicidal effect of some benzoylphenyl ureas on the grape berry moth *Lobesia botrana*. Phytoparasitica 20: 81

Baumhover AH, Graham AJ, Bitter BA, Hopkins DE, New WD, Dudley FH, Bushland RC (1955) Screwworm control through release of sterilized flies. J Econ Entomol 48: 462–466

Becher HM, Becker P, Procic-Immel R, Wirtz W (1983) CME 134, a new chitin synthesis inhibiting insecticide. 10th Int Congr Pl Prot, Brighton, vol 1, pp 408–415

Beers WH, Reich E (1970) Structure and activity of acetylcholine. Nature 228: 917–922

Begley MJ, Crombie L, Hadi HB, Josephs JL (1990) Synthesis and comparison of unnatural *trans*-B/C rotenoids with the natural *cis*-stereoisomers. In: Casida JE (ed) Pesticides and Alternatives. Elsevier, Amsterdam, pp 311–315

Bell J (1972) The acute toxicity of four common poisoning to the opossum, *Trichosurus vulpecula*. N Z Vet J 20: 212–214

Bentley EW, Greaves JH (1960) Some properties of fluoroacetamide as a rodenticide. J Hyg (Camb) 58: 125–132

Bergamsco R, Horn DHS (1980) The biological activities of ecdysteroids and ecdysteroid analogues. In: Hoffman JA (ed) Progress in ecdysone research. Elsevier, Amsterdam, pp 299–324

Beroza M (1976) Pest management with insect sex attractants. ACS symposium series no 23. American Chemical Society, Washington, DC

Berteau PE, Spath DP (1986) Toxicological and epidemiological effects of pesticide contamination in California groundwater. In: Garner WY, Honeycutt RC, Nigg HN (eds) Evaluation of pesticides in ground water. ACS symposium series. American Chemical Society, Washington DC, pp 423–435

Bierl BA, Beroza M, Collier CW (1970) Potent sex attraction of the gypsy moth; its isolation, identification, and synthesis. Science 170: 17–89

Bigley BA, Plapp FW (1978) Metabolism of *cis*- and *trans*-(^{14}C) permethrin by the tobacco budworm and the bollworm. J Agric Food Chem 26: 1128–1134

Birch MC (1974) Pheromones. American Elsevier, New York, 495 pp

Black BC, Hollingworth RM, Ahammadsahib KI, Kuhel CD, Donovan S. (1994) Insecticidal action and mitochondrial uncoupling activity of AC-303, 630 and related halogenated pyrroles. Pestic Biochem Physiol 50(2): 115–128

Blum M (1978) Biochemical defenses of insects. In: Rockstein M (ed) Biochemistry of insects. Academic Press, New York, pp 466–513

Bond EJ, de Matteis F (1969) Biochemical changes in rat liver after administration of CS_2 on the liver of rats. Br J Ind Med 26: 335–337

Bond EJ, Butler WH, de Matteis F, Barnes JM (1969) Effects of carbon disulfide on the liver of rats. Br J Ind Med 26: 335–337

Bowers WS (1976) Discovery of insect antiallatotropins. In: Gilbert LI (ed) The juvenile hormones. Plenum Press, New York, pp 398–408

Bowers WS, (1982) Toxicology of precocenes. In: Coats JR (ed) Insecticidal mode of action. Academic Press, New York, pp 403–427

Bowers WS, Martinez-Pardo R (1977) Antiallatotropins: inhibition of *Corpus allatum* development. Science 197: 1369–1371

Bowers WS, Fales HM, Thompson MJ, Uebel EC (1966) Juvenile hormone: identification of an active compound from balsam fir. Science 154: 1020–1021

Bowers WS, Ohta T, Cheere JS, Marsella PA (1976) Discovery of insect anti-juvenile hormone in plants. Science 193: 542–547

Brooks GT (1972) Pathways of enzymatic degradation of pesticides. In: Coulston F, Korte F (eds) Environmental quality and safety, vol 1. Thieme, Stuttgart

Brown AWA (1958) Nature and characterization of resistance. In: Insecticide resistance in arthropods. Monograph series, vol 38. WHO Geneva 240 pp

Brown AWA (1971) Pest resistance to insecticides. In: White-Stevens R (ed) Pesticides in the environment, vol 1, part II. Dekker, New York, pp 457–552

Brown AWA (ed) (1978) Insecticides and terrestrial vertebrates: insecticides and birds. Ecology of pesticides. Wiley, New York, 525 pp

Brown AWA, Pal R (1971) Insecticide resistance in arthropods. Monograph series, vol, 38, WHO, Geneva, pp 255–442

Brown C, Nestler C (1985) Catecholamines and indolalkylamines. Compr Insect Physiol Biochem Pharmacol 11: 435–497

Brown JJ (1994) Effect of nonsteroidal ecdysone agonist, tebufenozide, on host parasitoid interaction. Arch. Insect Biochem. Physiol. 26: 235–248

Browne LB (1977) Host-related responses and their suppression: some behavioral considerations. In: Shorey HH, McKelvey JJ Jr (eds) Chemical control of insect behavior—theory and application. Wiley, New York, pp 117–127

Brusick D (ed) (1980) The consequences of genotoxic effects in human and other mammals. Principles of genetic toxicology. Plenum Press, New York, 279 pp

Büchel KH (ed) (1983) Chemistry of pesticides. Wiley, New York, 518 pp

Buck JB, Keister ML (1949) Respiration and water loss under DDT poisoning on Phormia. Biol Bull 97: 64–81

Buckle JJ, Heap R, Saunders BC (1949) Toxic fluorine compounds containing the C-F link. III. Fluoroacetamide and related compounds. J Chem Soc 1949: 912–916

Bushland RC, Hopkins DE (1953) Sterilization of screwworm flies with X-rays and gamma rays. J Econ Entomol 46: 648–565

Busvine JR (1954) Houseflies resistant to a group of chlorinated hydrocarbon insecticides. *Nature* 174: 783–785

Busvine JR (1964) The insecticidal potency of γ-BHC and the chlorinated cyclodiene compounds and the significance to them. Bull Entomol Res 55: 271–288

Busvine JR (1971) A critical review on the techniques for testing insecticides. Commonwealth Agricultural Bureau, Slough, UK, 345 pp

Carlson DA, Mayer MS, Silhacek DL, James JD, Beroza M, Bierl BA (1971) Sex attractant pheromone of the housefly: isolation, identification, and synthesis. Science 174: 76–78

Carlson SA, Nelson DR, Langley PA, Coates TW, Davis TL, Leegwater-van den Linden ME. (1984) Contact sex pheromone in the tsetse fly *Glossina pallidipes* (Austen). Identification and synthesis. J Chem Ecol 10: 429–450

Carruthers RI, Hural K (1990) Fungi as naturally occurring entomopathogens. In: Baker RR, Dunns PE (eds) New directions in biological control. Alternatives for suppressing agricultural pests and diseases. Liss, New York pp 115–138

Carson R (1962) Silent spring. Houghton-Mifflin, Boston, 368 pp

Casarett LJ, Doull J (1975) Toxicology—the basic science of poisons. Macmillan, New York, 768 pp

Casida JE (1970) Mixed-function oxidase involvement in the biochemistry of insecticide synergists. J Agric Food Chem 18: 753–772

Casida JE (ed) (1973) Pyrethrum, the natural insecticide. Academic Press, New York, 329 pp

Casida JE (1993) Insecticide action at the GABA-gated chloride channel: recognition, progress and prospects. Arch. Insect Biochem Physiol 22: 13–23

Casida JE, Quistad GB (eds) (1995a) Pyrethrum flowers: production chemistry, toxicology, and uses. Oxford University Press, New York, 356 pp

Casida JE, Quistad GB (1995b) Metabolism and synergism of pyrethrins. In: Casida JE, Quistad GB (eds) Pyrethrum flowers: production, chemistry, toxicology, and uses. Oxford University Press, New York, pp 258–276

Casida JE, Kimmel EC, Elliott M, Janes NF (1971) Oxidative metabolism of pyrethrins in mammals. Nature 230: 326–327

Chang CK, Jordan, TW (1982) Penetration and metabolism of topically applied permethrin and cypermethrin in pyrethroid-tolerant *Wiseana cervinata* larvae. Pestic Biochem Physiol 17: 196–204

Chapman HC (1974) Biological control of mosquito larvae. Annu Rev Entomol 19: 33–59

Charles JF, de Barjac H (1981) Histopathologie de l'action de la delta-endotoxine de *Bacillus thuringienses* var. *israelensis* sur les larves d'*Aedes aegypti*. Entomophaga 26: 203–212

Chenoweth MB (1949) Monofluoroacetic acid and related compounds. Pharmacol Rev 1: 383–424

Chenoweth MB, Kandel A, Johnson LB, Bennett DR (1951) Factors influencing fluoroacetate poisoning. Practical treatment with glycerol monoacetate. J Pharmacol Exp Ther 102: 31–49

Cohen DB (1986) Groundwater contamination by toxic substances: a California assessment. In: Garner WY, Honeycutt RC, Nigg HN (eds) Evaluation of pesticides in groundwater. ACS symposium on evaluation of pesticides in groundwater. American Chemical Society, Washington DC, pp 499–529

Cohen E (1987) Chitin biochemistry: synthesis and inhibition. Annu Rev Entomol 32: 71–93

Cohen E (1990) Extracellular biopolymers as targets for pest control. In: Casida JE (ed) Pesticides and alternatives: innovative chemical and biochemical approaches to pest control. Elsevier, New York, pp 23–32

Cohen E, Casida JE (1980a) Properties of *Tribolium* gut chitin synthetase. Pestic Biochem Physiol 13: 121–128

Cohen E, Casida JE (1980b) Inhibition of *Tribolium* gut chitin synthetase. Pestic Biochem Physiol 13: 129–136

Cohen E, Casida JE (1983) Insect chitin synthetase as a biochemical probe for insecticidal compounds. In: Miyamoto J, Kearney PC (eds), Pesticide chemistry, human welfare and the environment. Pergamon Press, Oxford, pp 25–32

Cole LM, Nicholson RA, Casida JE (1993) Action of phenylpyrazole insecticides at the GABA-gated chloride channel. Pestic Biochem Physiol 46(1): 47–54

Conn EE, Stumpf PK, Bruening G, Doi RH (1987) Outlines of biochemistry, 5th edn Wiley, New York, 704 pp

Copplestone JF (1977) A global view of pesticide safety. In: Watson PL, Brown AWA (eds) Pesticide management and insecticide resistance. Academic Press, New York, pp 147–155

Corbett JR, Wright K, Baillie AC (1984) Insecticides acting elsewhere in the nervous system. In: The biochemical mode of action of pesticides, 2nd edn. Academic Press, London, pp 141–178

Croft BA (1977) Resistance in arthropod predators and parasites. In: Watson DL, Brown AWA (eds) Pesticide management and insecticide resistance. Academic Press, New York pp 377–393

Croft BA, Brown AWA (1975) Responses of arthropod natural enemies to insecticides. Annu Rev Entomol 20: 285–335

Croft BA, Strickler K (1983) Natural enemy resistance to pesticides: documentation, characterization, theory and application. In: Georghiou GP, Saito T (eds). Pest resistance to pesticides. Plenum Press, New York, pp 669–702

Cushing EC (1957) History of entomology in World War II. Smithsonian Institute Publications, no 4294. Lord Baltimore Press, Baltimore 117 pp

Dahm PA, Kearns CW (1951) A study of certain metabolic intermediates in the normal and DDT-poisoned housefly adult. Ann Entomol Soc Am 44: 573–580

Dähn U, Hagenmaier H, Höhne H, König WA, Wolf GA, Zähner H (1976) Metabolic products of microorganisms: 154. Nikkomycin, a new inhibitor of fungal chitin synthesis. Arch Microbiol 107: 143–160

Damluji SF, Tikrity S (1972) Mercury poisoning from wheat. Br Med J 1(803): 804

Danse LHJC, Van Velsen FL, van der Heijden CA (1984) Methyl bromide: carcinogenic effects in the rat forestomach. Toxicol Appl Pharmacol 72: 262–271

Darvas B, Polgár L, Tag El-din MH, Erós K, Wing KD (1992) Developmental disturbances in different insect orders caused by an ecdysteroid agonist RH-5849. J Econ Entomol 85: 2107–2112

Davich TB, Lindquist AW (1962) Exploratory studies on gamma-radiation for the sterilization of the boll weevil. J Econ Entomol 55: 164–167

David WAL (1975) The status of viruses pathogenic for insects and mites. Annu Rev Entomol 20: 97–117

Davidson EW (1990) Microbial control of vector insects. In: Baker RR, Dunn PE (eds) New directions in biological control. Alternatives for suppressing agricultural pests and diseases. Liss, New York, pp 199–212

Davies JH (1985) The pyrethroids: a historical introduction. In: Leahey JP (ed) The pyrethroid insecticides. Taylor and Francis, London, pp 1–41

De Barjac H, Coz J (1979) Sensibilité comparée de six espèces differentes de moustiques à *Bacillus thuringiensis* var. *israelensis*. Bull WHO 57: 139–141

De Cock A, Ishaaya I, Degheele D, Veierov D (1990) Vapor toxicity and concentration-dependent persistence of buprofezine applied to cotton foliage for controlling the sweetpotato whitefly (Homoptera: Aleyrodidae). J Econ Entomol 83: 1254–1260

Dekeyser MA, Downer RGH (1994) Biochemical and physiological targets for miticides. Pestic. Sci. 40: 85–101

Dethier VG (1947) Nature of chemical attractants. In: Dethier VG (ed) Chemical insect attractants and repellents. Blakiston, Philadelphia, pp 12–38

Dethier VG (1956) Repellents. Annu Rev Entomol 1: 181–202

Dethier VG, Browne LB, Smith CN (1960) The designation of chemicals in terms of the responses they elicit from insects. J Econ Entomol 53: 134–136

Devonshire AL, Field LM (1991) Gene amplifications and insecticide resistance. Annu Rev Entomol 36: 1–24

Dewitt JB, George JL (1960) US Fish & Wildlife Service, circ 84. US Government Printing Office, Washington DC

Dieke SH, Richter CP (1946) Comparative assays of rodenticides on wild Norway rats. Public Health Rep 61: 672–679

Dillwith JW, Blomquist GJ, Nelson DR (1981) Biosynthesis for the hydrocarbon components of the sex pheromone of the housefly *Musca domestica* L. Insect Biochem 11: 247–253

Drummond RC (1977) Resistance in ticks and insects of veterinary importance. In: Watson DL, Brown AWA (eds) Pesticide management and insecticide resistance. Academic Press, New York, pp 303–319

Dubitskii AM (1985) The biological regulation of *Culicidae* in the Soviet Union. In: Laird M, Miles JW (eds) Integrated mosquito control methodologies, vol 2. Biocontrol and other innovative components, and future directions. Academic Press, Orlando, pp 323–360

Dulmage HT (1981) Insecticidal activity of isolates of *Bacillus thuringiensis* and their potential for pest control. In: Burges HD (ed) Microbial control of pest and plant disease. 1970-1980 Academic Press London, pp 193–222

Edwards CA (ed) (1973) Environmental pollution by pesticides. Plenum Press, London, 542 pp

Edwards CA (1975). Persistent pesticides in the environment 2nd edn. CRC Press, Cleveland 170 pp

Edwards CA (1977) Nature and origins of pollution of aquatic systems by pesticides. In: Khan MAQ (ed) Pesticides in aquatic environments. Plenum Press, New York, pp 11–38

Egyed MN (1971) Experimental acute fluoroacetamide poisoning in sheep. III. Therapy. Refuah Vet 28: 70–73

Egyed MN (1973) Clinical, pathological, diagnostic and therapeutic aspects of fluoroacetate research in animals. Fluoride 6: 215–224

Egyed MN, Brisk Y (1965) Experimental fluoroacetamide poisoning in mice, rats and sheep. Refuah Vet 22: 273–274

Egyed MN, Miller GW (1971) Experimental acute fluoroacetamide poisoning in guinea pig and sheep. Fluoride 4: 137–142

Egyed MN, Shlosberg A (1973) Diagnosis of field cases of NaF acetate and F-acetamide poisoning in animals. Refuah Vet 30: 112–115

Egyed MN, Shlosberg A (1977) The efficiency of acetamide in prevention and treatment of F-acetamide poisoning in chickens. Fluoride 10: 34–37

Eisler R (1970) Acute toxicities of organochlorine and organophosphorus insecticides to estuarine fishes. US Dept. Interior Bureau Sport Fisheries and Wildlife, Tech Pap 46, Washington, DC, 12 pp

Eldefrawi ME, Abalis IM, Filbin MT, Eldefrawi AT (1985) Glutamate and GABA receptors of insect muscles: biochemical identification and interaction with insecticides. In: von Keyserlingk HC, Jaeger A, von Szczepanski C (eds) Approaches to new leads for insecticides. Springer Berlin Heidelberg New York, pp 101–116

Elliott M (1977) Synthetic pyrethroids. In: Elliott M (ed) Synthetic pyrethroids. ACS Symposium Series, no 42. American Chemical Society, Washington, DC, pp 1–28

Elliott M, Janes NF, Kimmel EC, Casida JE (1972) Metabolic fate of pyrethrin I, pyrethrin II, and allethrin administered orally to rats. J Agric Food Chem 20: 300–313

Elliott M, Farnham AW, Janes NF, Needham PH, Pulman DA, Stevenson JH (1973a) Potent Pyrethroid insecticides from modified cyclopropane acids. Nature 244: 456–457

Elliott M, Farnham AW, Janes NF, Needham PH, Pulman DA, Stevenson JH (1973b) A photostable pyrethroid. Nature 246: 169–170

Elliott M, Janes NF, Potter C (1978) The future of pyrethroids in insect control. Annu Rev Entomol 23: 443–469

Ellis-Pratt G (1983) The mode of action of pro-allatocidins. In: Whitehead DL, Bowers WS (eds) Natural products for innovative pest management. Pergamon Press, Oxford, pp 323–355

Endo H, Kakiki K, Misato T (1970) Mechanism of action of the anti-fungal agent polyoxin D. J Bacteriol 104: 189–196

Eto M (1969) Specificity and mechanism in the action of saligenin cyclic phosphorus esters. Residue Rev 25: 187–200

Eto M (1974) Organophosphorus pesticides: organic and biological chemistry. CRC Press, Boca Raton, 387 pp

Eto M (1979) Insecticides. In: Yamashita K, Mizutani J, Fujita T, Marumo S, Eto M, Takahashi N (eds) Textbook of pesticides Buneido Tokyo, pp 131–207

Eto M (1981) Structure and avian teratogenicity of organophosphorous compounds. J Pestic Sci 6: 95–106

Eto M, Casida JE (eds) (1995) Progress and prospects of organophosphorus agrochemicals. Kyushu University Press, Fukuoka, 190 pp

Eto M, Casida JE, Eto T (1962) Hydroxylation and cyclization involved in the metabolism of tri-o-cresyl phosphate. Biochem Pharmacol 11: 337–352

Eto M, Kinoshita Y, Kato I, Oshima Y (1963) Saligenin cyclic methyl phosphate and its thiono analogue: new insecticide related to the active metabolite of tri-o-cresyl phosphate. Nature 200: 171–172

Eto M, Okabe S, Ozoe Y, Maekawa K (1977) Oxidative activation of O.S-dimethyl phosphoramidothiolate. Pestic Biochem Physiol 7: 367–377

Evans PD (1978) Octopamine distribution in the insect nervous system. J Neurochem 30: 1009–1013

Falcon LA (1976) Problems associated with the use of arthropod viruses in pest control. Annu Rev Entomol 21: 305–324

Fang X-P, Reiser MJ, Gu ZM, Zhao G-X, McLaughlin JL (1993) Annonaceous acetogenins: an updated review. Phytochem Anal 4: 27–48

FAO (1977) Evaluation of some pesticide residues in food. Results of the deliberations of the Joint Meeting of the FAO Panel of Experts on pesticide residues and the environment and the WHO Expert Group on pesticide residues, Rome, Nov 22–30, 1976. Food and Agricultural Organization of the United Nations, Rome

FAO (1980) Pesticide residues in food—1979 evaluations. FAO plant production and protection, Pap 20 Suppl. Food and Agricultural Organization of the United Nations, Rome
FAO (1982) Pesticide residues in food—1981 evaluations. FAO plant production and protection, Pap 42. Food and Agricultural Organization of the United Nations, Rome
FAO (1983) Pesticide residues in food—1982 report. FAO plant production and protection, Pap 46. Food and Agricultural Organization of the United Nations, Rome
FAO/WHO (1981) Pesticide residues in food. Report of the 1980 Joint Meeting of the FAO Panel of Experts on pesticide residues in food and the environment and the WHO Expert Group on pesticide residues. FAO Plant Production, and Protection Pap 28 (Suppl). Food and Agricultural Organization of the United Nations Rome
Fashing NJ, Sagan H (1979) Effect of the JH analog methoprene on *Nasonia vitripennis* when administered via a host, *Sarcophaga bullata*. Environ Entomøl 8: 816–818
Fay RW, Perry AS (1965) Laboratory studies of ovipositional preferences of *Aedes aegypti*. Mosq News 25(3): 276–281
Federal Register (1982) Tolerances and exemptions from tolerances for pesticide chemicals in or on raw agricultural commodities; permethrin. Fed Register 47: 45008–45010
Feitelston JS, Quick TC, Gaertner F (1990) Alternative host for *Bacillus thuringiensis* δ-endotoxin genes. In: Baker RR, Dunn PE (eds) New direction in biological control. Alternatives for suppressing agricultural pests and diseases. Liss, New York, pp 561–572
Fest C, Schmidt K-J (1973) The chemistry of organophosphorus pesticides. Reactivity, synthesis, mode of action, toxicology. Springer, Berlin Heidelberg New York, 339 pp
Finney DJ (1949) The adjustment for a natural response rate in probit analysis. Ann Appl Biol 36: 187–195
Finney DJ (1952) Probit analysis, 2nd edn. Cambridge University Press, Cambridge, 318 pp
Fitzhugh OG (1965) Response of mammalian species to pesticides. In: Chichester CO (ed) Research in pesticides. Academic Press, New York, pp 59–67
Fletcher WW (1974) Pesticides and the environment. The pest war. Blackwell, Oxford, pp 146–173
Forbes SA (1915) The insect, the farmer, the teacher, the citizen, and the State Bull Ill State Lab Natural History
Forsberg CW, Henderson M, Henry E, Roberts JR (1976) *Bacillus thuringiensis*: its effect on environmental quality. Environmental Secretariat, National Research Council, Ottawa, Canada, publ no 15385, 135 pp
Fournier D, Bride JM, Mouchès C, Raymond M, Magnin M, Berge JB, Pasteur N, Georghiou GP (1987) Biochemical characterization of the esterases A_1 and B_1 associated with organophosphate resistance in the *Culex pipiens* L. complex. Pestic Biochem Physiol 27: 211–217
Franke FE, Thomas JE (1933) A study of the cause of death in experimental nicotine poisoning in dogs. J Pharmacol Ther 48: 199–208
Franke FE, Thomas JE (1936) Artificial respiration for nicotine poisoning. JAMA 106: 507–512
Fukami J (1956) Effect of some insecticides on the respiration of insect organs, with special reference to the effects of rotenone. Botyu Kagaku 21: 122–128
Fukami J (1961) Effects of rotenone on the respiration enzyme system of insect muscle. *Bull. Natl. Inst.* Agric Sci Ser C (Plant Pathol Entomol) 13: 33–45
Fukami H, Nakajima M (1971) Rotenone and the rotenoids. In: Jacobson M, Crosby DG (eds) Naturally occurring insecticides. Dekker, New York, pp 71–97
Fukami J, Shishido T (1966) Nature of a soluble glutathione-dependent enzyme system active in cleavage of methyl parathion to desmethyl parathion. J Econ Entomol 59: 1338–1346
Fukami H, Yamamoto I, Casida JE (1967) Metabolism of rotenone in vitro by tissue homogenates from mammals and insects. Science 155: 713–716
Fukuto TR, Sims JJ (1971) Metabolism of insecticides and fungicides. In: White-Stevens R (ed) Pesticides in the environment vol 1, part 1. Dekker, New York, pp 145–236
Fuxa JR (1990) New directions for insect control with baculoviruses. In: Baker RR, Dunn PE (eds) New directions in biological control. Alternatives for suppressing agricultural pests and diseases. Liss, New York, pp 97–113

Gage JC (1970) Methods in toxicology. Blackwell, Oxford 258 pp
Gaines TB (1960) The acute toxicity of pesticides to rats. Toxicol Appl Pharmacol 2: 88–99
Gaines TB (1969) Acute toxicity of pesticides. Toxicol Appl Pharmacol 14: 515–534
Garner WY, Honeycutt RC, Nigg H (1986) Evaluation of pesticides in groundwater. ACS symposium series no 315. American Chemical Society, Washington, DC, 573 pp
Garrido A, Beitia F, Gruenholz P (1984) Effects of PP 618 on immature stages of *Encarsia formosa* and *Cales noaki* (Hymenoptera: Aphelinidae). British Crop Protection Conf, Pests and diseases Brighton, pp 305–310
Gasith A, Perry AS (1980) Rate of parathion in a fish pond ecosystem and its impact on food-chain organisms. In: IAEA (ed) Agrochemical residue–biota interactions in soil and aquatic ecosystems. International Atomic Energy Agency, Vienna, pp 125–151
Georghiou GP (1983) Management of resistance in arthropods. In: Georghiou GP, Saito T (eds) Pest resistance to pesticides. Plenum Press, New York, pp 769–792
Georghiou GP (1986) The magnitude of the resistance problem. In: Pesticide resistance: strategies and tactics for management. National Academy of Sciences Press Washington DC 471 pp
Georghiou GP (1987) Insecticides and pest resistance: the consequences of abuse. 36th Annual Faculty Research Lecture University of California, Riverside
Georghiou GP (1990) Overview of insecticide resistance. In: Managing resistance to agrochemicals. From fundamental research to practical strategies. ACS symposium series no 421. American Chemical Society, Washington, DC
Georghiou GP, Pasteur N (1978) Electrophoretic esterase patterns in insecticide-R and S mosquitoes. J Econ Entomol 71: 201–205
Georghiou GP, Saito I. (eds) (1983) Pest resistance to pesticides. Plenum Press New York, 809 pp
Georghiou GP, Pasteur N, Hawley MK (1980) Linkage relationship between OP resistant and a highly active esterase-B in *Culex quinquefaciatus*, California. J Econ Entomol 73: 301–305
Gerolt P (1972) Mode of entry of oxime carbamates into insects. *Pestic. Sci.* 3: 43–55
Gerschenfeld HM (1973) Chemical transmission in invertebrate central nervous system. *Physiol Rev* 53: 1–119
Gerstl Z, Chen Y, Mingelgrin U, Yaron B (1989) Toxic organic chemicals in porous media. Springer, Berlin, Heidelberg New York, 343 pp
Gilbert LI, Bollenbacher WE, Goodman W, Smith SL, Agui N, Grander N, Sedlak BJ (1980) Hormones controlling insect metamorphosis. Recent Prog Horm Res 36: 401–449
Goldberg LG, Margalit J (1977) A bacterial spore demonstrating rapid larvicidal activity against *Anopheles sergentii, Uranotaenia unquiculata, Culex univitattas, Aedes aegypti* and *Culex pipiens.* Mosq News 37: 355–358
Granett J, Weseloh RM, Helgert E (1975) Activity of JH analogs on hymenopterous parasitoids of the gypsy moth. Entomol Exp Appl 18: 377–383
Grosscurt AC (1978) Diflubenzuron—some aspects of its ovicidal and larvicidal mode of action and on evaluation of its practical possibilities. Pestic Sci 9: 373–386
Grosscurt AC, Stoker A (1991) Resistance to diflubenzuron in insects. Meded Fac Laudbouwwet Rijksuniv Gent 56: 1151–1159
Grossman SJ, Hsia MTS (1982) Metabolic fate of the hepatotoxic anti-JH precocene II in the rat. Toxicology 25: 293–298
Guthrie FE (1950) Effect of temperature in toxicity of certain organic insecticides. J Econ Entomol 43: 559–560
Haag, HB (1931) Toxicological studies of *Derris elliptica* and its constituents, Rotenone. Pharmacol Exp Ther 43: 193–208
Hackler RE, Suhr RG, Sheets JJ, Hatton CJ, Johnson PL, Davis LN, Edie RG, Kaster SV, Jourdan GP, Jackson JL, Krumkalns EV (1994) Chemistry and miticidal activity of fused pyrimidine derivatives of fenazaquin. In: Briggs GG (ed) Advances in chemistry of insect control III. Royal Society of Chemistry, London, pp 70–84

Haga T, Tobi T, Koyanagi T, Nishiyama R (1982) Structure activity relationships of a series of benzoyl-pyridyloxyphenyl-urea derivatives. Abstr 5th Int Congr Pestic Chem, IUPAC, Kyoto, p IID-7

Hajjar NP, Casida JE, (1979) Structure activity relationships of benzoylphenyl ureas as toxicants and chitin synthesis inhibitors in *Oncopeltus fasciatus*. Pestic Biochem Physiol 11: 33–45

Hall RJ (1987) Impact of pesticides on bird populations. In: Marco GJ, Hollingworth RM, Durham W (eds) Silent spring revisited. American Chemical Society, Washington, DC, pp 85–111

Hamaguchi H, Kajihara O, Katoh M (1995) Development of a new acaricide, fenpyroximate. J. Pestic Sci 20: 173–175, 203–212

Hammock BD (1993) Recombinant baculoviruses as biological insecticides. In: Lumsden RD, Vaughn JL (eds) Pest management: biologically based technologies. ACS symposium series. American Chemical Society, Washington, DC, pp 313–325

Hargreaves PA, Wainwright OH, Swaine G, Corcoran RJ (1978) Residual ethylene dibromide and inorganic bromide levels in some fruit and vegetables after fumigation with ethylene dibromide or methyl bromide. Aust J Exp Agric Anim Husb 18: 586–590

Harvey GT, Brown AWA (1951) The effect of insecticides on oxygen consumption in *Blattella*. Can J Zool 29: 42–53

Hartwell WV, Hayes GR, Funckes AJ (1964) Respiratory exposure of volunteers to parathion. Arch Environ Health 8: 820–826

Havron A (1983) Studies toward selection of *Aphytis* wasps for pesticide resistance. PhD Thesis, Hebrew University of Jerusalem

Hayes GR, Funckes AJ, Hartwell WV (1964) Dermal exposure of human volunteers to parathion. Arch. Environ. Health 8: 829–833

Hayes WJJr (1982) Pesticides studies in man. Williams and Wilkins, Baltimore 672 pp

Hayes WJJr, Laws ER (1991) Handbook of pesticide toxicology. vol I, vol II, vol III. Academic Press New York, 1576 pp

Henderson C, Pickering QH, Tarzwell CM (1959) Relative toxicity of ten chlorinated hydrocarbon insecticides to four species of fish. Trans Am Fish Soc 88: 23–32

Henrick CA (1982) JH-analogs: structure–activity relationships. In: Coats JR (ed) Insecticidal mode of action. Academic Press, New York, pp 315–402

Henry JE (1990) Control of insects by protozoa. In: Baker RR, Dunn PE (eds) New direction in biological control. Alternatives for suppressing agricultural pests and diseases. Liss, New York, pp 161–176

Hervé JJ (1985) Agricultural, public health and animal usage. In: Leahey JP (ed) The pyrethroid insecticides. Taylor and Francis, London, pp 343–417

Hill EF, Dale WE, Miles JW (1971) DDT intoxication in birds: subchronic effects and brain residues. Toxicol Appl Pharmacol 20: 502–514

Hill IR (1985) Effects on non-target organisms in terrestrial and aquatic environments. In: Leahey JP (ed) The pyrethroid insecticides. Taylor and Fancis, London, pp 151–238

Hirata K, Kawamura Y, Kudo M, Igarashi H (1995) Development of a new acaricide, pyridaben. J Pestic Sci 20: 213–221

Hobbiger F (1955) Effect of nicotinhydroxamic acid methoxide on human plasma cholinesterase inhibited by organophosphates containing a dialkylphosphate group. Br J Pharmacol Chemother 10: 356–362

Hodgson E, Tate LG (1976) Cytochrome P450 interactions. In: Wilkinson CF (ed) Insecticide biochemistry and physiology. Plenum Press, New York, pp 115–148

Hollingshaus JG (1987) Inhibition of mitochondrial electron transport by hydramethylnon: a new amidinohydrazone insecticide. Pestic Biochem Physiol 27: 61–70

Hollingsworth RM, Lund AE (1982) Biological and neurotoxic effect of formamidine pesticides. In: Coats JR (ed) Insecticide mode of action. Academic Press, New York, pp 189–227

Hollingsworth RM, Ahammadsahib KI, Gadelhak G, McLaughlin JL (1994) New inhibitors of complex I of the mitochondrial electron transport chain with activity as pesticides. Biochem Soc. Trans 22: 230–233

Holmstead, RL, Casida JE, Ruzo LO (1977) Photochemical reactions of pyrethroid insecticides. In: Elliott M (ed) Synthetic pyrethroids. ACS symposium series, no 42. American Chemical Society, Washingtan DC pp 137–146

Holst H (1975) Die Fertilitätsbeeinflussende Wirkung des neuen Insektizids PDD 60-40 bei *Epilachna varivestis* Muls. (Col. Coccinellidae). Z Pflanzenkr Pflanzenschutz 82: 1–7

Horn DHS, Galbraith MN, Kelly BA, Kinnear JF, Martin MD, Middleton EJ, Virgonia CTF (1981) Moulting hormones LIII. The synthesis and biological activity of some ecdysone analogues. Aust J Chem 34: 2607–2618

Horowitz AR, Ishaaya I (1992) Susceptibility of the sweet potato whitefly (Homoptera: Aleyrodidae) to buprofezine during the cotton season. J Econ Entomol 85: 318–324

Horowitz AR, Ishaaya I (1994) Managing resistance to insect growth regulators in the sweet potato whitefly (Homoptera: Aleyrodidae). J Econ Entomol 87: 866–871

Horowitz AR, Toscano NC, Youngman RR, Georghiou GP (1988) Synergism of insecticides with DEF in sweet potato whitefly (Homoptera: Aleyrodidae). J Econ Entomol 81: 110–114

Horowitz AR, Seligman IM, Forer G, Bar D, Ishaaya I (1993) Preventive insecticide resistance strategy in *Helicoverpa (Heliothis) armigera* (Lepidoptera: Noctuidae) in Israeli cotton. J Econ Entomol 86: 205–212

Horowitz AR, Forer G, Ishaaya I (1994) Managing resistance in *Bemisia tabaci* in Israel with emphasis on cotton. Pestic Sci 42: 113–122

Howarth FG (1991) Environmental impacts of classical biology control. Annu Rev Entomol 36: 485–509

Hoy MA, Cave FE (1988) Guthion resistant strain of walnut aphid parasite. Calif Agric 42(4): 4–5

Hoy MA, Conley J (1987) Toxicology of pesticides to western predatory mite. Calif Agric 4 (7–8): 12–14

Hoyle G (1953) Potassium ions and insect nerve muscle. J Exp Biol, 30: 121–125

Hsia MTS, Grossman S, Schrankel KR (1981) Hepatotoxicity in the anti-JH precocene II and the generation of dehydrodiol metabolites. Chem Biol Interact 37(3): 265–277

Hudson RH, Tucker RK, Haegele MA (1972) Effect of age on sensitivity: acute oral toxicity of 14 pesticides to mallard ducks of several ages. Toxicol Appl Pharmacol 22: 556–561

Hutchens JO, Wagner H, Podolsky B, McMahon TM (1949) The effect of ethanol and various metabolites in fluoroacetate poisoning. J Pharmacol Exp Ther 95: 62–70

International Agency for Research on Cancer (1976) On the evaluation of carcinogenic risks of chemicals to man. IARC Monogr 11: 17, 159

International Agency for Research on Cancer (1977) 2,4-D and esters. In: IARC monographs on the evaluation of the carcinogenic risk of chemicals to man: some fumigants, the herbicides, 2,4-D and 2,4,5-T, chlorinated dibenzodioxins and miscellaneous industrial chemicals. IARC Monogr 15: 111–138

International Agency for Research on Cancer (1979) Some halogenated hydrocarbons. IARC Monogr 20: 371–392

Ishaaya I (1990a) Buprofezin and other insect growth regulators for controlling cotton pests. Pestic Outlook 1(2): 30–33

Ishaaya I (1990b) Benzoylphenyl ureas and other selective insect control agents. Mechanism and application. In: Casida JE (ed) Pesticides and alternatives. Elsevier, New York, pp 365–376

Ishaaya I (1992) Insect resistance to benzoylphenyl ureas, and other insect growth regulators, mechanisms and countermeasures. In: Mullin CA, Scott JG (eds) Molecular mechanisms of insecticide resistance. ACS series, no 505. American Chemical Society, Washington, DC, pp 213–246

Ishaaya I (1993) Insect detoxifying enzymes: their importance in pesticide synergism and resistance. Arch Insect Biochem Physiol 22: 263–276

Ishaaya I, Blumberg D, Yarom I (1989) Buprofezin—a novel IGR for controlling whiteflies and scale insects. Meded Fac Laudbouwwet Rijksuniv Gent 54: 1003–1008

Ishaaya I, Casida JE (1974) Dietary TH 6040 alters composition and enzyme activity of house fly larval cuticle. Pestic Biochem Physiol 4: 484–490

Ishaaya I, Casida JE (1975) Phenyltin compounds inhibit digestive enzymes of *Tribolium confusum* larvae. Pestic Biochem Physiol 5: 350-358

Ishaaya I, Casida JE (1980) Properties and toxicological significance of esterases hydrolyzing permethrin and cypermethrin in *Trichoplusia ni* larval gut and integument. Pestic Biochem Physiol 14: 178–184

Ishaaya I, Casida JE (1981) Pyrethroid esterase(s) may contribute to natural pyrethroid tolerance of the larvae of the cotton green lacewing. Environ Entomol 10: 681–684

Ishaaya I, De Cock A, Degheele D (1994) Pyriproxyfen, a potent suppressor of egg hatch and adult formation of the greenhouse whitefly (Homoptera: Aleyrodidae). J Econ Entomol 87: 1185–1189

Ishaaya I, Horowitz AR (1992) Novel phenoxy juvenile hormone analog (pyriproxyfen) suppresses embryogenesis and adult formation of sweet potato whitefly (Homoptera: Aleyrodidae). J Econ Entomol 85: 2113–2117

Ishaaya I, Horowitz AR (1995) Pyriproxyfen, a novel insect growth regulator for controlling whiteflies: mechanism and resistance management. Pestic Sci 43: 227–232

Ishaaya I, Klein M (1990) Response of laboratory (susceptible) and field resistant strains of *Spodoptera littoralis* (Lepidoptera: Nectuidae) to teflubenzuron. J Econ Entomol 83: 59–62

Ishaaya I, Elsner A, Ascher KRS, Casida JE (1983) Synthetic pyrethroids: toxicity and synergism on dietary exposure of *Tribolium castaneum* (Herbst) larvae. Pestic Sci 14: 367–372

Ishaaya I, Yablonski S, Ascher KRS, Casida JE (1984) Pyrethroid synergism and prevention of emergence in *Tribolium castaneum* and *Musca domestica vicima* by the insect growth regulator ROB-J223. Phytoparasitica 12: 99–108

Ishaaya I, Navon A, Gurevitz E (1986) Comparative toxicity of chlorfluazuron (IKI-7894) and cypermethrin to *Spodoptera littoralis*, *Lobesia botrana* and *Drosophila melanogaster*. Crop Protect 5: 385–388

Ishaaya I, Mendelson Z, Ascher KRS, Casida JE (1987) Cypermethrin synergism by pyrethroid esterase inhibitors in adult of the whitefly *Bemisia tabaci*. Pestic Biochem Physiol 28: 155–162

Ishaaya I, Mendelson Z, Melamed-Madjar V (1988) Effect of buprofezine on embryogenesis and progeny formation of sweet potato whitefly (Homoptera: Aleyrodidae). J Econ Entomol 81: 781–784

Ishaaya I, Yablonski S, Horowitz AR (1995) Comparative toxicity of two ecdysteroid agonists, RH-2485 and RH-5992, on susceptible and pyrethroid-resistant strains of the Egyptian cotton leafworm, *Spodoptera littoralis*. Phytoparasitica 23: 139–145

Ishaaya I, Yablonski S, Mendelson Z, Mansour Y, Horowitz AR (1996) Novaluron (MCW-275), a novel benzoylphenyl urea, suppressing developing stages of lepidopteran, whitefly and agromyzid leafminer pests. In: Brighton Crop Protection Conference, Pests and diseases. British Crop Protection Council, Farnham, p. 1013–1020.

Ishida T, Suzuki J, Tsukidate Y (1994) YI-5301, a novel oxazoline acaricide. In: Brighton Crop Protection Conference, Pests and diseases, vol 1. British Crop Protection Council, Farnham, pp 37–44

Isono K, Suzuki S (1968) The structure of polyoxins A and B. Tetrahedron Lett 9: 1133–1137

Itaya N (1987) Insect juvenile hormone analogues as an insect growth regulator. Sumitomo Pyrethroid World 8: 2–4

Izawa Y, Uchida M, Sugimoto T, Asai T (1985) Inhibition of chitin biosynthesis by buprofezin analogs in relation to their activity controlling *Nilaparvata lugens* Stal. Pestic Biochem Physiol 24: 343–347

Jacobson M (1965) Insect sex attractants. Wiley, New York, 154 pp

Jalili MA, Abbasi AH (1961) Poisoning by ethyl mercury toluene sulphonanilide. Br J Ind Med 18: 303–308

Jao LT, Casida JE (1974) Insect pyrethroid-hydrolyzing esterases. Pestic Biochem Physiol 4: 465–472

Johansen C (1963) The effect of pesticides on the alfalfa leafcutting bee *Megachile rotundata*. Washington State Univ Agric Exp Stn Circ 418, 12 pp

Johansen C, Eves J (1967) Toxicity of insecticides to the alkali bee and the alfalfa leafcutting bee. Washington State Univ Agric Exp Stn Circ 475, 15 pp
Johnson MK (1982) The target for initiation of delayed neurotoxicity by organophosphorus esters: biochemical studies and toxicological application. Rev Biochem Toxicol 4: 141–212
Jones KC (ed) (1991) Organic contaminants in the environment. Elsevier, London
Jutson AR (1989) Insect pheromones in plant protection. Wiley, New York
Kanno H, Ikeda K, Asai T, Maekawa S (1981) 2-tert-Butylimino-3-isopropyl-5-phenylperhydro-1,3,5 thiadiazin-4-one (MNI 750), a new insecticide. In: British Crop Protection Conf. Pests and diseases. British Crop Protection Council, Farnham, pp 56–69
Katz M (1961) Acute toxicity of some organic insecticides to three species of salmonids and to the three-spine stickleback. Trans Am Fish Soc 40: 264–268
Kavlock R, Chernoff N, Baron R, Linder R, Rogers E, Carver B, Dilley J, Simmon V (1979) Toxicity studies with decamethrin synthetic pyrethroid insecticide. J Environ Pathol Toxicol 2: 751–765
Kawada H (1988) An insect growth regulator against cockroaches. Sumitomo Pyrethroid World 11: 2–4
Kaya HK (1990) Entomopathogenic nematodes in biological control of insects. In: Baker RR, Dunn PE (eds) New directions in biological control. Alternatives for suppressing agricultural pests and diseases. Liss, New York, pp 189–198
Kehoe J, Marder E (1976) Identification and effects of neural transmitters in invertebrates. Annu Rev Pharmacol Toxicol 10: 245–268
Keiding J (1977) Resistance in the housefly in Denmark and elsewhere. In: Watson DL, Brown AWA (eds) Pesticide management and insecticide resistance. Academic Press, New York, pp 261–302
Kimura S, Toeda K, Miyamoto T, Yamamoto I (1984) Activation and detoxication of S-alkyl phosphorothiolate insecticides. J Pestic Sci 9: 137–142
Knipling EF (1953) The greater hazard—insects or insecticides? J. Econ Entomol 46: 1–7
Knipling EF (1955) Possibilities of insect control or eradication through the use of sexually sterile males. J Econ Entomol 48: 459–462
Knipling EF (1959) Sterile male method of population control. Science 139 (3380): 902–904
Knipling EF (1960) Eradication of the screwworm fly. Sci Amer 203(4): 54–61
Knipling EF (1968) The potential role of sterility for pest control. In: LaBreque GC, Smith C (eds) Principle of insect chemosterilization. Appleton-Century-Crofts, New York, pp 7–39
Kobinata K, Uramoto M, Nishii M, Kusakabe H, Nakamura G, Isono K (1980) Neopolyoxins A, B and C. New chitin synthetase inhibitors. Agric Biol Chem 44: 1709–1711
Koehler PG, Patterson RJ (1991) Incorporation of pyriproxyfen in a German cockroach (Dictyoptera: Blattellidae) management program. J Econ Entomol 84: 917–921
Koolman J, Karlson P (1985) Regulation of ecdysteroid titer: degradation. In: Kerkut GA, Gilbert LI (eds) Comprehensive insect physiology, biochemistry and pharmacology, vol 7 Pergamon Press, Oxford, pp 343–361
Kristinsson H (1994) Pymetrozine, a new insecticide. In: Briggs GG (ed) Advances in the chemistry of insect control III. Royal Society of Chemistry, Cambridge, pp 85–102
Kuhr RJ, Dorough HW (1976) Carbamate insecticides: chemistry, biochemistry and toxicology. CRC Press, Cleveland, 300 pp
Kydonieus AF, Beroza M (1982) Insect suppression with controlled release pheromone systems, vols I, II. CRC Press, Boca Raton 312 pp
LaBreque GC (1968) Laboratory procedures. In: LaBreque GC, Smith, CN (eds) Principles of insect chemosterilants. Appleton-Century-Croft, New York, pp 42–98
LaBreque GC, Smith CN, Meifert DW (1962) A field experiment in the control of housefly with chemosterilant baits. J Econ Entomol 55: 449–451
Lacey LA, Undeen AH (1986) Microbial control of black flies and mosquitoes. Annu Rev Entomol 31: 265–296
Lacey LA, Urbina MJ, Heitzman CM (1984) Sustained release formulations of *Bacillus sphaericus* and *Bacillus thuringiensis* (H-14) for control of container breeding *Culex quinquefasciatus*. Mosq News 44: 26–32

Lagunes AT (1980) Impact of the use of mixtures and sequences tissues in the evaluation of resistance in *Culex quinquefasciatus*. PhD Thesis, University of California, Riverside

Langley PA (1990) Control of the tsetse fly using a juvenile hormone mimic, pyriproxyfen. Sumitomo Pyrethroid World 15: 2–5

Langley PA, Coates TW, Carlson DA (1982) Sex recognition pheromone in the tsetse fly *Glossina pallidipes* Austen. Experientia 38: 473–474

Lasota JA, Dybas RA (1991) Avermectins, a novel class of compounds. Implications for use in arthropod pest control. Annu Rev Entomol 36: 91–117

Laws ERJr, Curley A, Biros FJ (1967) Men with intensive occupational exposure to DDT. A clinical and chemical study. Arch Environ Health 15: 766–775

Leahey JP (1985) (ed) Metabolism and environmental degradation. In: Leahey JP(ed) The pyrethroid insecticides. Taylor and Francis, London pp 153, 263–342

Lee HL, Cheong WH (1987) Field evaluation of the efficacy of B.t. for the control of *Aedes albopictus*. Mosq Borne Dis Bull 3: 57–63

Leemans J, Reynaerts A, Höfte H, Peferoen M, Van Mellaert H, Jooz H (1990) Insecticidal crystal proteins from *Bacillus thuringiensis* and their use in transgenic crops. In: Baker RR, Dunn PE (eds) New direction in biological control. Alternatives for suppressing agricultural pests and diseases. Liss, New York, pp 573–581

Lehman AJ (1951–1952) Chemicals in foods: a report to the Association of Food and Drug Officials on current developments. Part II. Pesticides. Section I. Introduction. Section II. Dermal toxicity. Section III. Subacute and chronic toxicity. Section IV. Biochemistry. Section V. Pathology. Association of Food and Drug Officials, US Q Bull: I 15: 122–133 (1951); II. 16:3–9 (1952); III. 16: 47–53 (1952); IV. 16: 85–91 (1952); V. 16: 126–132 (1952)

Lehman AJD, Fassett DW, Gerarde HW, Stakinger HE, Zapp JW (1964) Principles for evaluating the toxicity of household substances. Public no 1138 National Academy of Sciences, National Research Council, Washington, DC

Lichtenstein EP, Fuhremann TW, Schulz KR (1971) Persistence and vertical distribution of DDT, lindane and aldrin residues 10 and 15 years after a single soil application. J Agric Food Chem 19: 718–772

Lindahl PE, Oberg KE (1961) The effect of rotenone on respiration and its points of attack. *Exp Cell Res* 23: 228–237

Lindquist, AW (1955) Possibilities in the use of gamma radiation for control or eradication of the screwworm. J Econ Entomol 48: 467–469

Lindquist AW (1961) New ways to control insects. Pest Control 29(6): 9, 11–12

Link KP (1944) The anticoagulant from spoiled sweet clover hay. Harvey Lect 39: 162–216

Lisella FS, Long KR, Scot HG (1971) Toxicology of rodenticides and their relation to human health. J Environ Health 33: 231–237, 361–365

Litchfield MH (1983) Characterization of the principal mammalian toxicological and biological actions of synthetic pyrethroids. In: Miyamoto J, Kearney PC (eds) Pesticide chemistry: human welfare and the environment vol II. Pergamon Press, Oxford, pp 207–211

Litchfield MH (1985) Toxicity of pyrethroids to mammals. In: Leahey JP(ed) The pyrethroid insecticides. Taylor and Francis, London, pp 99–150

Liu MY, Lanford J, Casida JE (1993) Relevance of tritiated imidacloprid binding site in house fly head acetylcholine receptor to insecticidal activity of 2-nitromethylene- and 2-nitroiminoimidazolidines. Pestic Biochem Physiol 46(3): 200–206

Love JL, Winchester RV, Keating DL (1979) Fumigant and contact insecticide residues on dried fruit, cereals, nuts and spices imported into New Zealand. NZJ Sci 22: 95–98

Ludwig D (1946) Effect of DDT on metabolism: Popillia. Ann Entomol Soc Am, 39: 496–509

Macek KJ, McAllister WA (1970) Insecticide susceptibility of some common fish families. Trans Am Fish Soc 99: 20–27

Mandava NB, Morgan ED (1988) Pheromones Parts A and B. Handbook of natural pesticides, vols 4, 5. CRC Press, Boca Raton. 291 pp 243 pp

Marco GJ, Hollingsworth RM, Durham W (1987) Silent spring revisited. American Chemical Society, Washington, DC, 214 pp

Marer PJ, Flint ML, Stimmann MW (1988) The safe and effective use of insecticides. Publ no 3324 University of California, Division of Agriculture and Natural Resources-Oakland

Margalit J, Bobrolgo H (1984) The effect of organic materials and solids in water on the persistence of *Bacillus thuringiensis* var. *israelensis* serotype H14. Z Angew Entomol 97: 516–520

Margalit J, Markus A, Pelah Z (1984) Effect of encapsulation on the persistence of *Bacillus thuringiensis* var. *israelensis* serotype H-14. Appl Microbiol Biotechnol 19: 382–382

Marks EP, Leighton T, Leighton F (1982) Mode of action of chitin-synthesis, inhibition. In: Coats JR (ed) Insecticide mode of action. Academic Press, New York, pp 283–313

Matsumura F (1972) Metabolism of insecticides in microorganisms and insects. In: Coulsten F, Korte F (eds) Environmental quality and safety, vol 1. Academic Press, New York, pp 96–106

Matsumura F (1975) Naturally occurring botanical insecticides. In: Matsumura F(ed) Toxicology of insecticides, 2nd edn. Plenum Press, New York, pp 134–141

Matsumura F (1985) Toxicology of insecticides 2 ed. Plenum Press, New York, 598 pp

Matsumura F, Brown AWA (1961) Biochemistry of malathion resistance in *Culex tarsalis*. J Econ Entomol 54: 1176–1185

Matsumura F, Clark JM (1982) ATP-utilizing system in the squid axons: a review on the biochemical aspects of ion-transport. Prog Neurobiol 18: 231–255

Matsumura F, O'Brien RD (1963) A comparative study of the modes of action of fluoroacetamide and fluoroacetate in the mouse and American cockroach. Biochem Pharmacol 12: 1201–1205

Mayer RT, Meola SM, Coppage DL, DeLoach JR (1980) Utilization of imaginal tissues from pupae of the stable fly for the study of chitin synthesis and screening of chitin synthesis inhibition. J Econ Entomol 73: 76–80

McBain JB, Yamamoto I, Casida JE (1971) Mechanism of activation and deactivation of Dyfonate (O-ethyl S-phenyl ethylphosphonodithioate) by rat liver microsomes. Life Sci 10: 947–954

McCoy CW (1990) Entomogenous fungi as microbial pesticides. In: Baker RR, Dunn PE(eds) New directions for Biological Control. Alternatives for Suppressing Agricultural Pests and Diseases. Liss, New York, pp 139–159

McGaughey WH (1990) Insect resistance to *Bacillus thuringiensis d*-endotoxin. In: Baker RR Dunn PE(eds) New directions in biological control. Alternatives for suppressing agricultural pests and diseases. Liss, New York, pp 583–598

Melander AL (1914) Can insects become resistant to sprays? J Econ Entomol 7: 167–172

Mendel Z, Blumberg D, Ishaaya I (1994) Effect of some insect growth regulators on natural enemies of scale insects (Hom.: Coccoidea). Entomophaga 39: 199–209

Mendelssohn H (1972) Effect of pesticides on bird life. The impact of pesticides on bird life in Israel. Int Counc Bird Preserv Bull 11: 75–104

Mendelssohn H, Paz U (1977) Mass mortality of birds of prey caused by Azodrin an organophosphorus insecticide. Biol Conserv 11: 163–170

Mendelssohn H, Schlueter P, Aderet Y (1979) Report on Azodrin poisoning of birds of prey in the Hula Valley in Israel. Int Counc Bird Preserv Bull 13: 124–129

Mengle DC, Casida JE (1958) Inhibition and recovery of brain cholinesterase activity in house flies poisoned with organophosphate and carbamate compounds. J Econ Entomol 51: 750–757

Metcalf CL, Flint RL, Metcalf RF (1962) Destructive and useful insects, 4th edn. McGraw-Hill, New York, 1087 pp

Metcalf MA, Metcalf RA (eds) (1993) Destructive and useful insects—their habits and control. 5th edn. McGraw-Hill, New York

Metcalf RL (1971) The chemistry and biology of pesticides. In:White-Stevens R (ed) Pesticides in the environment vol 1, Part 1. Dekker, New York, pp 1–144

Metker LW (1980) Subchronic inhalation toxicity of 3-(phenoxyphenyl) methyl (± *cis*, *trans*-3-(2,2-dichloroethenyl)-2,2-dimethylcyclopropane carboxylate (permethrin). Report no 75-51-0026-80. US Army Environmental Hygiene Agency, Aberdeen Proving Ground, MD

Milani R, Travaglino A (1957) Ricerche genetiche sulla resistenza al DDT in *Musca domestica* concatenazione del gene kdr (knockdown resistance) on due mutanti morfologigi. Riv Parasitol 18: 199–202

Miller LK, Lingg AJ, Bulla LA Jr (1983) Bacterial, viral and fungal insecticides. Science 219: 715–721

Miyamoto J (1976) Degradation, metabolism and toxicity of synthetic pyrethroids. Environ Health Perspect 14: 15–28

Miyamoto T, Yamamoto I (1977) Mechanism of phosphinyl disulfide formation from a phosphorodithioate ester. J Pestic Sci 2: 303–310

Miyamoto T, Yamamoto I (1980) Reaction of phosphinyl and phosphinothioyl disulfides with diazomethane. Agric Biol Chem 44: 2581–2586

Monthean C, Potter DE (1992) Effects of RH-5849, a novel insect growth regulator, on Japanese beetle (Coleoptera: Scarabaeidae) and fall armyworm (Lepidoptera: Noctuidae) in turfgrass. J Econ Entomol 85: 507–513

Moriarty F (1975) Organochlorine insecticides: persistent organic pollutants. Academic Press, New York, 302 pp

Motoyama N, Dauterman WC (1972) In vitro metabolism of azinphosmethyl in susceptible and resistant houseflies. Pestic Biochem Physiol 2: 113–122

Motoyama N, Rock GC, Dauterman WC (1971) Studies on the mechanism of azinphosmethyl resistance in the predaceous mite, *Neoseiulus (T.) fallacis* (family: Phytoseiidae). Pestic Biochem Physiol 1: 205–215

Motoyama N, Nishizawa Y, Nagakura A, Takemasa T, Dauterman WC (1991) Selective inhibition of the cytochrome P450-dependent monooxydases from the rat liver and housefly abdomen. In: Hodgson E, Roe RM, Motoyama N (eds) Pesticides and the future: toxicological studies of risks and benefits. North Carolina State University at Raleigh, pp 173–184

Mouchès C, Pasteur N, Berge JB, Hyrien O, Raymond M, de Saint Vincet BR, de Silvestri M, Georghiou GP (1986) Amplification of an esterase gene is responsible for insecticide resistance in a *Culex* mosquito from California. Science, 233: 778–780

Mouchès C, Magnin M, Berge JB, de Silvestri M, Beyssat N, Pasteur N, Georghiou GP (1987) Overproduction of detoxifying esterases in OP-R *Culex* mosquitoes and their presence in other insects. Proc Natl Acad Sci USA 84: 2113–2116

Murphy SD (1975) Pesticides. In: Casarett LJ, Doull J (eds) Toxicology—the basic science of poisons. Macmillan, New York, pp 408–453

Muszkat L (1994) Groundwater contamination by organic pollutants. In: Zoller U (ed) Groundwater contamination and control. Dekker, New York, pp 271–289

Muszkat L, Lahav D, Ronen D, Magaritz M (1993a) Penetration of pesticides and industrial organics deep into soil and into groundwater. Arch Insect Biochem Physiol 22: 487–499

Muszkat L, Magaritz M, Ronen L (1988) Large scale contamination of deep groundwaters by organic pollutants. In: Longevialle P (ed) Advances in mass spectrometry, vol 11B. Heyden, Gir Sur Yvette pp 1128–1129

Muszkat L, Raucher D, Magaritz M, Ronen D, Amiel A (1993b) Unsaturated zone groundwater contamination by organic pollutants in a sewage-effluent irrigated site. Ground Water 31: 556–565

Muszkat L, Rosenthal E, Ronen D, Magaritz M (1989) Organic contaminants in the Israeli coastal aquifer. In: Luria M, Steinberger Y, Spanier E (eds) Environmental quality and ecosystem stability, vol. IV A—environmental quality. ISEEQS Publishers, Jerusalem, pp 471–477

Muzzarelli RAA (1977) Chitin. Pergamon Press, Oxford 309 pp

Nachtomi E (1970) The metabolism of ethylene dibromide in the rat; the enzymic reaction with glutathione in vitro and in vivo. ***Biochem. Pharmacol.*** 19: 2853–2859

Nachtomi E, Sarma DSR (1977) Repair of rat liver DNA *in vivo* damaged by ethylene dibromide. Biochem Pharmacol 26: 1941–1945

Nagata T (1986) Timing of buprofezine application for control of the brown planthopper, *Nilaparvata lugens* Stal. (Homoptera: Delphacidae). Appl. Entomol Zool 21: 357–362

Nakatsugawa T, Dahm PA (1967) Microsomal metabolism of parathion. Biochem Pharmacol 16: 25–38

Nassis M, Kamel O (1977) A field trial with permethrin against body lice, ***Pediculus humanus***, in Egypt (1976). Pestic Sci 8: 301–304
National Research Council (1992) Neem—a tree for solving global problems. National Academic Press, Washington, DC, 141 pp
Nauman K (1990) Synthetic pyrethroid insecticides: structure and properties. Springer, Berlin Heidelberg New York, 241 pp
Navon A (1993) Control of lepidopteran pests with ***Bacillus thuringiensis***. In: Entwistle PF Cory JS Bailey MJ Higgs S (eds) ***Bacillus thuringiensis*** an environmental biopesticide: theory and practice. Wiley, Chichester, pp 125–146
Navon A, Klein M, Braun S (1990) ***Bacillus thuringiensis*** potency bioassays against ***Heliothis armigera***, ***Earias insultana*** and ***Spodoptera littoralis*** larvae based on standardized diets. J Invert Pathol 33: 387–393
Nishino C, Manabe S, Kuwahara K, Kimura R, Takayanagi H (1983) Isolation of sex pheromones of the American cockroach by monitoring with electroantennogram responses. Insect Biochem 13: 65–70
O'Brien RD (1967) Insecticides: action and metabolism. Academic Press, New York, 332 pp
Ohana B, Margalit J, Barak Z (1987) Fate of ***Bacillus thuringiensis*** subsp. ***israelensis*** under field simulated conditions. Appl Environ Microbiol 53: 828–831
Ohkawa H, Ohkawa R, Yamamoto I, Casida JE (1972) Enzymatic mechanism and toxicological significance of hydrogen cyanide liberation from various organothiocyanates and organonitrile in mice and houseflies. Pestic Biochem Physiol 2: 95–112
Ohkawa H, Kikuchi R, Miyamoto J (1980) Bioaccumulation and biodegradation of the (S)-acid isomer of fenvalerate (sumicidin) in an aquatic model ecosystem. J Pestic Sci 5: 11–22
Oppenoorth FJ (1965) Biochemical genetics of insecticide resistance. Annu Rev Entomol 10: 185–206
Oppenoorth FJ (1985) Biochemistry and genetics of insect resistance. In: Kerkut GS, Gilbert LI (eds) Insect control, vol 12. Pergamon Press New York, pp 731–773
Oppenoorth FJ, Welling W (1976) Biochemistry and physiology resistance. In: Wilkinson CF (ed) Insecticide biochemistry and physiology. Plenum Press, New York, pp 507–544
Page GW (1981) Comparison of groundwater and surface water for patterns and levels of contamination by toxic substances. Environ Sci Technol 14: 1475–1481
Peleg BA (1983) Effect of a new insect growth regulator RC-13-5223 on hymenopterous parasites of scaly insects. Entomophaga 28: 367–372
Peleg BA (1988) Effect of a new phenoxy juvenile hormone analog on California red scale (Homoptera: Diaspididae), Florida wax scale (Homoptera: Coccidae) and the ectoparasite ***Aphytis holoxanthus*** DeBache (Hymenoptera: Aphelinidae). J Econ Entomol 81: 88–92
Peleg BA, Gothilf S (1980) Effect of juvenoid Altosid in the development of three hymenopterous parasites. Entomophaga 25(3): 323–328
Pener MP, Orshan L (1977) Morphogenic effect of precocene 2, an anti-juvenile hormone in the African migratory locust, ***Locusta migratoria migratorioides***. Isr J Zool 26: 262–263
Pener MP, Orshan L, de Wilde J (1977) Precocenes II causes atrophy of corpora allata in Locusta migration. Nature 272: 350–353
Pener MP, Troetschler R, Fridman-Cohen S, Zeldes I (1981) Comparative studies of the effect of precocenes in various grasshoppers and locust species. In: Sehnal F, Zabza A, Menn JJ, Cymborowski B (eds) Int Conf Regulation of insect development and behavior, Karpaz, Poland 23-28 June 1980, Technical University Press, Wroclaw, pp 357–375
Perring FH, Mellanby K (eds) (1977) Ecological effects of pesticides. Academic Press, New York 193 pp
Perry AS (1974) Insecticide resistance in insects and its ecological and economic thrust. In: Khan MAQ, Bederka JP Jr (eds) Survival in toxic environments. Academic Press, New York, pp 399–446
Perry AS, Agosin M (1974) The physiology of insecticide resistance by insects. In: Rockstein M(ed) The physiology of insecta vol IV 2nd edn. Academic Press, New York, pp 3–121
Perry AS, Buckner AJ (1958) Biochemical investigations of DDT-resistance in the human body louse ***Pediculus humanus humanus***. Am J Trop Med Hyg 7(6): 620–626

Perry AS, Fay RW (1967) Correlation of chemical constitution and physical properties of fatty acid esters with oviposition response of *Aedes aegypti*. Mosq News 27(2): 175–183

Perry AS, Mattson AM, Buckner AJ (1958) The mechanism of synergistic action of DMC with DDT against resistant houseflies. Biol Bull 104: 426–438

Perry AS, Gasith A, Mozel Y (1983) Pesticide residues in fish and aquatic invertebrates. Arch Toxicol Suppl 6: 199–204

Perry AS, Muszkat L, Perry RY (1989) Pollution hazards from toxic organic chemicals. In: Gerstl Z, Chen Y, Mingelgrin U, Yaron B (eds) Toxic organic chemicals in porous media. Springer, Berlin Heidelberg New York, pp 16–33

Perry AS, Sidis I, Zemach A (1990) Organochlorine insecticide residues in birds and bird eggs in the coastal plain of Israel. Bull Environ Contam Toxicol 45: 523–530

Persoons CJ, Ritter FJ, Lichtendonk WJ (1974) Sex pheromones of the American cockroach *Periplaneta americana*. Isolation and partial identification of two excitants. Proc K Ned Acad Wet Ser C Biol Med Sci 77: 201–204

Persoons CJ, Verwiel PEJ, Talman E, Ritter FJ (1979) Sex pheromone of the American cockroach *Periplaneta americana*. Isolation and structure elucidation of Periplanone-B. J Chem Ecol 5: 221–236

Peters RA (1948) Therapy of acute fluoride poisoning. Am. J. Med. Sci. 216: 278–285

Peters RA (1963a) Organofluorine compounds present in certain plants and their effect on animals. Biochem J 88: 55 p

Peters RA (1963b) Lethal synthesis and carbone-fluorine compounds. In: Peter RA (ed) Biochemical lesions and lethal synthesis. Macmillan New York, pp 88–142

Phillips MA, Worden AN (1957) The mammalian oral toxicity of fluoroacetamide. J Sc Food Agric 8: 653–657

Pickering QH, Henderson C, Lemke EA (1962) The toxicity of organic phosphorus insecticides to different species of warm water fishes. Trans. Am Fish Soc 91: 175–184

Pillmore RB (1973) Toxicity of pyrethrum to fish and wild-life. In: Casida JE(ed) Pyrethrum. Academic Press, New York, pp 143–165

Pimentel D (1971) Ecological effects of pesticides on non-target species. US Government Printing Office Washington, DC, 220 pp

Pimentel D (1990) Pest management in agriculture, 2nd edn. CRC Press, New York

Pimentel, D, Levitan L (1986) Pesticides: amounts applied and amounts reaching pests. Bioscience 36: 86–91

Pimentel D, Andow D, Dyson-Hudson R, Gallahan D, Jacobson S, Irish M, Kroop S, Moss A, Schreiner I, Shepard M, Thompson T, Vinzant B (1980) Environmental and social costs of pesticides: a preliminary assessment. OIKOS 34: 126–140

Plapp FW (1976) Biochemical genetics of insecticide resistance. Annu Rev Entomol 21: 179–198

Porter AG, Davidson EW, Liu JW (1993) Mosquitocidal toxins of bacilli and their genetic maipulation for effective biological control of mosquitoes. Microbiol Rev 57: 838–861

Post G, Schroeder TR (1971) The toxicity of four insecticides to four salmonid species. Bull Environ Contam Toxicol 6: 144–155

Post LC, de Jong BJ, Vincent WR (1974) 1-(2,6-Disubstituted benzoyl)-3-phenylurea insecticides: inhibitors of chitin synthesis. Pestic Biochem Physiol 4: 473–483

Prestwick GD, Blomquist GJ (1987) Pheromone biochemistry. Academic Press, New York, 565 pp

Raasch MS (1958) 5-Fluoronorvaline and 6-fluoronorleucine. J Org Chem 23: 1567–1568

Rashed SS, Mulla MS (1989) Factors influencing ingestion of particulate materials by mosquito larvae (Diptera: Culicidae). J Med Entomol 26: 210–216

Richards AG (1951) The chemical components and other properties, the anatomy and development and the permeability. In: The integument of arthropods. University of Minnesota Press, Minneapolis, 911 pp

Richards AG (1978) The chemistry of insect cuticle. In: Reckstein M(ed) Biochemistry of insects. Academic Press, New York, pp 205–232

Richards AG, Cutkomp LK (1945) Cholinesterase of nerve: apis and periplaneta. J Cell Comp Physiol 26: 57–61

Richards G, Hoffman JA (1987) Introduction to the insect neuroendocrine system. In: Morgan ED, Mandava NB (eds) CRC handbook of natural pesticides, vol III. Insect growth regulators part A. CRC Press, Boca Raton pp 1–14
Richardson CH (1925) The oviposition response of insects. USDA Bull 1324: 1–17
Robbins WE, Kaplanis JN, Thompson MJ, Shortino TJ, Joyner SC (1970) Ecdysone and synthetic analogs: molting hormone activity and inhibitive effects on insect growth, metamorphosis and reproduction. Steroids 16: 105–125
Robinson WH (1970) Acute toxicity of sodium monofluoroacetate to cattle. J Wildl Manage 34: 647–648
Roelofs WL (1975) Manipulating sex pheromones for insect suppression. Environ Lett 8: 41
Roelofs WL (1978) Chemical control of insects by pheromones. In: Reckstein M(ed) Biochemistry of insects. Academic Press, New York, pp 419–464
Roelofs WL, Cardé RT, Taschenberg EF, Weires RW Jr (1976) Pheromone research for the control of lepidopterous pests in New York. In: Beroza M (ed) Pest management with insect sex attractants. ACS symposium series no 23. American Chemical Society, Washington, DC, pp 75–87
Rosenheim JA, Hoy MA, Gordon J, Stewart JR (1989) Selecting for insecticide resistance in a California red scale parasitoid. Calif Agric 43(1): 17–18
Rothnie H, Geiser M (1990) Differential expression of the 3-δ-endotoxin genes in ***Bacillus thuringiensis*** subsp. ***Kurstaki*** HD-1. In: Baker RR, Dunn PE (eds) New directions in biological control. Alternatives for suppressing agricultural pests and diseases. Liss, New York, pp 599–608
Ruder FJ, Kayser H (1992) The carbodiimide product of diafenthiuron reacts covalently with two mitochondrial proteins, the FO-proteolipid and porin, and inhibits mitochondrial ATPase in vitro. Pestic Biochem Physiol 42(3): 248–261
Rupprecht JK, Hui Y-H, McLaughlin JL (1990) Annonaceous acetogenins: a review. J Nat Prod 53: 237–278
Saleh MA, Casida JE (1979) Toxaphene composition, structure–toxicity relations and metabolism. In: Geissbuehler H (ed) Advances in pesticide science part 3. Pergamon Press, Oxford, pp 562–566
Sandhu SS, Warren WJ, Nelson P (1978) Pesticidal residue in rural potable water. J Am Water Works Assoc 70: 41–45
Sarasua MJ, Santiago-Alvarez C (1983) Effect of diflubenzuron on the fecundity of ***Ceratitis capitata***. Entomol Exp Appl 33: 223–225
Savit J, Kollros J, Tobias J (1946) Measured dose of gamma-hexachlorocyclohexane required to kill flies and cockroaches, and a comparison with DDT. Proc Soc Exp Biol Med 62: 44–48
Sbragia RJ, Bisarbi-Ershadi B, Rigterink RH, Clifford DP, Dutton R (1983) XRD-473, a new acylurea insecticide effective against ***Heliothis***. In: 10th Int. Congr on Plant Protection, Brighton, vol 1 pp 417–424
Schechter MS, Green N, LaForge FB (1949) Constituents of pyrethrum flowers, XXIII cinerolone and the synthesis of related cyclopentenolones. J Am Chem Soc 71: 3165–3173
Schmeltz I (1971) Nicotine and other tobacco alkaloids. In: Jacobson M, Crosby DG (eds) Naturally occurring insecticides. Dekker, New York, pp 99–136
Schmutterer H (1990) Properties and potential of natural pesticides from the neem tree ***Azadirachta indica***. Annu Rev Entomol 35: 271–297
Schmutterer H (ed) (1995) The neem tree. VCH, Weinheim, 696 pp
Schnepf, EH (1995) ***Bacillus thuringiensis*** toxins: regulation, activities and structural diversity. Curr Opin Biotech 6: 305–312
Schnitzerling HJ, Nolan J, Hughes S (1983) Toxicology and metabolism of some synthetic pyrethroids in larvae of susceptible and resistant strains of the cattle tick ***Boophilus microplus*** (can). Pestic Sci 14: 64–72
Schoenig GP (1995) Mammalian toxicology of pyrethrum extract. In: Casida JE(ed) Pyrethrum, the natural insecticide. Academic Press, New York, pp 249–257
Schräder G (1963) Die Entwicklung neuer insektizider Phosphorsäure-Ester. Verlag Chemie, Weinheim, 444 pp

Schrankel KR, Grossman SJ, Hsia MTS (1982) Precocene II nephrotoxicity in the rat. Toxicol Lett 12: 95–100

Schreck CE (1977) Techniques for the evaluation of insect repellents: a critical review. *Annu. Rev. Entomol.* 22: 101–119

Seifert J, Casida JE (1981) Mechanism of teratogenesis induced by organophosphorus and methylcarbamate insecticides. *Prog Pest Biochem* 1: 219–246

Sekizawa J (ed) (1997) A compendium of pesticide safety data. LIC Inc, Tokyo

Sharratt M (1974) The effect on man of the use of chemicals in agriculture. In: Irvine DEG, Knights B (eds) Pollution and the use of chemicals in agriculture. Butterworths, London, pp 76–88

Shepard HH (ed) (1951) Plant products: nicotine and other alkaloids. Pyrethrins, rotenone, and miscellaneous plant extractives. The chemistry and action of insecticides. McGraw Hill, New York, 504 pp

Shishido T, Fukami J (1963) Studies on the selective toxicities of organic phosphorus insecticides (III). The charactors of enzyme system in cleavage of methyl parathion to desmethyl parathion in the supernatant of several species of homogenates (Part 1). Botyu–Kagaku 28: 77–81

Shono T, Casida JE (1978) Species-specificity in enzymatic oxidation of pyrethroid insecticides: 3-phenoxybenzyl and a-cyano-3-phenoxybenzyl-3-(2,2,-dihalovinyl)-2,2-dimethylcyclopropanecarboxylates. J Pestic Sci 3: 165–168

Shono T, Ohsawa K, Casida JE (1979) Metabolism of *trans*- and *cis*-permethrin, *trans*- and *cis*-cypermethrin and decamethrin by microsomal enzymes. *J. Agric. Food Chem.* 27(2): 316–325

Shorey H, McKelvey J (1977) Chemical control of insect behavior: theory and application. Wiley, New York, 414 pp

Silhacek DL, Oberlander H, Porcheron P (1990) Action of RH-5849, a non-steroidal ecdysteroid mimic, on *Plodia interpunctella* (Hübner) in vivo and in vitro Arch Insect Biochem Physiol 15: 201–212

Simkin J, Galun R (1983) Microencapsulated natural pyrethrum—an improved insect repellent. ICIPE, International center of insect physiology and ecology, Nairobi, annual report, pp 151–161

Simmons SW (1959) The use of DDT insecticides in human medicine. In: Müller PH (ed) DDT—the insecticide dichlorodiphenyltrichloroethane and its significance, vol II. Human and veterinary medicine. Birkhauser, Basel pp 251–502

Slama K, Williams CM (1965) Juvenile hormone activity for the bug *Pyrrhocoris apterus*. Proc Natl Acad Sci USA 54: 411–414

Smagghe G, Degheele D (1992a) Effects of RH-5849, the first nonsteroidal ecdysteroid agonist, on larvae of *Spodoptera littoralis* (Boisd.) (Lepidoptera: Noctuidae). Arch Insect Biochem Physiol 21: 119–128

Smagghe G, Degheele D (1992b) Effect of nonsteroidal ecdysteroid agonist RH-5849 on reproduction of *Spodoptera littoralis* (Boisd.) (Lepidoptera: Noctuidae). Parasitica 48: 23–29

Smagghe G, Degheele D (1993) Metabolism, pharmacokinetics, and toxicity of the first nonsteroidal ecdysteroid agonist RH-5849 to *Spodoptera exempta* (Walker), *Spodoptera exigua* (Hübner), and *Leptinotarsa decemlineata* (Say). Pestic Biochem Physiol 46: 149–160

Smagghe G, Degheele D (1994a) Action of the nonsteroidal ecdysteroid mimic RH-5849 on larval development and adult reproduction of insects of different orders. Invert Rep Dev 25: 227–236

Smagghe G, Degheele D (1994b) Action of a novel nonsteroidal ecdysteroid mimic, tebufenozide (RH-5492), on insects of different orders. Pestic Sci 42: 85–92

Smilowitz Z, Martinka CA, Jowyk EA (1976) The influence of a juvenile hormone mimic (JHM) on the growth and development of the cabbage looper and the endoparasite *Hyposoter exiguae*. Environ Entomol 5: 1178–1182

Smissaert HR (1964) Cholinesterase inhibition in red spider mites *Tetranychus urticae* susceptible and resistant to organophosphate. Science 143: 129–131

Smith CN, LaBreque GC, Bonkovec AB (1964) Insect chemosterilants. Annu Rev Entomol 9: 269–284
Smyth HF, Weil CS, West JS, Carpenter CP (1970) An exploration of joint toxic action II. Equitoxic versus equivolume mixtures. Toxicol Appl Pharmacol 17: 498–503
Soderlund DM, Sanborn JR, Lee PW (1983) Metabolism of pyrethrins and pyrethroids in insects. In: Hutson DH, Roberts TR (eds) Progress in pesticide biochemistry vol 3. Wiley Chichester, pp 401–435
Sollmann T (1957) A manual of pharmacology and its application to toxicology and therapeutics 8th edn. Saunders, Philadelphia
Song JH, Narahashi T (1996) Modulation of sodium channels of rat cerebellar Purkinje neurons by the pyrethroid treatment J Pharmacol Exp Ther 277: 445–453
Sovocool GW, Lewis RG, Harless RL, Wilson NK, Zehr RD (1977) Analysis of technical chlordane by gas chromotography/mass spectrometry. Anal Chem 49: 734–740
Sowa BA, Marks EP (1975) An in vitro system for the quantitative measurement of chitin synthesis in the cockroach: inhibition by TH 6040 and polyoxin D. Insect Biochem 5: 858–859
Staal GB (1986) Anti-JH agents. Annu Rev Entomol 31: 391–429
Starratt AN, Bond EJ (1981) Metabolism of methyl bromide by susceptible and resistant strains of the granary weevil, *Sitophilus granarius* (L). Pestic Biochem Physiol15: 275–281
Staudinger H, Ruzicka L (1924) Insekttentotende Stoffe I, V and X Helv Chem Acta 7: 177–259, 377–458
Steinhaus EA (1956) Microbial control—the emergence of an idea. Hilgardia 26: 107–160
Steiner LF, Rohwer GG, Ayers EL, Christenson LD (1961) The role of attractants in the recent Mediterranean fruit fly eradication program in Florida. J Econ Entomol 54: 30–35
Steiner LF, Mitchell WC, Harris EJ, Kozuma TT, Fujimoto MS (1965) Oriental fruit fly eradication by male annihilation. J Econ Entomol 58(5): 961-964
Stewart RD, Boettner EA, Southworth RS, Erney JC (1963) Acute CCl_4 intoxication. JAMA 183: 994–997
Stickel LF (1973) Pesticide residues in birds and mammals. In: Edwards CA (ed) Environmental pollution by pesticides Plenum Press, New York, pp 254–312
Storey BT (1981) Inhibitions of energy-coupling site 1 of the mitochondrial respiratory chain In: Erecinsca M, Wilson DF (eds) Inhibitors of mitochondrial functions. International Encyclopaedia of Pharmacological Therapeutics, sect 107 Pergamon Press, New York, pp 101–108
Sussman JL, Silman I (1992) Acetylcholinesterase: structure and use as a model for specific cation-protein interactions. Curr Opin Struct Biol 2: 721–729
Suzuki S, Isono K, Nagatsu J, Mizutani T, Koseki K, Kodama H, Suzuki A (1965) A new antibiotic, polyoxin A. J Antibiot Ser A, 18: 131
Swaroop S, Uemura K (1956) Probit analysis. Mimeograph Malaria no 178. WHO, Geneva
Takahashi Y, Kyomura N, Yamamoto I (1977) Synergistic insecticidal action of *N*-methyl- and *N*-propylcarbamates to the green rice leafhopper, *Nephotettix cincticeps*, resistant to aryl *N*-methylcarbamates. J Pestic Sci 2: 467-470
Tanaka K, Ohsawa K, Honda H, Yamamoto I (1981) Copulation release pheromone, erectin, from the azuki bean weevil (*C. chinensis* L). J Pestic Sci 6: 75–82
Tatebayashi H, Narahashi T (1994) Differential mechanism of action of the pyrethroid tetramethrin on tetrodoxin-sensitive and tetradoxin-resistant sodium channels. J Pharmacol Exp Ther 270: 595–603
Thomas W, Ellar DJ (1983) Mechanism of action of *Bacillus thuringiensis* var *israelensis* insecticidal δ-endotoxin. FEBS Lett 154(2): 362-368
Tobias JM, Kollros JJ (1946) Loci of action of DDT in the cockroach (*Periplaneta americana*). Biol Bull 91: 247–255
Tomizawa M, Yamamoto I (1993) Structure–activity relationships of nicotinoids and imidacloprid analogs. J Pesticide Sci 18: 91–98
Torkelson R, Oyen F (1977) The toxicology of 1,3-dichloropropene as determined by repeated exposure of laboratory animals. Am Ind Hyg Assoc J **38**: 217–223

Tourtellotte WW, Coon JM (1951) Treatment of fluoroacetate poisoning in mice and dogs. J Pharmacol Exp Ther 101: 82–91

Treherne JE, Pichon Y (1972) The insect blood–brain barrier Adv Insect Physiol 9: 257–311

Trevan JW (1922) An apparatus for the measurement of small quantities of fluid. Lancet 202: 786

Tucker RK, Crabtree DG (1970) Handbook of toxicity of pesticides to wildlife. Resource publication, no 84. Bureau of Sport Fisheries and Wildlife, Denver Wildlife Research Center Denver, 131 pp

Tucker RK, Haegele MA (1971) Comparative acute oral toxicity of pesticides to six species of birds. Toxicol Appl Pharmacol 20: 57–95

Turner RB (1968) Chemistry of insect chemosterilants In: LaBreque GC, Smith CM (eds) Principles of insect chemosterilization. Appleton-Century-Crofts, New York, pp 159–274

Uchida M, Izawa Y, Sugimoto T (1987) Inhibition of prostaglandin biosynthesis and oviposition by an insect growth regulator, buprofezin, in *Nilaparvata lugens* Stål. Pestic Biochem Physiol 27: 71–75

Unai T, Yamamoto I (1973) Synthesis of the stereoisomers of natural rotenone. J Pestic Sci 37: 897–901

Van Daalen JJ, Meltzer J, Mulder R, Wellinga K (1972) A selective insecticide with a novel mode of action. Naturwissenschaften 59: 312–313

Verloop A, Ferrell CD (1977) Benzoylphenyl ureas—a new group of larvicides interfering with chitin deposition. In: Plimmer JR (ed) Pesticide chemistry in the 20th century. American Chemical Society, Washington, DC, pp 237–270

Vinson SB (1974) Effect of an insect growth regulator on two parasitoids developing from treated tobacco budworm larvae. J Econ Entomol 67: 335–336

Ward JC, Spencer DA (1947) Notes on the pharmacology of sodium fluoroacetate-compound 1080. J Am Pharm Assoc Sci Ed 36: 59–62

Ware GW (1983) Pesticides, theory and application. Freeman, San Francisco, 308 pp

Warthen JD Jr, Uebel EC, Lusby WR, Adler VE (1983) Investigations of a sex pheromone for the oriental cockroach, *Blatta orientalis*. J Insect Physiol 29: 605–609

Weber WJ (ed) (1979) Health hazards from pigeons, starlings and English sparrows. Thompson Fresno Ca, 138 pp

West TF, Campbell GA (1950) DDT-synthetic insecticide which stopped a typhus epidemic. DDT and newer persistent insecticides, 2nd edn. Chapman and Hall, London, 632 pp

Whittle AJ (1994) 3-Aryl pyrimidinones: a novel class of potent insecticide. In: Briggs GG (ed) Advances in the chemistry of insect control III. Royal Society of Chemistry, Cambridge, pp 156–170

WHO (1975) Recommended classification of pesticides by hazards. WHO Chron 29: 397–401

WHO (1979) DDT and its derivatives. Environmental Health Criteria, no 9 WHO, Geneva

WHO (1986) Organophosphorus insecticides: a general introduction. Environmental Health Criteria no 63: WHO, Geneva 181 pp

WHO (1988) The WHO recommended classification of pesticides by hazards and guidelines to classification 1988–1989 WHO/VBC/88.953 WHO, Geneva

Whorton D, Krauss RM, Marshall S, Milby TH (1977) Infertility in male pesticide workers. Lancet 2: 1259–1261

Whorton D, Milby TH, Krauss RM, Stuffs HA (1979) Testicular function in DBCP exposed pesticide workers J Occup Med 21: 1161–1166

Wilkinson CF (1971) Effects of synergists on the metabolism and toxicity of anticholinesterases Bull WHO 44: 171–190

Wilkinson CF (1976a) Insecticide synergists and their mode of action. In: Tahori AS (ed) Pesticide chemistry. Proc 2nd Int Congr on Pesticides and Chemistry, 1971, vol 2. Gordon and Breach, New York, pp 117–159

Wilkinson CF (1976b) Insecticide synergism. In: Metcalf RL, McKelvey JJ Jr (eds) The future for insecticides: needs and prospects vol 6. Wiley, New York, pp 195–218

Williams CM (1946) Anaesthesia by carbon dioxide Science 103: 57

Williams CM (1956) The juvenile hormone of insects. Nature. 178: 212–213

Williams CM (1967) Third-generation pesticides. Sci Am 217(1): 13–17
Williams RT (ed) (1959) The metabolism and detoxication of drugs, toxic substances and other organic compounds. Detoxication mechanisms, 2nd edn. Wiley, New York, 288 pp
Wilson D, Anema BP (1988) Development of buprofezine for control of whitefly *Trialeurodes vaporariorum* and *Bemisia tabaci* on glasshouse crops in the Netherlands and the UK. In: British Crop Protection Conf, Pests and diseases, Brighton. British Crop Protection Council, Farnham, pp 175–180
Wilson JG (1972) Environmental effects on development-teratology. In: Assali NS (ed) Pathophysiology of gestation vol 2. Academic Press, New York, pp 269–320
Wilson NK, Sovocool GW (1977) Spectroscopic characterization of some chlorinated tricyclic hydrocarbons of the C_{10} series: chlordanes, chlordenes and nonaclors. Org Magn Reson 9: 536–542
Wing KD (1988) RH-5849, a nonsteroidal ecdysone agonist: effects on a *Drosophila* cell line Science 241: 467–469
Wing KD, Aller HE (1990) Ecdysteroid agonists as novel insect growth regulators. In: Casida JE (ed) Pesticides and alternatives: innovative chemical and biological approaches to pest control. Elsevier, Amsterdam, pp 251–257
Wing KD, Glickman AH, Casida JE (1982) Oxidative bioactivation of S-alkyl phosphorothiolate pesticides: stereospecificity of profenofos insecticide activation. *Science* 219: 63–65
Wing KD, Slawecki RA, Carlson GR (1988) RH-5849, a nonsteroidal ecdysone agonist: effects on larval Lepidoptera. Science 241: 470–472
Wood DL, Silverstein RM, Nakajima M (1970) Control of insect behavior by natural products. Academic Press, New York
Woolson EA (ed) (1975) Arsenical pesticides ACS symposium series, no 7. American Chemical Society, Washington, DC, 176 pp
Wright JE, Harris RL (1976) Ovicidal activity of Thompson-Hayward TH 6040 in the stable fly and horn fly after surface contact by adults. J Econ Entomol 69: 728–730
Wright JE, Spates GE (1972) A new approach in integrated control: insect JH plus a hymenopteran parasite against the stable fly. Science 178: 1279–1293
Wright JW, Fritz RF, Haworth J (1972) Changing concepts of vector control in malaria eradication. Annu Rev Entomol 17: 75–102
Wustner DA, Desmarchelier J, Fukuto TR (1972) Structure for the oxygenated product of peracid oxidation of Dyfonate insecticide (*o*-ethyl *s*-phenylethyl phospho-nodithioate). Life Sci 11: 583
Yamagishi M, Kakinohana H, Kuba H, Kohama T, Nakamoto Y, Sokei Y, Kinjo K (1993) Eradication of the melon fly from Okinawa, Japan, by means of the sterile insect technique. In: Proc Int Symp on Management of insect pests: nuclear and related molecular and genetic techniques IAEA, Vienna, pp 49–60
Yamamoto I (1965) Nicotinoids as insecticides. In: Metcalf RL (ed) Advances in pest control research. Interscience, New York, pp 231–260
Yamamoto I (1970) Mode of action of pyrethroids, nicotinoids, and rotenoids. *Annu Rev Entomol* 15: 257–272
Yamamoto I (1973) Mode of action of synergists in enhancing the insecticidal activity of pyrethrum and pyrethroids. In: Casida JE (ed) Pyrethrum, the natural insecticide. Academic Press, New York pp 195–210
Yamamoto I (1990) Chemical ecology of Bruchids In: Fujii K, Gatehouse AMR, Johnson CD, Mitchel R, Yoshida T (eds) Bruchids and legumes: economics, ecology and coevolution. Kluwer, Dordrecht, pp 53–62
Yamamoto I, Casida JE (1966) 0-Demethyl pyrethrin II analogs from oxidation of pyrethrin I, allethrin, dimethrin and phthalthrin by a house fly enzyme system. J Econ Entomol 59: 1542–1543
Yamamoto I, Kamimura H, Yamamoto R, Sakai S, Goda M (1962) Studies on nicotinoids as an insecticide—relation of structure to toxicity. Agric Biol Chem 26: 709–716
Yamamoto I, Soeda Y, Kamimura H, Yamamoto R (1968) Cholinesterase inhibition by nicotinoids and pyridylalkylamines—its significance to mode of action. Agric Biol Chem 32: 1341–1348

Yamamoto I, Kimmel EC, Casida JE (1969) Oxidative metabolism of pyrethroids in houseflies. J Agric Food Chem 17: 1227–1236

Yamamoto I, Elliot M, Casida JE (1971a) The metabolic fate of pyrethrin I, pyrethrin II, and allethrin. Bull WHO 44: 347–348

Yamamoto I, Takahashi Y, Kyomura N (1983) Suppression of altered acetylcholinesterase of the green rice leafhopper by N-propyl and N-methyl carbamate combinations. In: Georghiou GP, Saito T (eds) Pest resistance to pesticides Plenum Press, New York, pp 579–594

Yamamoto I, Kyomura N, Takahashi Y (1993) Negatively correlated cross resistance: combinations of N-methylcarbamate with N-propylcarbamate or oxadiazolone for green rice leafhopper. Arch Insect Biochem 22: 277–288

Yamamoto I, Unai T, Ohkawa H, Casida JE (1971b) Stereochemical considerations in the formation and biological activity of the rotenone metabolites. Pestic Biochem Physiol 1: 143–150

Yamamoto I, Yabuta G, Tomizawa M, Saito T, Miyamoto T, Kagabu S (1995) Molecular mechanism for selective toxicity of nicotinoids and neonicotinoids. J Pestic Sci 20: 33–40

Yamamoto I, Honda H, Miyamoto T, Ohsawa K (1996) Chemosensory and tactile regulation of copulation and oviposition by *C chenensis* L and *C maculatus* Fabricius (Coleoptera: Bruchidae) 20th Int Congr on Entomology, Florence, Aug 25–31 562 pp

Yamamoto T, Powell G (1993) *Bacillus thuringiensis* crystal proteins: recent advances in understanding its insecticidal activity In: Kim L (ed) Advanced engineered pesticides Dekker, New York, pp 3–42

Yarom I, Blumberg D, Ishaaya I (1988) Effect of buprofezine on California red scale (Homoptera: Diaspididae) and Mediterranean black scale (Homoptera: Coccidae) J Econ Entomol 81: 1581–1585

Yasui M, Fukada M, Maekawa S (1985) Effect of buprofezine on different developmental stages of the greenhouse whiteflies, *Trialeurodes vaporariorum* (Westwood) (Homoptera: Aleyrodidae). Appl Entomol Zool 20: 340–347

Yasui M, Fukada M, Maekawa S (1987) Effect of buprofezine on reproduction of the greenhouse whiteflies, *Trialeurodes vaporariorum* (Westwood) (Homoptera: Aleyrodidae). Appl Entomol Zool 22: 266–271

Yiallouros M, Storch V, Becker N (1995) Ultrastructural midgut events in *Chironomus thummi thummi* larvae after the exposure to B.t.i. and B.s. 9th Eur Meet Society for Vector Ecology, 4–7 Sept 1995, Prague

Yousten AA (1984) *Bacillus sphaericus*: microbiological factors related to its potential as a mosquito larvicide Adv Biotechnol Processes 3: 315–343

Yunowitz H, Yawets A (1988) Interaction between δ-endotoxin produced by *Bt* spp *entomocidus* and liposomes FEBs Lett 230 (1,2): 105–108

Yunowitz H, Yawets A (1989) The interaction of the δ-endotoxin purified from *B thur* subsp *entomocidus* with lipocoms and midgut epithelium membranes from two insect species. Isr J Entomol 23: 255–258

Zahavi M, Tahori A (1970a) Difference in acetylcholinesterase sensitivity to phosphamidone in Mediterranean fruit fly strains. Isr J Entomol, 5: 185–192

Zahavi M, Tahori A (1970b) Sensitivity of acetylcholinesterase in spider mites to organophosphorous compounds. Biochem Pharmacol 19: 219–225

Zlotkin E (1991) Venon neurotoxins—models for selective insecticides Phytoparasitica 19: 177–182

The Contribution of Albert S. Perry to Insect Resistance and Environmental Toxicology

A. Yawetz[1] and A.S. Tahori[2]

Professor Albert S. Perry died suddenly on February 18, 1992. Al, as he was known among his friends, was born in Salonica, Greece, in 1915. His family emigrated to Israel (then Palestine), when he was 6 years old. He subsequently excelled in his scholastic record at an agricultural school, which led him to continue in academic studies at the University of California, Berkeley, where he obtained a B.S. degree in Entomology in 1941. When World War II broke out, he volunteered for the US Army and served as parachutist, participating in the invasion of Normandy, where he was decorated for bravery. After his return from the army he enrolled in graduate studies at the University of California. In 1950 he submitted his doctoral thesis on "Detoxification of DDT in Resistant Houseflies and the Inhibiting Effect of Synergists", the result of research conducted under the supervision of Prof. W.M. Hoskins, then the uncrowned Dean of Insect Toxicologists. After obtaining his Ph.D. degree, in Entomology and Agricultural Chemistry, he moved to the Communicable Disease Center (CDC), USPHS, at Savannah, Georgia, where he remained until 1974.

Al's main interests and scientific contribution were in the fields of insect resistance to insecticides and the impact of pesticide residues on the environment and especially on wildlife.

In the classic review on "The physiology of insecticide resistance by insects", which Prof. Perry published with Prof. M. Agosin in 1974, they proposed that resistance is biphasic. Phase I is due to selection of variants in the population attaining a resistance level commensurable with the gene pool for resistance initially present in the population. It is assumed that resistance is due in most instances to enzymatic detoxication of the chemical in which genetic selection results in increased enzyme activity. However, in phase II, acceleration of resistance takes place by induction of preexisting detoxifying enzymes toward enhanced activity, resulting in faster breakdown of the chemical. The inducer may be the insecticide itself, which is being used for the control of the particular insect.

Al studied the metabolism of DDT in susceptible and DDT-resistant body lice, *Pediculus humanus humanus.* The resistant (Korean) strain tolerated blood concentrations of 200 ppm, whereas the susceptible (Orlando) strain succumbed

[1]Institute for Nature Conservation Research, Faculty of Life Sciences, Tel-Aviv University, Israel.
[2]Ramat Hasharon, Israel

to 10–20 ppm DDT. He later found that homogenates of both strains metabolized DDT in vitro to form *p, p'*-dichlorodiphenyl-acetic acid (DDA), like the metabolism of DDT in the mammalian host, rather than *p, p'*-dichlorodiphenyldichloroethylene (DDE), the normal metabolite of DDT in insects.

Al worked also on cross-resistance, a unique phenomenon in insect toxicology among both closely related compounds and those relatively distant in structure. For example, he found that 1,1,-bis(4-chlorophenyl-2-nitrobutane (Prolan)-resistant flies metabolized Prolan to neutral and acidic derivatives, but they also dehydrochlorinated DDT via a different mechanism.

Later, Al investigated insecticide resistance in anopeline mosquitoes. He found that field-collected adults of *Anopheles atroparvus, A. maculipenis, A. labranchiae, A. superpictus* and *A. claviger* from Italy exhibited normal susceptibility to DDT and little conversion of DDT and DDE. Laboratory colonies of DDT-resistant *Anopheles atroparvus* and *A. stephensi* produced little DDE in proportion to their resistance, and showed no significant difference in DDT metabolism from their susceptible counterparts. Only in various strains of *A. sacharovi* did he recover substantial quantities of DDE, so that he could correlate the resistance mechanism with DDT dehydrochlorination.

In another study Al found that the enzymatic degradation of DDT in the body louse yielded three metabolites, DDE, DDA, and 4,4-dichlorobenzophenone. Three or more enzymes were involved in these biotransformations, two of which he isolated and partially purified.

He then examined the metabolic fate of the cyclodiene insecticides. Early investigations on the fate of these compounds in insects demonstrated the biotransformation of heptachlor, aldrin, and isodrin to their epoxides, i.e., heptachlor epoxide, dieldrin, and endrin, respectively. He found that the onset of symptoms of heptachlor poisoning, after a latent period of 1–2 h, conincided with the appearance of heptachlor epoxide in the tissues. This observation led him to infer that transformation to the parent compound is an activation process, not involving a detoxification mechanism.

In his work on synergism, Al established that 1,1-bis(*p*-chlorophenyl) ethanol (DMC) acts as a competetive substrate for DDT dehydrochlorination, blocking thereby DDT detoxification. He pointed out the similarity between DDT dehydrochlorination and DMC dehydration. Al's studies showed that DMC was metabolized by the housefly to 1,1-bis (*p*-chlorophenyl) ethylene and then to DDA. However, he noted a lack of synergism by DMC in DDT-resistant *P. humanus humanus,* and DMC proved to have only a mild or no synergistic effect with *o*-chloro-DDT.

Al was among the pioneer scientists who discovered the fundamental role of microsomal monooxygenation in insecticide metabolism in insects. Insects have been shownd to catalyze a series of oxidative reactions which include incorporation of half of the dissolved oxygen molecule into hydrophobic insecticide while the other oxygen atom is reduced to water. In both insect and mammalian systems, NADPH is the preferred electron donor. Al demonstrated the induction of cytochrome P_{450}, the main component of the microsomal monooxygenase system, in a diazinon-resistant housefly strain, compared with a susceptible strain. He noticed that age and diet modified the levels of cytochrome P_{450}. At that early stage of cytochrome P_{450} research, the question of the multiplicity of cytochrome P_{450} was not yet solved, and it was not known whether to attribute the variety of monooxygenation reactions to a single or several hemoproteins. He contributed significantly to the development of the multienzyme concept with his finding that microsomes prepared from a susceptible strain or from strains resistant to DDT, dieldrin, prolan, and malathion, had a carbon-monoxide difference spectrum with a peak at 452 nm, while a diazinon-resistant strain showed absorbance maxima at 448 nm, indicating the presence of a hemoprotein with different spectrophotometric properties in the resistant strain. He contributed also to the understanding of the mechanism of action of piperonyl butoxide (PB) as a synergist to various insecticides by showing that PB inhibits cytochrome P_{450} activity in insects.

Al returned to Israel in 1974, to the Institute for Nature Conservation Research, The George S. Wise Faculty of Life Sciences, at Tel-Aviv University. He was concerned mainly with the deleterious effects of pesticides and industrial wastes on the quality of the environment in Israel. The increase in environmental pollutants in Israel affected wildlife by changing species composition and population dynamics and decreasing the abundance of various species. Al and his students investigated the biochemical mechanism by which wildlife resist the destruction of their habitat by toxic chemicals. Those mechanisms were related largely to the hepatic microsomal monooxygenase system with cytochrome P_{450} being the final oxidase. Polychlorinated biphenyls (PCBs) were found to be the major chemical contaminants of industrial origin, with direct effects on cytochrome P_{450} content and composition. The impact on monooxyganase systems, in both the liver and internal secretion glands, was correlated with pathological changes of reproduction physiology and behavior.

Al participated in the October 1991 United States/Israeli BARD Symposium on "New Targets for Insect Management in Crop Protection", at Shoresh, Israel. All participants were impressed and stirred by the sagacity and sound judgement he demonstrated in his remarks as a member of the final discussion panel which was held on this occasion.

Notwithstanding his international reputation as a scientist who contributed to the very foundations of modern environmental toxicology, he was very modest and shy, never trying to impose his opinions, and always ready to listen and examine every idea of his associates and students. In the hard, very demanding, and competitive academic world, Al was a unique phenomenon, as he was appreciated and loved by all who knew him. Al was loved because he was good-hearted and his friendship was totally altruistic, given without any consideration of interest or gain. The cautious and careful way typical of his scientific work was exercised in all his human relations.

Albert Perry: Relevant Publications

I. Chapters in Books

1. Perry AS (1964) The physiology of insecticide resistance by insects. In: Rockstein M (ed). The physiology of Insecta vol III. Academic Press, New York, pp 285–378
2. Perry AS (1968) The physiology of insecticide resistance by insects. In: Rockstein M (ed) The physiology of Insecta, vol III. Academic Press, New York, pp 286–368
3. Perry AS, Agostin, M (1974) The physiology of insecticide resistance by insects. In: Rockstein, M (ed) The physiology of Insecta vol VI, 2nd edn. Academic Press, New York, pp 3–102
4. Agostin M, Perry AS (1974) Microsomal mixed-function oxidases In: Rockstein M (ed). The physiology of Insecta, vol V, 2nd edn. Academic Press, New York pp 538–590
5. Perry, AS (1974) Insecticide resistance in insects and its ecological and economic thrust. In: Khan MAQ, Bederka JP Jr (eds) Survival in toxic environments. Academic Press, New York
6. Perry AS (1976) Microsomal oxidations in insects and role of cytochrome P-450 In: Gunther, FA (ed) Residue reviews. Springer, Berlin Heidelberg New York
7. Perry A, Perry RY (1989) Effects in arid regions. In: Bourdeau P, Haines JA, Klein W, Krishna Murti CR (eds) Ecotoxicology and climate. 1989 SCOPE. John Wiley & Sons, New York, pp 155–193
8. Perry A, Muszkat L, Perry RY (1989) Pollution hazards from toxic organic chemicals. In: Gerstl Z, Chen Y, Mingelgrin U, Yaron B (eds) Toxic organic chemicals in porous media. Ecological studies 73. Springer, Berlin Heidelberg New York, pp 16–33

II. Scientific Articles

1. Perry AS, Hoskins WM (1950) The detoxification of DDT by resistant houseflies and inhibition of this process by piperonyl cyclonene. Science 111: 600–601
2. Perry AS, Hoskins WM (1951) Detoxification of DDT as a factor in the resistance of houseflies. J Econ Entomol 44: 850–857
3. Perry AS, Hoskins WM (1951) Synergistic action with DDT toward resistant houseflies. J Econ Entomol 44: 839–850
4. Perry AS, Fay RW, and Buckner AJ (1953) Dehydrochlorination as a measure of DDT resistance in houseflies. J Econ Entomol 46: 972–976
5. Perry AS, Mattson AM, Buckner AJ (1953) The mechanism of synergistic action of DMC with DDT against resistant houseflies. Biol Bull 104: 426–438
6. Perry AS, Sacktor B (1955) Detoxification of DDT in relation to cytochrome oxidase activity in resistant and susceptible houseflies. Ann Entomol Soc Am 48: 329–333
7. Perry AS, Jensen JA, Pearce GW (1955) Colorimetric and radiometric determinations of DDT and its metabolites in resistant houseflies. Agric Food Chem 3: 1008–1011
8. Brown AWA, Perry AS (1956) Dehydrochlorination of DDT by resistant houseflies and mosquitoes. Nature 178: 368–369
9. Perry AS (1958) Factors associated with DDT resistance in the housefly *Musca domestica* L Proc 10th Int Congr Entomol 1956 2: 157–172
10. Perry AS, Buckner AJ (1958) Biochemical investigations on DDT resistance in the human body louse *Pediculus humanus humanus* Am J Trop Med Hyg 7: 620–626
11. Perry AS, Mattson AM, Buckner AJ (1958) The metabolism of heptachlor by resistant and susceptible houseflies. J Econ Entomol 51: 346–351
12. Jones JC, Perry AS (1959) A nib-point microloop and its calibration. Mosquito News 19: 26–27
13. Perry AS and Buckner AJ (1959) The metabolic fate of Prolan in a Dilan-resistant strain of houseflies J Econ Entomol 52: 997–1002
14. Perry AS (1960) Investigations on the mechanism of DDT resistance in certain anopheline mosquitoes. Bull Wildl Health Organi 22: 743–756

15. Perry AS (1960) Metabolism of insecticides by various insect species. Agric Food Chem 8: 266–272
16. Perry AS (1960) Biochemical aspects of insect resistance to the chlorinated hydrocarbon insecticides. Misc Pub Entomol Soc Am 2: 119–137
17. Perry AS, Miller S, Buckner AJ (1963) The enzymatic in vitro degradation of DDT by susceptible and DDT-resistant body lice. Agric Food Chem 11: 457–462
18. Miller S, Perry AS (1964) Separation and purification of DDT-degrading enzymes from the human body louse. Agric Food Chem 12: 167–169
19. Perry AS, Pearce GW, Buckner AJ (1964) The absorption, distribution, and fate of C^{14}-Aldrin and C^{14}-Dieldrin by susceptible and resistant houseflies. J Econ Entomol 57: 867–872
20. Perry AS, Miller S (1965) The essential role of folic acid and the effect of antimetabolites on growth and metamorphosis of housefly larvae *Musca domestica* L. J Insect Physiol 11: 1277–1287
21. Miller S, Perry AS (1965) Isolation, purification, and characterization of N^5-formyl-tetrahydrofolic acid (folinic acid) from the housefly. Life Sci 4: 1573–1580
22. Fay RW, Perry AS (1965) Laboratory studies of ovipositional preferences of *Aedes aegypti* Mosquito News 25: 276–281
23. Perry AS (1966) Biochemistry of insecticide resistance in mosquitoes Mosquito News 26: 301–309
24. Perry AS, Hennessy DJ, Miles JW (1967) Comparative toxicity and metabolism of *p,p'*-DDT and various substituted DDT-derivatives by susceptible and resistant house flies. J Econ Entomol 60: 568–673
25. Perry AS, Fay RW (1967) Correlation of chemical constitution and physical properties of fatty acid esters with oviposition response of *Aedes aegypti* Mosquito News 27: 175–183
26. Perry AS (1967) Current status of the biochemistry of insecticide resistance. Information circular on insecticide resistance, insect behaviour and vector genetics. World Health Organi 62: 40–55
27. Kalra RL, Perry AS, Miles JW (1967) Studies on the mechanism of DDT resistance in *Culex pipiens fatigans* Bull Wildl Health Organi 37: 651–656
28. Miller S, Perry AS (1968) RNA synthesis in the developing ovary of the housefly: incorporation of C^{14}-formate. J Insect Physiol 14: 581–589
29. Kalra RL, Perry AS, Miles JW (1968) Toxicity, pick-up, and metabolism of *p,p'*-DDT and related compounds in *Culex pipiens fatigans* and *Aedes aegypti* Indian J Exp Biol 6: 37
30. Keirans JE, Perry AS, Fay RW (1968) Preliminary observations on the effect of alcohol and oil films on the oviposition of *Culex pipiens quinquefasciatus* (Say) Mosquito News 28: 71–74
31. Perry AS (1968) An apparatus for dispensing housefly eggs mechanically. J Econ Entomol 61: 1112–1113
32. Perry AS (1970) Studies on microsomal cytochrome P-450 in resistant and susceptible houseflies. Life Sci 9 (LL): 335–350
33. Perry AS, Dale WE, Buckner AJ (1971) Induction and repression of microsomal mixed-function oxidases and cytochrome P-450 in resistant and susceptible house flies. Pestic Biochem Physiol 1: 131–142
34. Capdevila J, Perry AS, Morello A, Agostin M (1973) Effect of phenobarbital and naphthalene on some components of the electron transport system and the hydroxylating activity of housefly microsomes. Biochemistry 12: 1445–1451
35. Capdevila J, Perry AS, Morello A, Agostin M (1973) Some spectral properties of cytochrome P-450 from microsomes isolated from control, phenobarbital- and naphthalene-treated house flies. Biochem Biophys Acta 314: 93–103
36. Capdevila J, Perry AS, Agostin M (1974) Spectral and catalytic properties of cytochrome P-450 from a diazinon-resistant housefly strain. Chem-Biol Interactions 9: 105–116
37. Yawetz A, Agostin M, Perry AS (1978) Components of the electron transport system in the mixed-function-oxidase system in wild birds. Pestic Biochem Physiol 8: 44–52
38. Yawetz A, Agostin M, Perry AS (1978) Spectral and catalytic properties of cytochrome P-450 from four wild birds species. Comp Biochem Physiol 59C: 45–49

39. Yawetz A, Agostin M, Perry AS (1979) Metabolism of Parathion and brain cholinesterase inhibition in four species of wild birds. Pestic Biochem Physiol 11: 294–300
40. Gasith A, Perry AS (1980) Fate of Parathion in a fish pond ecosystem and its impact on food-chain organisms. Agrochemical residue—biota interactions in soil and aquatic ecosystems. International Atomic Energy Agency, Vienna, pp 125–151
41. Rinsky A, Perry AS (1981) Induction of the mixed-function oxidase system in the liver of the barn owl *Tyto alba* by PCBs. Pestic Biochem Physiol 16: 72–78
42. Manulis S, Ishaaya I, Perry AS (1981) Acetylcholinesterase of *Aphis citricola*: properties and significance for determining toxicity of systemic organophosphorus and carbamate toxicity. Pestic Biochem Physiol 15: 267–274
43. Yawetz A, Perry AS, Freeman A, Katchalski-Katzir E (1984) Monooxygenase activity of rat liver microsomes immobilized by entrapment in a crosslinked prepolimerized polyacrylamide hydrazide. Biochem Biophys Acta 798: 204–209
44. Koren B, Yawetz A, Perry AS (1984) Biochemical properties characterizing the development of tolerance to malathion in *Ceratitis capitata* Wiedemann (Diptera: Tephritidae). J Econ Entomol 77(4): 864–867
45. Gazit Y, Ishaaya I, Perry AS (1989) Detoxification and synergism of diflubenzuron and chlorfluazuron in the red flour bettle *Tribolium castaneum* Pestic Biochem Physiol 34: 103–110
46. Perry AS, Şidis I, Zemach A (1990) Organochlorine insecticide residues in birds and bird eggs in the coastal plain of Israel Bull Environ Contam Toxicol 45: 523–530
47. Nir R, Gasith A, Perry AS (1990) Cadmium uptake and toxicity to water hyacinth: effect of repeated exposures under controlled conditions. Bull Environ Contam Toxicol 44: 149–157
48. Fishelson L, Yawetz A, Perry AS, Zuk-Rimon Z, Manelis R, Dotan A (1994) The environmental health profile (EHP) for the Acre valley: xenobiotics in animals and physiological evidence of stress. Sci Total Environ 144: 33–45

Subject Index

MIX
Papier aus verantwortungsvollen Quellen
Paper from responsible sources
FSC® C105338

If you have any concerns about our products,
you can contact us on
ProductSafety@springernature.com

In case Publisher is established outside the EU,
the EU authorized representative is:
Springer Nature Customer Service Center GmbH
Europaplatz 3, 69115 Heidelberg, Germany

Printed by Libri Plureos GmbH
in Hamburg, Germany